# Atlas of Medical Parasitology

# Atlas of Medical Parasitology

An Atlas of Important Protozoa, Helminths and Arthropods, Mostly in Colour

Viqar Zaman
D.Sc., M.R.C. Path., M.B.B.S., D.T.M. & H.
Professor and Head of the Department of Microbiology
Faculty of Medicine
University of Singapore

430 Illustrations — 318 in Colour;
33 Life Cycle Diagrams

Springer-Science+Business Media, B.V.

# Atlas of Medical Parasitology

First published by Adis Press
Published in the United States of America and Japan
by Lea & Febiger

ISBN 978-94-017-5414-9 ISBN 978-94-017-5412-5 (eBook)
DOI 10.1007/978-94-017-5412-5

First Printing


Originally published by ADIS Press Australasia Pty Ltd in 1979
Softcover reprint of the hardcover 1st edition 1979



Colour separation by Hong Kong Graphic Arts Service Centre.

To Rosy, Sara and Fara

# Preface

The main purpose of this atlas is to provide a visual aid to the morphology, life cycle and clinical aspects of medically important parasites. The descriptive text is brief and is meant to act as a guide to the illustrative material. More detailed information should be obtained from other major publications. Emphasis is placed on features which are useful for diagnostic work, both in the laboratory and in clinical practice. It is, therefore, hoped that the atlas will prove useful to both students of parasitology and of tropical medicine.

The micro photographic techniques involved a liberal use of interference and phase contrast, as these methods reveal structural details better than other forms of light microscopy. In the case of protozoa, information on the ultrastructure is also provided as this helps to correlate the structural organisation with the functional activity of the cell. The magnification given with each illustration refers to instrumental reading and not to the final enlargement. Size relationships, therefore, should not be made unless so indicated in the text.

During the preparation of this book I received parasitological material from many colleagues. In this respect I would like to thank in particular Mrs K.M.G. Adam, Drs Paul C. Beaver, A.A. Buck, A.S. Dissanaike, T. Dondero, Ho Beng-Chuan, S. Inoki, Ian Jack, Mak Joon-Wah, R.A. Neal, Tom Orihel, S. Ramalingam, John Schacher, K. Shanmugaratnam, R. Sinniah, P. Wenk and M. Yokogawa.

Photographic material that was received is separately acknowledged in the text. My thanks are also due to Cambridge Instruments, U.K. for conducting the scanning electronmicroscopy on Schistosomes.

All the helminth life cycle drawings, with the exception of *Angiostrongylus cantonensis* and *Wuchereria bancrofti,* were redrawn from the charts of the Center for Disease Control (C.D.C.), Atlanta, Georgia, U.S.A. I wish to express my deep gratitude to the Parasitology Section of the Center and Dr Mae Melvin for giving me permission to do so. My thanks are due to W.H.O. and C.D.C. for permission to use the drawings of adult mosquito, flea and louse from 'Vector Control in International Health' by W.H.O., Geneva, 1972.

Finally, my thanks are also due to Loh Ah Keong who made all the line drawings, Evelyn Goh who assisted me in the photography and to Graeme Avery and Dennis Nickless of ADIS Press for editorial advice.

*Viqar Zaman*

# Contents


## Part I: Protozoa

1. Classification of Protozoa 1
2. Amoebae Infecting Man 3
3. Intestinal and Genital Flagellates 35
4. Haemoflagellates 45
5. Balantidium coli 73
6. Malaria Parasites 79
7. Toxoplasma gondii 99
8. Sarcocystis and Isospora 121
9. Pneumocystis 129

## Part II: Helminths

10. Cestodes 131
11. Trematodes 163
12. Nematodes 195

## Part III: Arthropods

13. Arthropods 259

Subject Index 283

# Part I: Protozoa

# Section 1 Classification of Protozoa of Medical Importance

Phylum Protozoa

Subphylum Sarcomastigophora (pseudopodia and flagella)
Ciliophora (cilia)
Apicomplexa (apical complex)

| Subphyllum | Superclass | Class | Genus |
|---|---|---|---|
| Sarcomastigophora | Sarcodina (pseudopodia for locomotion) | Rhizopodea | *Entamoeba*<br>*Endolimax*<br>*Iodamoeba*<br>*Dientamoeba*<br>*Acanthamoeba*<br>*Naegleria* |
| | Mastigophora (flagellum for locomotion) | Zoomastigophorea | *Giardia*<br>*Chilomastix*<br>*Trichomonas*<br>*Leishmania*<br>*Trypanosoma* |
| | Opalinata (no medical importance) | | |
| Ciliophora | | Ciliatea | *Balantidium* |
| Apicomplexa | | Sporozoasidea | *Plasmodia*<br>coccidia including<br>*Toxoplasma*<br>*Sarcocystis*<br>*Isospora* |

# Section 2
# Amoebae Infecting Man

## Terms Used in Relation to Amoebae

**Trophozoite** — Literally means any stage in the life cycle of a protozoan which can ingest food. In practice it refers to the motile form, which in the case of *Entamoeba histolytica,* is capable of invading the tissues.

**Cyst** — Is the non-motile form, which is protected by a distinct membrane or a cyst wall. This is the infective stage of the parasite with the exception of *Dientamoeba fragilis* and *Entamoeba gingivalis.*

**Pre-cyst** — Is the rounded form of trophozoite which precedes the cystic stage. It differs from the cyst in not having a cyst wall.

**Excystation** — Is the process of emergence of the trophozoite from the cyst.

**Encystation** — Is the process of formation of the cyst from the trophozoite.

**Metacyst** — Is the trophozoite which emerges from the cyst.

**Ectoplasm** — Is the external hyaline portion of the cytoplasm which is generally visible in a moving trophozoite.

**Endoplasm** — Is the internal granular portion of the cytoplasm which contains various food inclusions.

**Chromatoid body** — Is the RNA-protein complex which stains deeply with basic dyes but not with iodine. It is found in the genus *Entamoeba.* In the electron microscope it reveals a crystalline structure resembling virus particles.

**Glycogen vacuole** — Is a glycogen reserve which stains deeply with iodine and is found mainly in the cysts.

**Pseudopod** — Literally means false foot. It refers to temporary cytoplasmic processes formed at the surface of the trophozoite.

**Food vacuole** — A membrane bound vesicle in the cytoplasm formed around an ingested food particle.

**Contractile vacuole** — A membrane bound vesicle in the cytoplasm which collects and expels water from the cell. This is found in the free-living amoebae.

**Chromatin** — Is the substance in the cell nucleus which stains with basic dyes.

## Classification

For convenience of description they can be classified as following:

Pathogenic intestinal amoeba
*Entamoeba histolytica*

Non-pathogenic intestinal amoebae
*Entamoeba hartmanni*
*Entamoeba coli*
*Iodamoeba butschlii*
*Endolimax nana*
*Dientamoeba fragilis*
*Entamoeba gingivalis*

Free-living amoebae
*Naegleria* spp
*Acanthamoeba (Hartmanella)* spp

# Pathogenic intestinal amoeba

## Entamoeba histolytica

This is a very common parasite in the tropics. The virulence of *E. histolytica* varies considerably. It may live as a commensal in the lumen of the large intestine or it may invade the host tissues. The factors which determine the virulence of the parasite are not clearly understood. It is, however, known that strains isolated from asymptomatic patients are usually less invasive or non-invasive to laboratory animals (Neal, 1966).

### *Life-cycle of E. histolytica* [1]

Infection occurs by the ingestion of a mature cyst of the parasite. If the trophozoite is ingested it disintegrates in the stomach without producing infection. Excystation occurs in the lower region of the small intestine and the metacyst rapidly divides to give rise to 8 small amoebae. These enter the large intestine and may: (1) invade the host tissues, (2) live in the lumen of the large intestine without invasion, or (3) undergo encystation.

Only the cysts are able to survive in the external environment for any length of time. Immature (single or binucleated) or mature (four nucleated) cysts may be passed in the faeces. Immature cysts are able to mature in the external environment and become infective. Trophozoites are unable to encyst outside the body and are not infective.

The tissue invasion will cause bleeding which will lead to ingestion of red cells by the trophozoites. The trophozoites invading the intestinal tissues destroy the epithelial lining of the large intestine by producing proteolytic enzymes. The ulcers resulting from epithelial destruction may be shallow, eroding only the mucosa, or they may be deep, having entered the submucosa. In the submucosa they rapidly multiply and spread laterally giving rise to broad based or 'flask-shaped' ulcers. Alternatively, the trophozoites may produce microabscesses in the submucosa, which eventually rupture

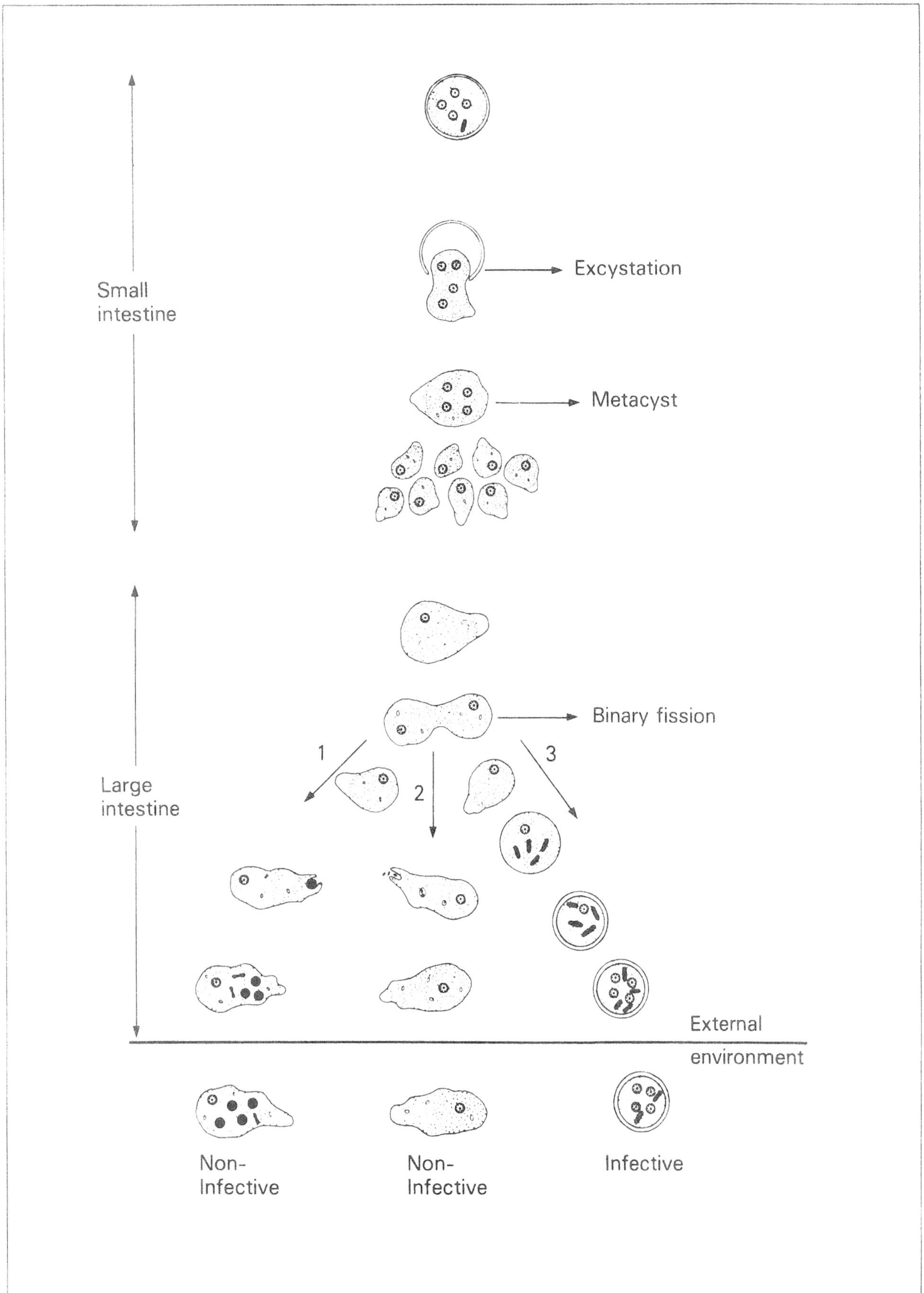
Small
intestine
Excystation
Metacyst
Large
intestine
Binary fission
1
2
3
External
environment
Non-
Infective
Non-
Infective
Infective

1

through the epithelium, also giving rise to 'flask-shaped' ulcers. The trophozoites from the intestinal tissues may be carried to the extra-intestinal organs via the portal circulation.

The most commonly involved extra-intestinal organ is the liver. In the liver the parasites may produce diffuse hepatitis or localised abscesses. The abscesses may spread to neighbouring organs by direct extension or the parasites from the abscesses may be carried to other organs via the circulatory system.

## *Clinical aspects*

### Acute amoebic dysentery

The presenting symptom is diarrhoea which is accompanied with blood, mucus and sometimes tenesmus. The condition may last for a few days or weeks, after which it disappears spontaneously or is transformed into chronic amoebiasis. As a complication, severe intestinal haemorrhage or rarely perforation may occur. During the acute stage trophozoites are passed in the stools.

### Chronic and asymptomatic amoebiasis

In chronic amoebiasis symptoms of diarrhoea alternate with constipation. Vague abdominal pain, dyspepsia and asthenia are common. As a complication amoeboma might develop. Some patients remain asymptomatic. Cysts are generally passed in the stools.

### Hepatic amoebiasis

This may manifest as amoebic hepatitis consisting of pain in the right hypochondrium with mild to low grade fever. In liver abscess there is hepatomegaly, marked tenderness of the liver, mild to high grade fever, leucocytosis and raised ESR. 'Anchovy-sauce' type pus (thick reddish-brown pus) is obtained by aspiration of the abscess.

## *Diagnosis*

Intestinal amoebiasis is diagnosed by the faecal examination which may reveal either the presence of trophozoites or cysts. If trophozoites containing red cells are detected, it will be safe to assume that ulceration of the large intestine is present. Charcot-leyden crystals, which are derived from eosinophils are often present in intestinal amoebiasis.

In extra-intestinal amoebiasis, cysts or trophozoites of *E. histolytica* may not be present in the faeces. A serological examination is, therefore, of value. The 4 serological tests most commonly employed are gel-diffusion, indirect haemagglutination, latex agglutination and the fluorescent-antibody test. These tests become positive in over 90% of proven cases of hepatic amoebiasis (Ambrose-Thomas and Truong, 1972).

Radiological examination may reveal a picture similar to that of ulcerative colitis if severe ulceration of the large intestine is present. If amoeboma has developed, a filling defect will become visible. In the case of liver involvement there is generally some basal pleuritis with pleural effusion. In the case of liver abscess good visualisation of the cavity is obtained after injecting air, radio-opaque dye or angiography.

*Table I.* Differential characteristics of intestinal amoebae (live specimens)

| | *E. histolytica* | *E. hartmanni* | *E. coli* | *E. polecki* | *E. nana* | *I. butschlii* | *D. fragilis* * |
|---|---|---|---|---|---|---|---|
| *Trophozoite* | | | | | | | |
| Movement | Active especially in acute dysentery | Sluggish | Sluggish | Sluggish | Sluggish | Active often | Active often multiple pseudopods |
| Inclusions | Red cells in case of tissue invasion | No red cells | No red cells | No red cells | No red cells | No red cells | No red cells |

* This is now regarded as a *Trichomonad* (Camp et al., 1974)

*Table II.* Differential characteristics of intestinal amoebae (stained specimens)

| | *E. histolytica* | *E. hartmanni* | *E. coli* | *E. polecki* | *E. nana* | *I. butschlii* | *D. fragilis* |
|---|---|---|---|---|---|---|---|
| *Trophozoite* (haematoxylin stain) | | | | | | | |
| No. of nuclei | 1 | 1 | 1 | 1 | 1 | 1 | 2 |
| Karyosome (nucleolus) | Generally central small | Generally central small | Generally eccentric large | Generally eccentric small | Large | Large generally irregular | Large may be fragmented |
| Peripheral chromatin | Generally symmetrical fine | Generally symmetrical fine | Generally asymmetrical fine | Generally symmetrical fine | Incon-spicuous | Incon-spicuous | Incon-spicuous |
| *Cyst* (mature) (haematoxylin stain) | | | | | | | |
| No. of nuclei | 4 | 4 | 8 | 1 | 1 | 1 | Cysts not found |
| Shape | Circular | Circular | Circular | Circular | Generally oval | Irregular | — |
| Size (diameter) | 12-20μm | 2-8μm | 15-25μm | 11-15μm | 8-12μm | 10-16μm | — |
| Chromatoid body | Generally large rounded ends | Generally large rounded ends | Generally small splintered ends | Variable shape | Absent | Absent | — |
| Glycogen vacuole (Iodine stain) | Diffuse | Diffuse | Diffuse | Diffuse | Absent | Large and sharply demarcated | — |
| Other inclusions | Nil | Nil | Nil | Usually large faintly staining mass | Nil | Sometimes minute rod like bodies | — |

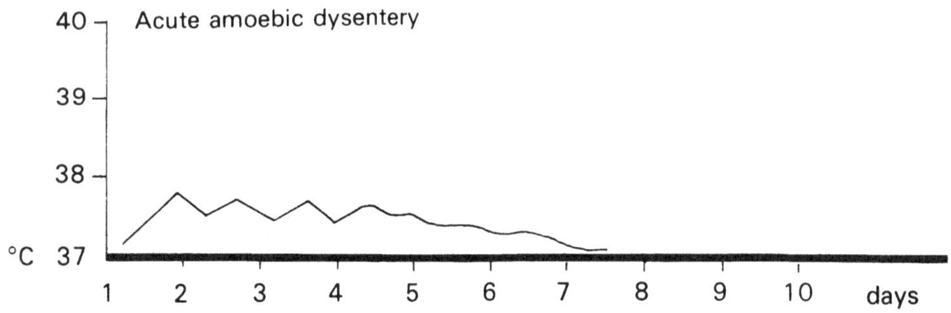


| *Symptoms* | *Signs* | *Stools* | *Serum antibodies* | *Course* |
|---|---|---|---|---|
| Bloody diarrhoea<br>Abdominal discomfort<br>Dyspepsia | Tenderness in colon on palpation | Trophozoites containing red cells | + or - | Spontaneous remission<br>Haemorrhage<br>Extra-intestinal involvement |

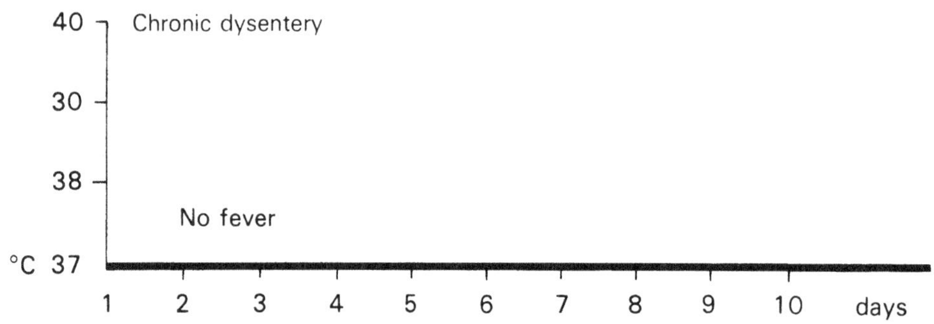


| *Symptoms* | *Signs* | *Stools* | *Serum antibodies* | *Course* |
|---|---|---|---|---|
| Irregular bowel movement<br>Dyspepsia<br>Asthenia | Tenderness in colon on deep palpation | Cysts with 4 nuclei | + or - | Extra-intestinal involvement<br>Ameoboma |

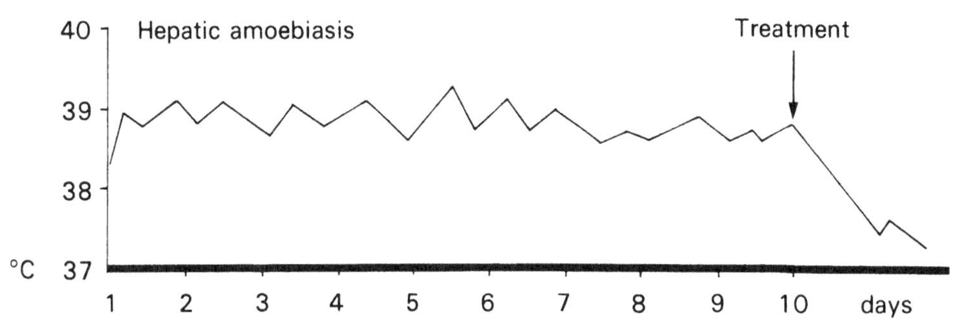


| *Symptoms* | *Signs* | *Stools* | *Serum antibodies* | *Blood* | *Course* |
|---|---|---|---|---|---|
| Pain in the right hypochondrium<br>Fever<br>Dyspepsia | Tenderness on pressure<br>Enlarged liver | + or - | + | Leucocytosis<br>ESR raised | Rupture into neighbouring organs<br>Chemotherapy plus aspiration<br>Good prognosis |

### *Complications of amoebiasis* [2]

The intestinal complications consist of haemorrhage, appendicitis, perforation and amoeboma. The main extra-intestinal complication is hepatic amoebiasis. This occurs as a result of trophozoites entering the portal circulation from the intestinal tissues and becoming lodged in the liver. The liver abscess once formed may rupture into any of the neighbouring organs. In addition, the amoebae may be carried by the systemic circulation from the infected tissues to settle in the lungs, brain and other organs of the body.

### References

Ambroise-Thomas, P. and Truong, T.K.: Fluorescent antibody test in amebiasis. Clinical applications. American Journal of Tropical Medicine and Hygiene 21: 907 (1972).

Camp, R.R.; Mattern, C.F.T. and Honigberg, B.M.: Study of *Dientamoeba fragilis* Jepps and Dobell. I. Electronmicroscopic observations of the binucleate stages. II. Taxonomic position and revision of the genus. Journal of Protozoology 21: 69 (1974).

Neal, R.A.: Experimental studies on *Entamoeba* with reference to speciation; in Dawes (Ed) Advances in Parasitology Vol. 4 (Academic Press, London and New York 1966).

2

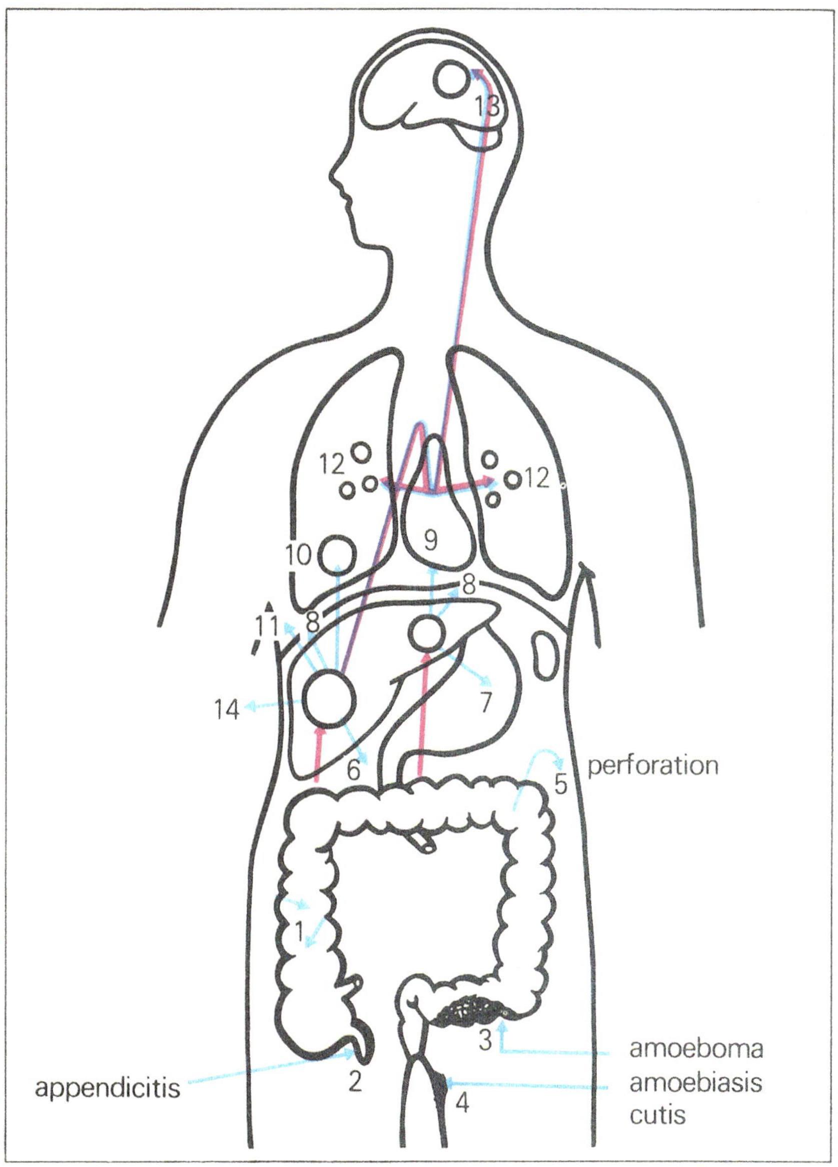


**2 Complications of amoebiasis**

1. Haemorrhage;
2. Appendicitis;
3. Amoeboma,
4. Amoebiasis cutis;
5. Perforation;
6. Rupture of liver abscess into peritoneal cavity;
7. Rupture into stomach;
8. Rupture into subdiaphragmmatic space;
9. Rupture into pericardium;
10. Rupture into lung parenchyma;
11. Rupture into pleura;
12. Blood borne spread to lung parenchyma;
13. Blood borne spread to brain;
14. Rupture into exterior. (Red line shows blood borne extension via portal circulation. Blue line shows direct extension. Red and blue show extension via systemic circulation).

**3 Starch ingestion by E. histolytica.** These photographs taken at an interval of 4 seconds illustrate stepwise the process of phagocytosis in *Entamoeba.* Phase contrast X1000.

**4 a) E. histolytica with starch crystals.** The scanning electron micrographs show the surface of *Entamoeba* and the starch crystals.
**b)** The lower electron micrograph shows a crystal being drawn into the cytoplasm. X6000 and X8000. st. = starch. A = amoeba.

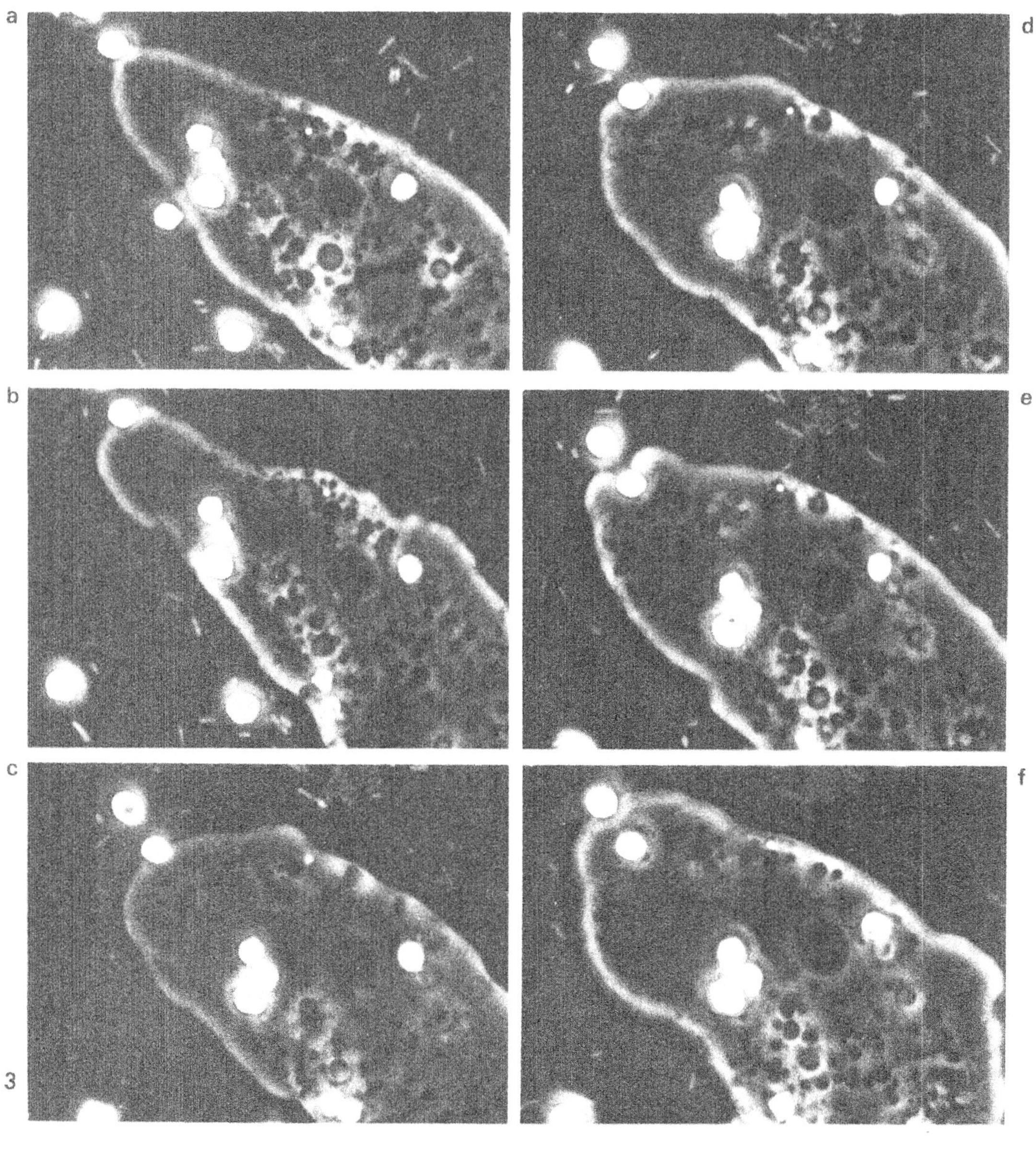

3

a

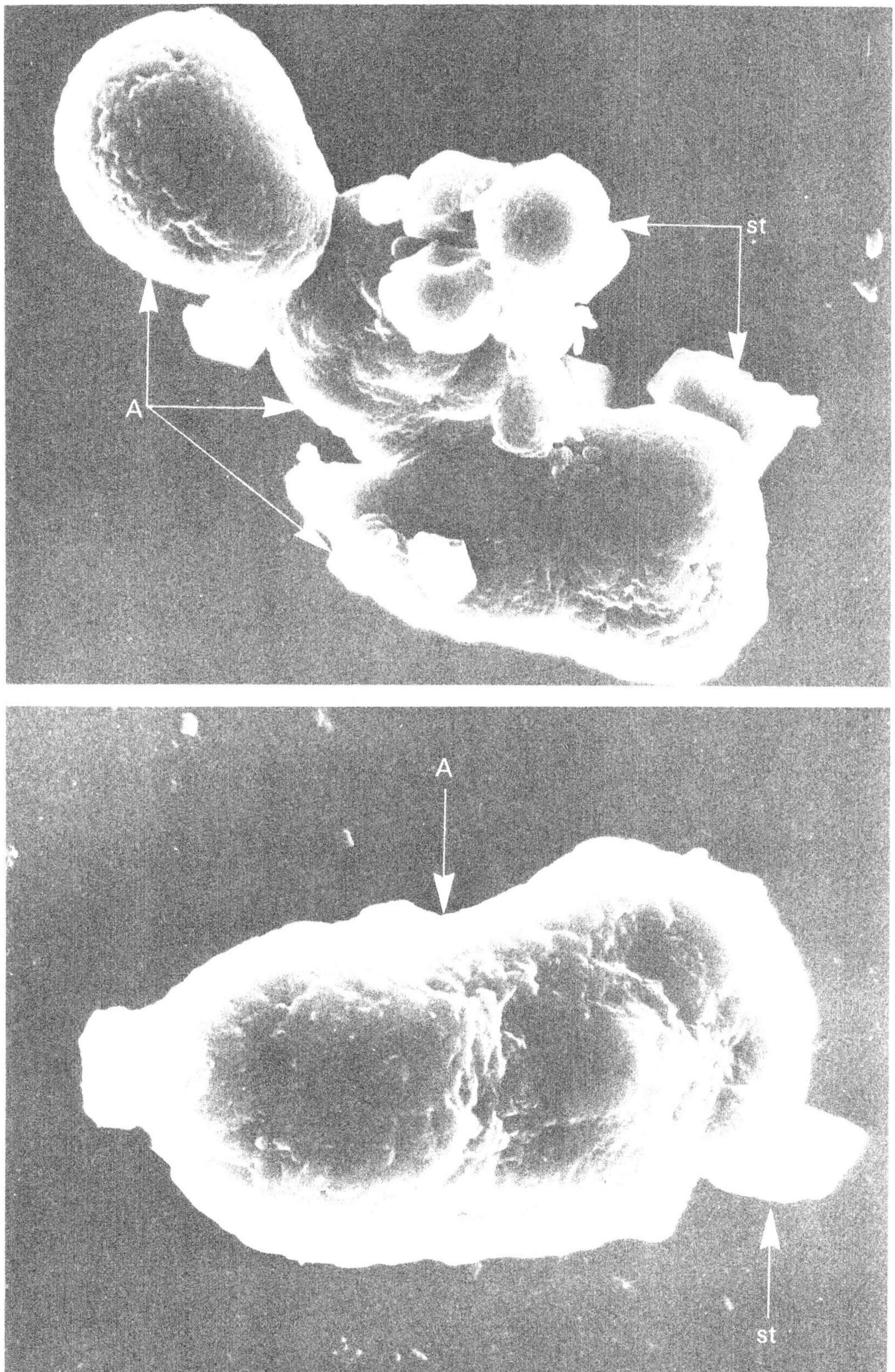


b

4

**5 a) Ingestion of bacteria by E. histolytica.** Bacteria are an important source of food for *Entamoeba* and ingestion occurs by the extension of pseudopods around the bacterial cell. **b)** All the ingested food particles are surrounded by plasmalemma which forms the wall of the food vacuole. Electron micrograph. X20,000. B = bacterium.

**6 Cyst of E. histolytica.** The crystalline structure is the chromatoid body cut longitudinally. Glycogen appears as an irregular mass lying adjacent to the chromatoid body. The food vacuoles contain amorphous structures as bacteria and other food matter is digested before encystation occurs. The smaller vacuoles (secretory vacuoles) probably carry enzymes used for excystation. Electron micrograph. X1600. FV = food vacuole; N = nucleus; G = glycogen mass; CB = chromatoid body; SV = secretory vacuole.

**7 Cyst of Naegleria fowleri.** The cytoplasm contains mitochondria which are absent from *Entamoeba.* The nucleolus is large and there is no chromatoid body or glycogen mass. Electron micrograph. X1600. M = mitochondria. N = nucleus. Nl = nucleolus. SV = secretory vacuole.

a

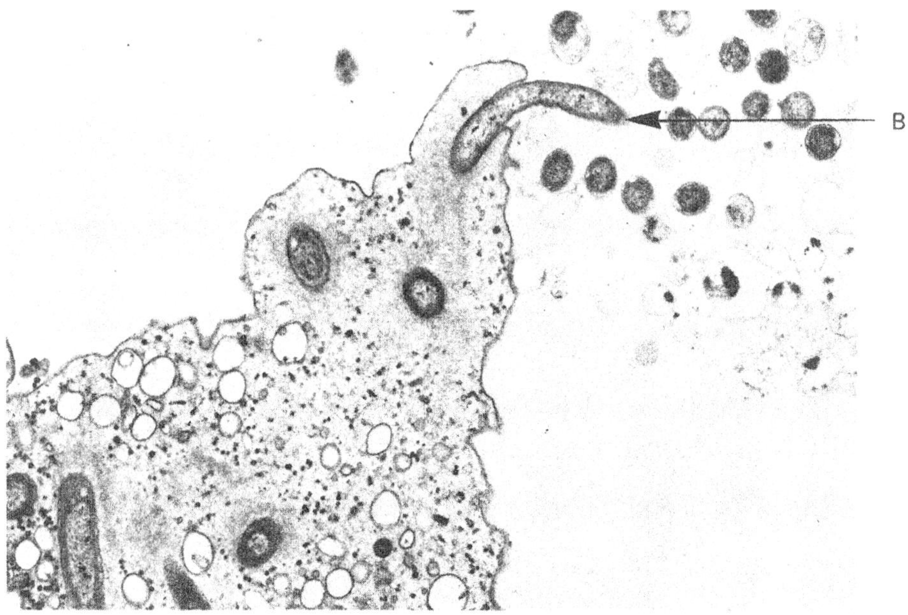


b

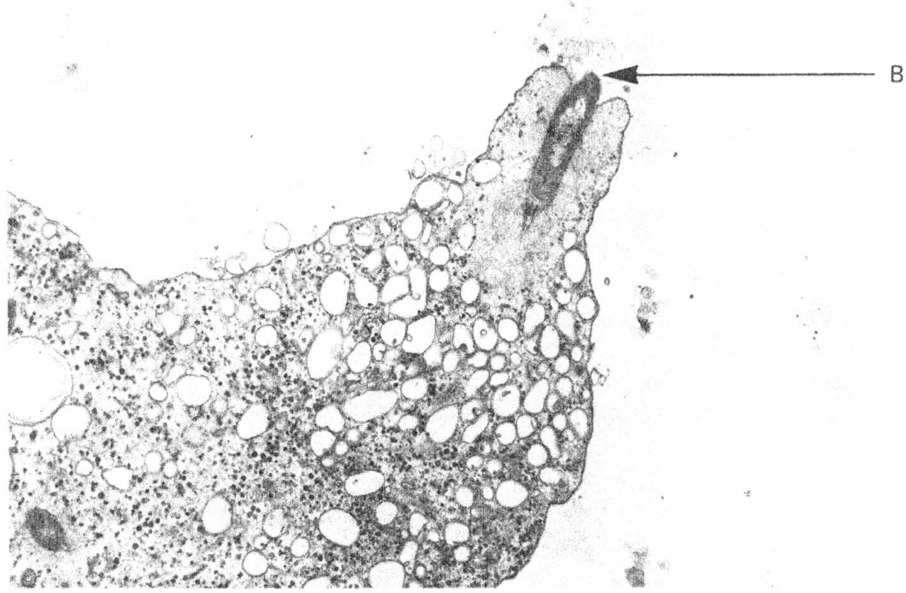


5

6

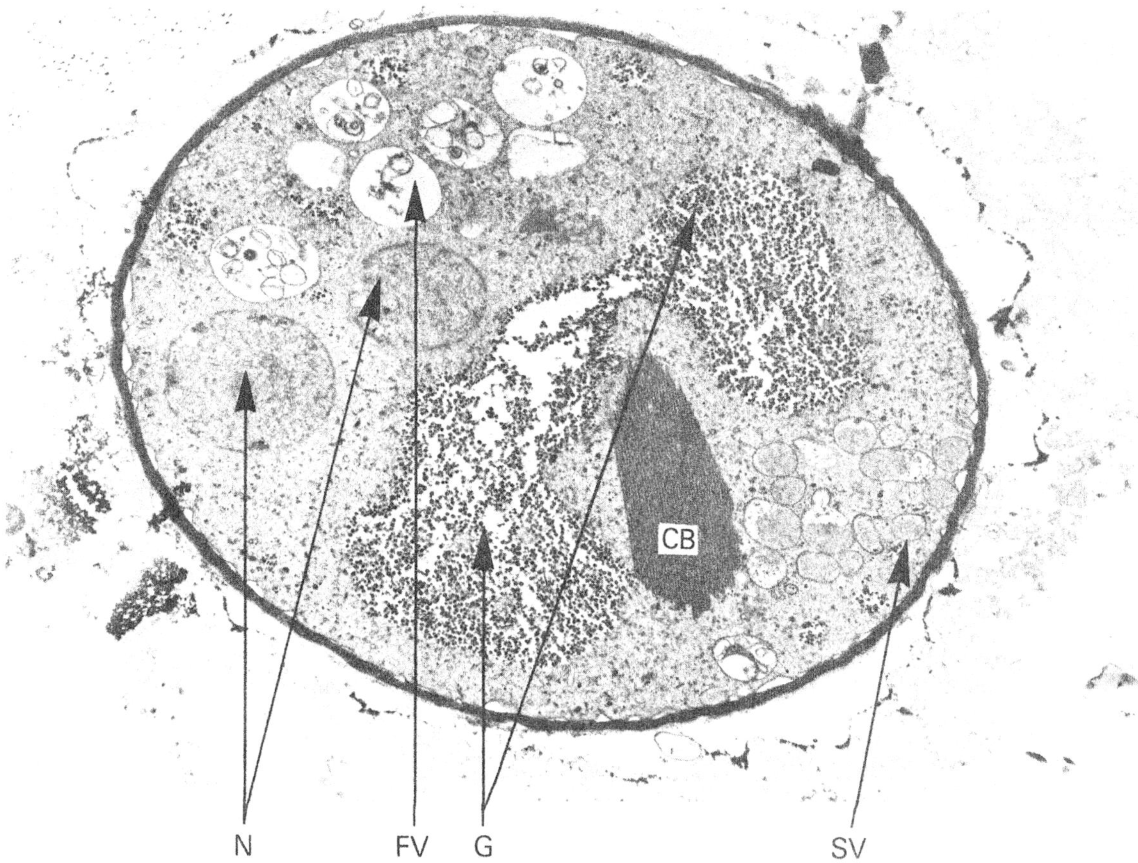


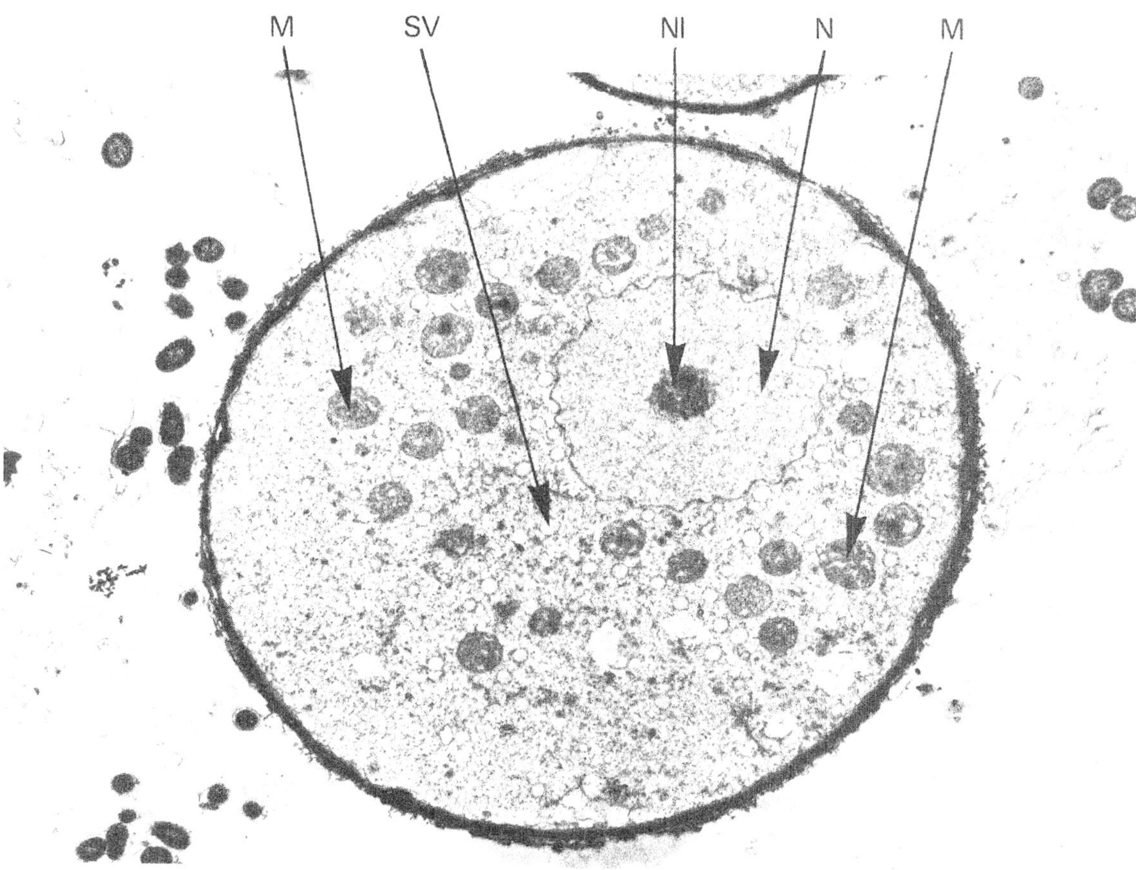


7

**8 Cross-section of chromatoid body of Entamoeba.** A fully formed body appears hexagonal in cross-section. It is made up of spherical bodies or particles measuring 25-30mμ. Electron micrograph. X100,000.

**9 X-ray of a large amoebic liver abscess.** A fluid level has formed after aspiration due to entry of air.

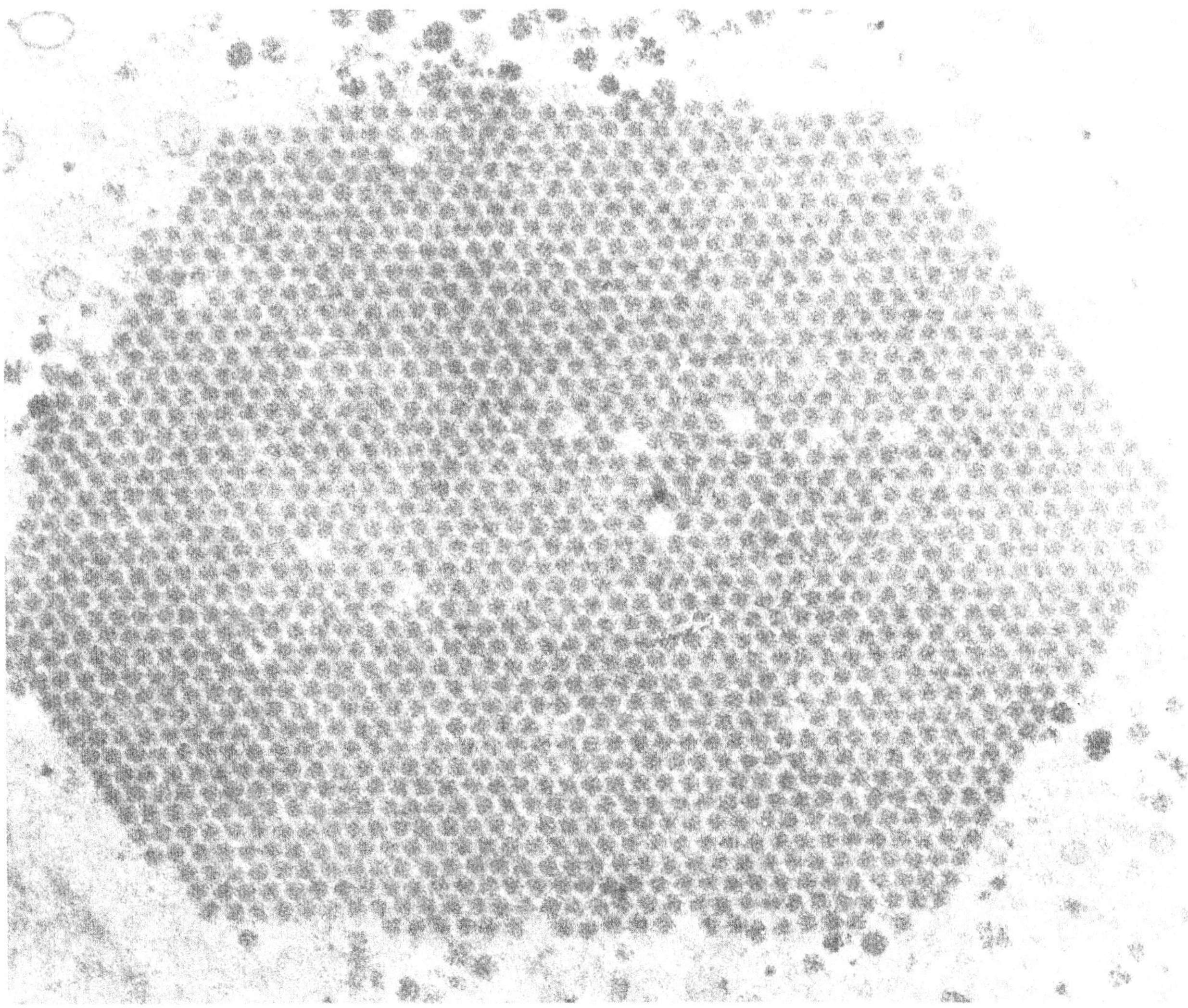

8

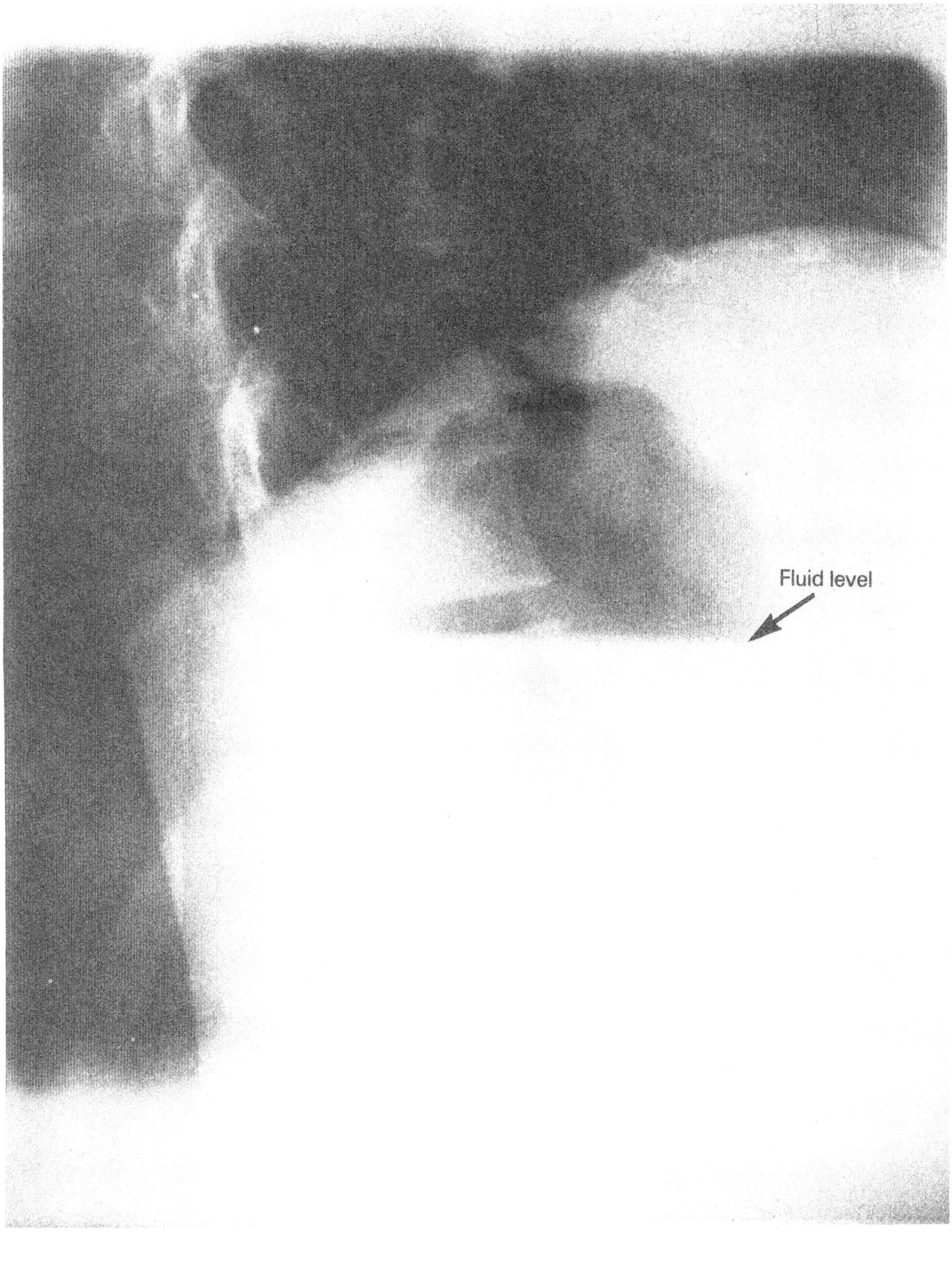


9

**10 Ultrastructure of 'Entamoeba' type nucleus.** All species of *Entamoeba* display certain common characteristics with regard to their nuclear morphology. The nuclear membrane is lined with peripheral chromatin which appears as electron dense masses, a circular nucleolus which is in the centre or eccentrically placed and the nucleoplasm which contains circular internuclear bodies. The function of these internuclear bodies is not known. NM = nuclear membrane; PC = peripheral chromatin; NI = nucleolus; INB = intranuclear bodies; FV = food vacuole. X20,000.

**11 E. histolytica** trophozoite from amoebic dysentery faeces. The circular bodies in the cytoplasm are red cells. The parasite is actively motile and generally moves in one direction at a time. The background shows bacteria, red cells and degenerated white cells. Interference contrast. X1000. Enlarged by 9.6.

**12 E. histolytica fed** *in vitro* **(cultures) with red cells.** The parasites have ingested a large number of red cells and have become sluggish due to distension of the plasmalemma. Interference contrast. X1000. Enlarged by 23.4.

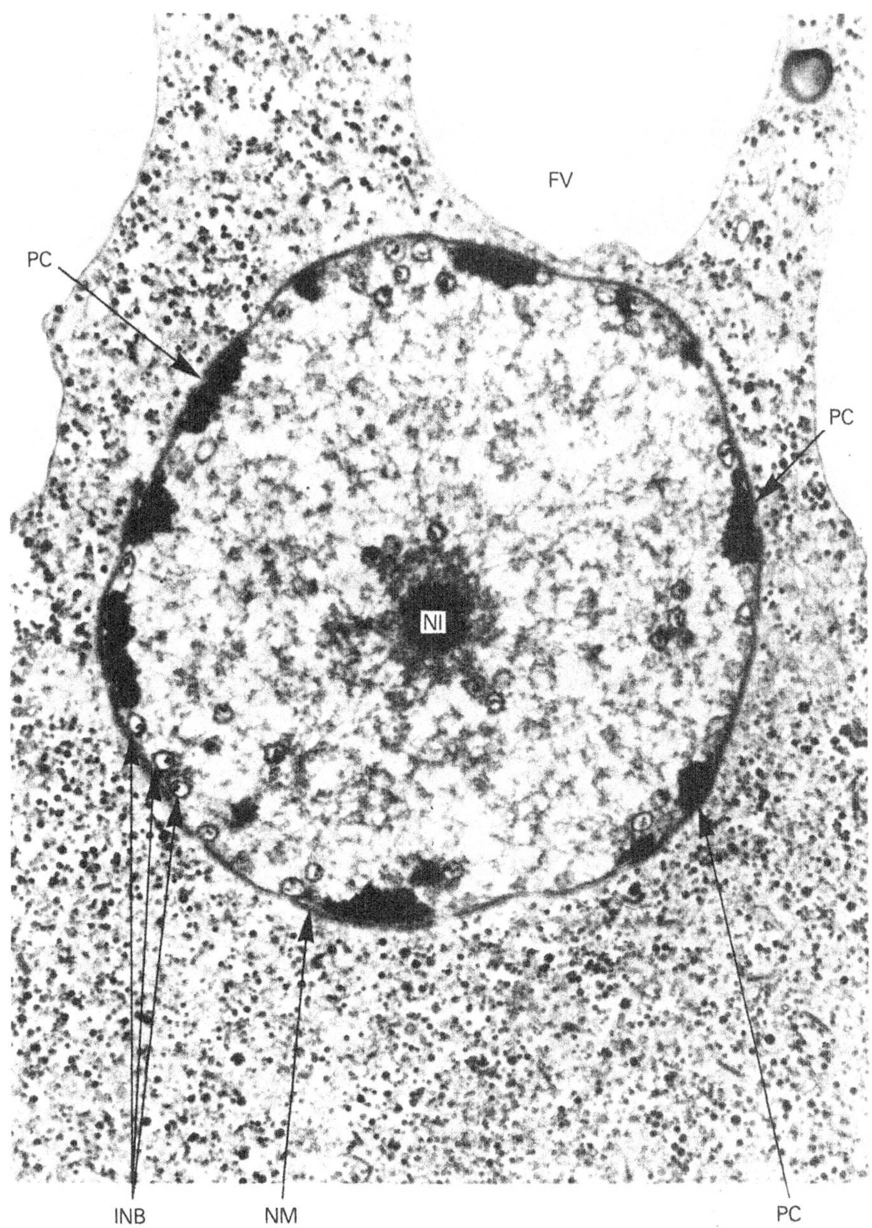


10

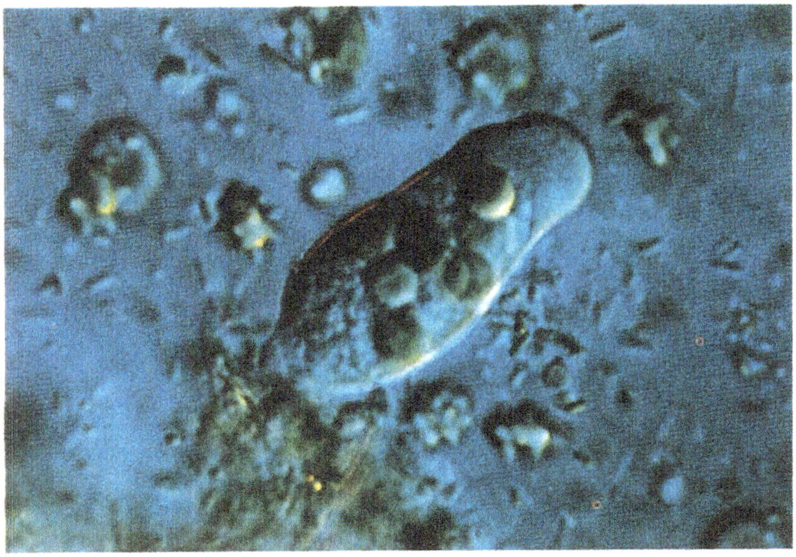

11

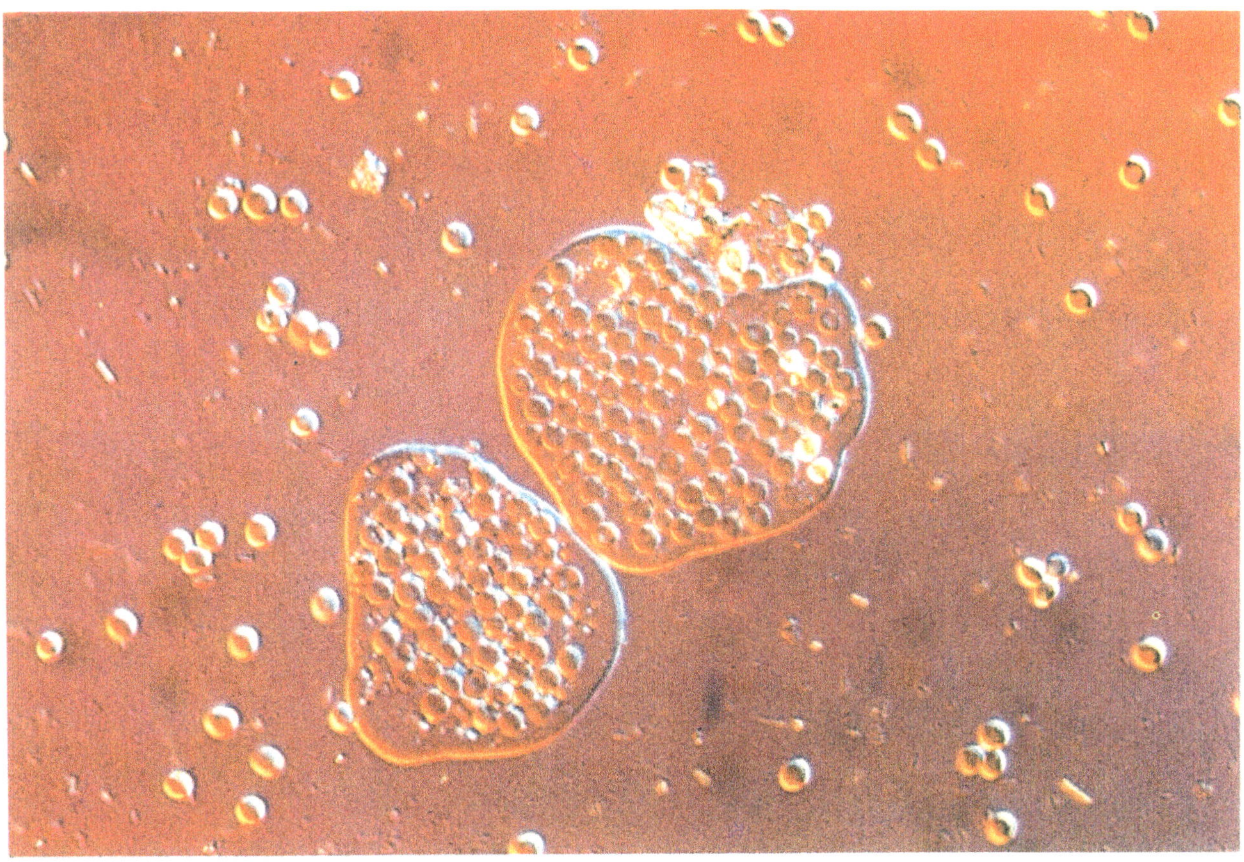

12

**13 Charcot Leyden crystal.** These diamond shaped crystals are often seen in amoebic dysentery faeces and may also be present in other parasitic infections. They are absent in bacillary dysentery. Interference contrast. X400. Enlarged by 9.6.

**14 E. histolytica trophozoite.** Showing the nuclear structure. Iron-haematoxylin. X1000. Enlarged by 9.6.

**15, 16 E. histolytica cyst.** Stained with iodine showing the presence of 4 nuclei. The nuclei do not lie in the same plain. A change of focus is required to count the total number. X1000. Enlarged by 6.5.

**17 E. histolytica, a uninucleated cyst.** The nucleolus becomes large before division. The chromatoid bodies are the dark structures in the cytoplasm. Iron-haematoxylin. X1000. Enlarged by 6.5.

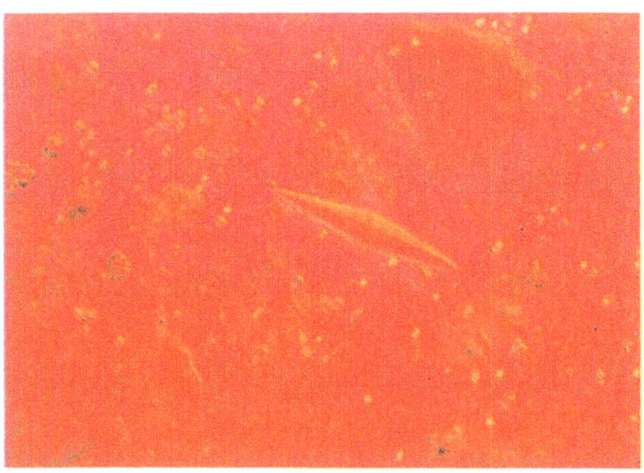

13

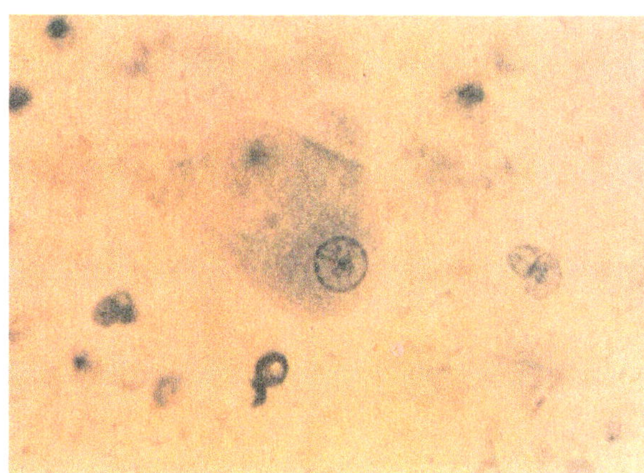

14

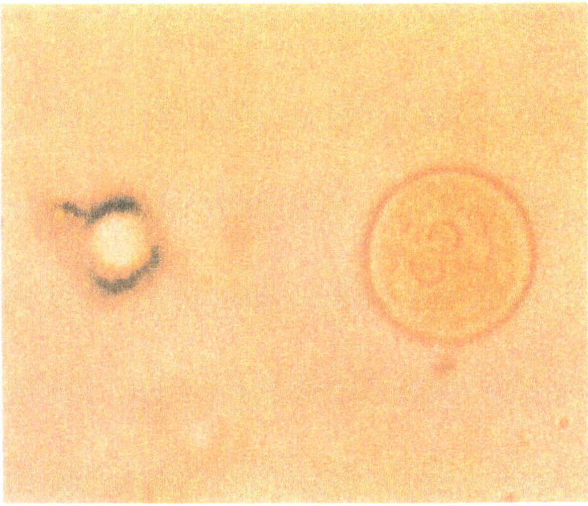

15

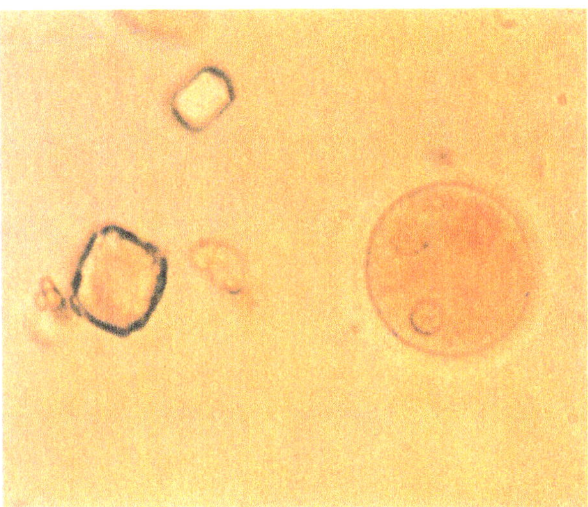

16

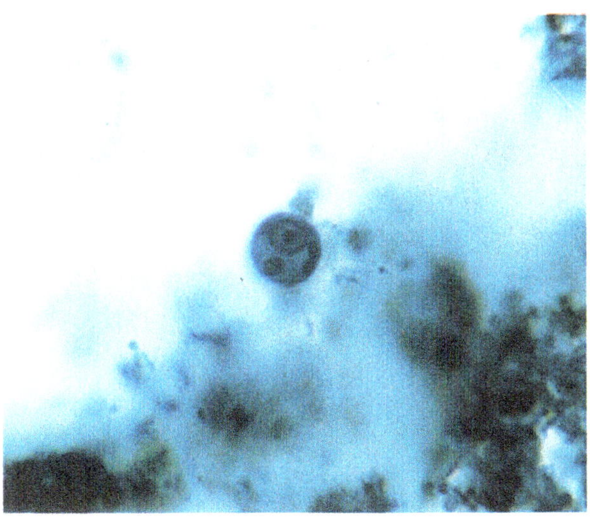

17

**18 E. histolytica, a binucleated cyst.** The central clear area is occupied by glycogen and is not stained by iron-haematoxylin. The amount of glycogen decreases as the cyst matures. X1000. Enlarged by 6.5.

**19 E. histolytica cyst.** Showing the bar shaped structure of chromatoid body. Iron-haematoxylin. X1000. Enlarged by 6.5.

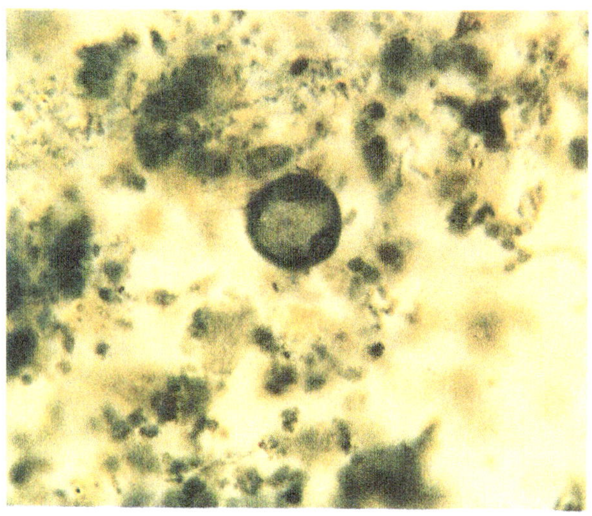

18

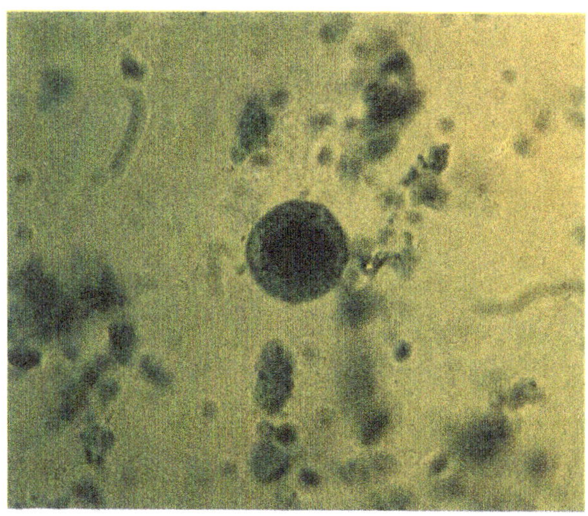

19

**20 E. coli trophozoite.** Showing a large eccentrically placed nucleolus and irregular chromatin. Iron-haematoxylin. X1000. Enlarged by 6.5.

**21, 22 E. coli cyst iodine stained.** A change of focus shows the presence of 8 nuclei. X1000. Enlarged by 6.5.

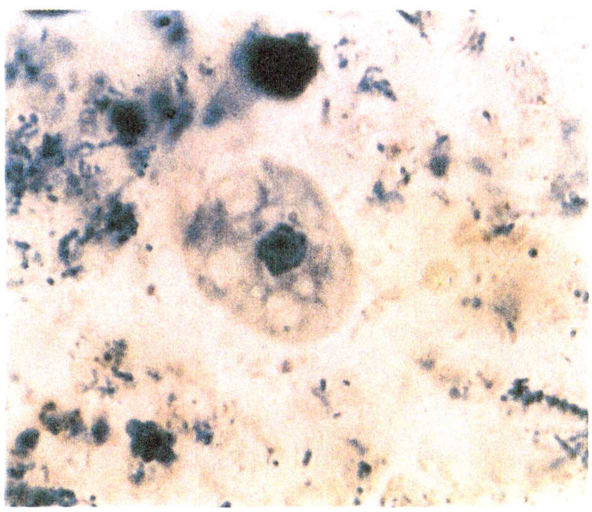

20

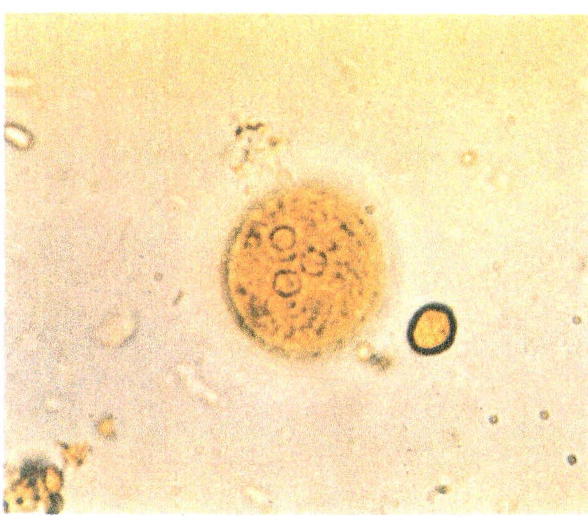

21

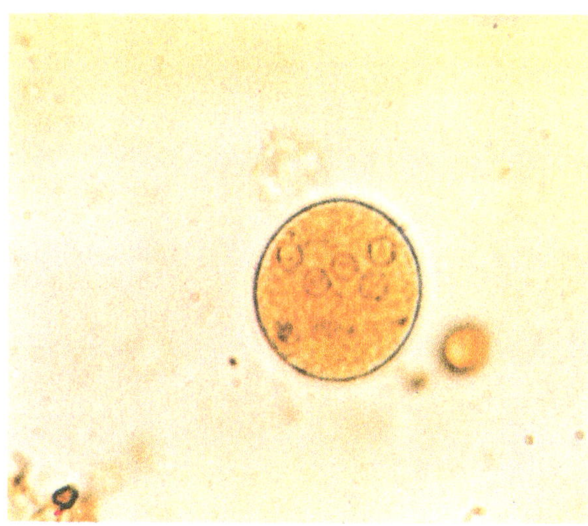

22

**23 Dientamoeba fragilis.** Trophozoite showing the 2 nucleoli. The peripheral chromatin is not present. Iron-haematoxylin. X1000. Enlarged by 6.5.

**24 Endolimax nana trophozoite.** Showing a single large nucleolus. The nuclear membrane is visible but the peripheral chromatin is not present. Iron-haematoxylin. X1000. Enlarged by 6.5.

**25 Endolimax nana cyst.** Showing 4 dot like nucleoli iodine stain. X1000. Enlarged by 6.5.

**26 Iodamoeba butschlii cyst.** Showing darkly staining glycogen vacuole. Iodine stain. Interference contrast. X1000. Enlarged by 9.6.

**27 Entamoeba polecki cyst.** Showing a single nucleus and a diffuse mass in the cytoplasm. The nature of the 'diffuse mass' is not known. Iron-haematoxylin. X1000. Enlarged by 9.6.

**28 E. hartmanni trophozoite.** Note the small size as compared to *E. histolytica.* A single nucleus is visible. Iron-haematoxylin. X1000. Enlarged by 9.6.

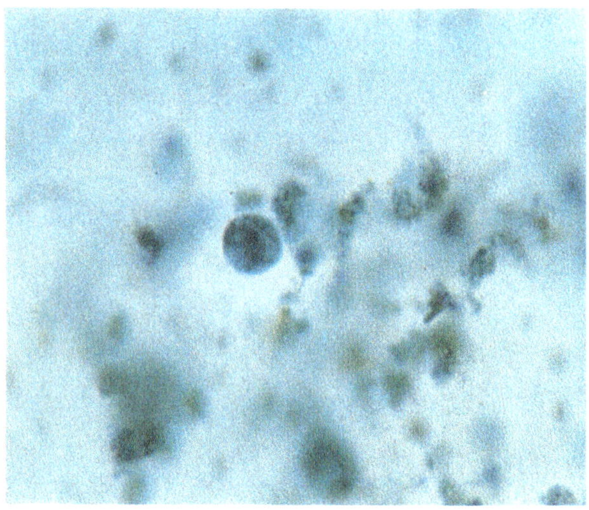

23

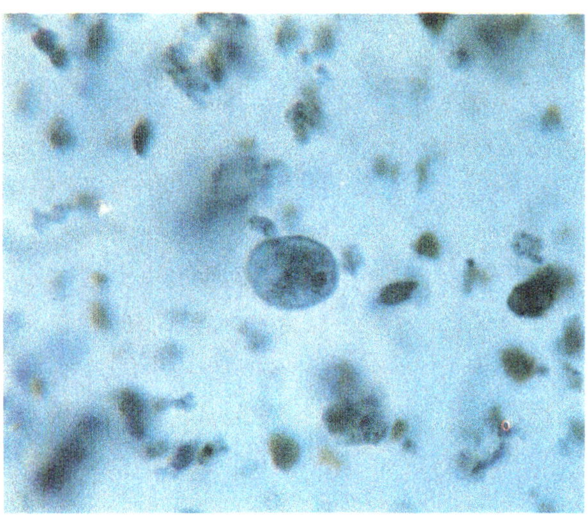

24

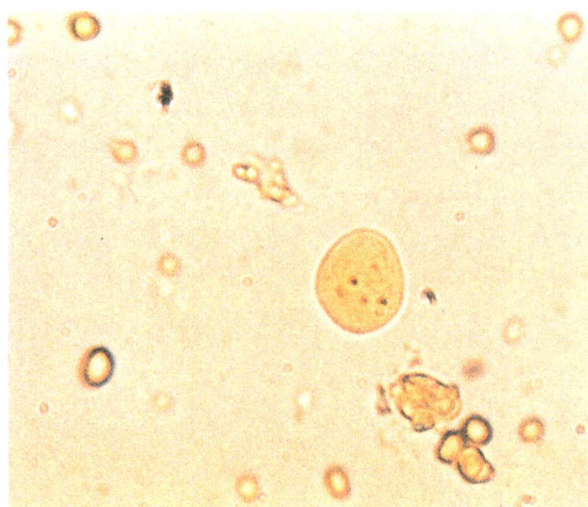

25

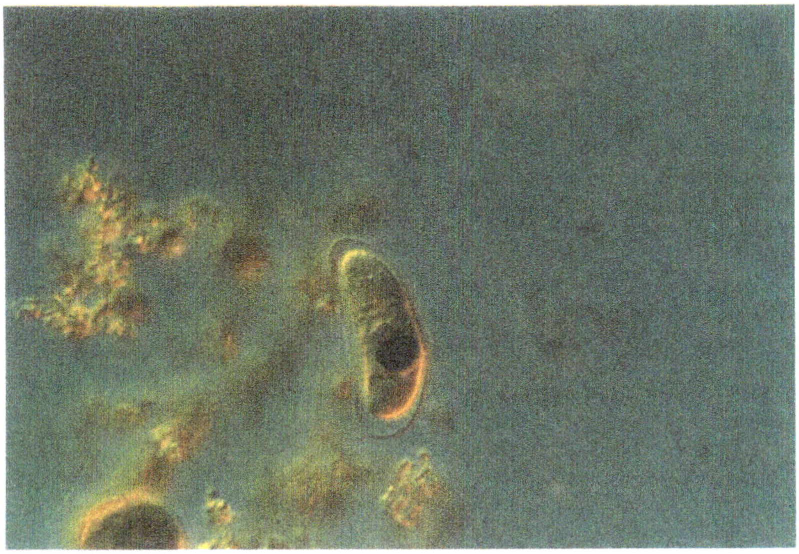

26

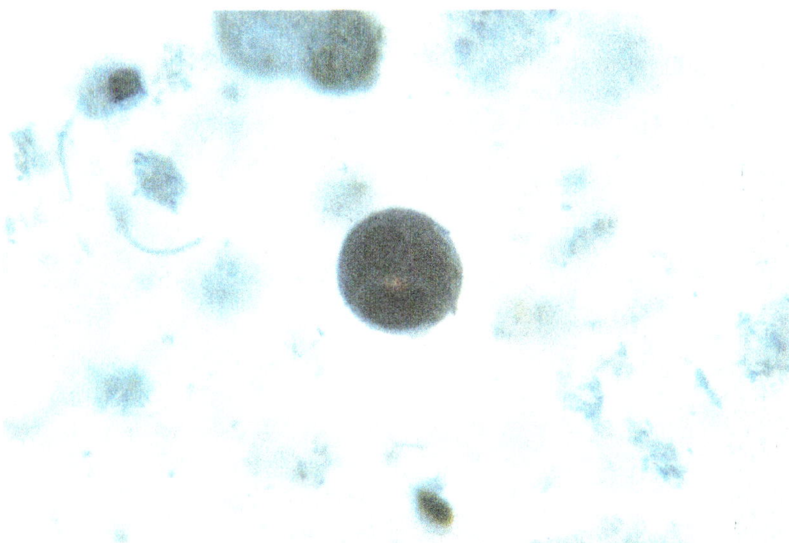

27

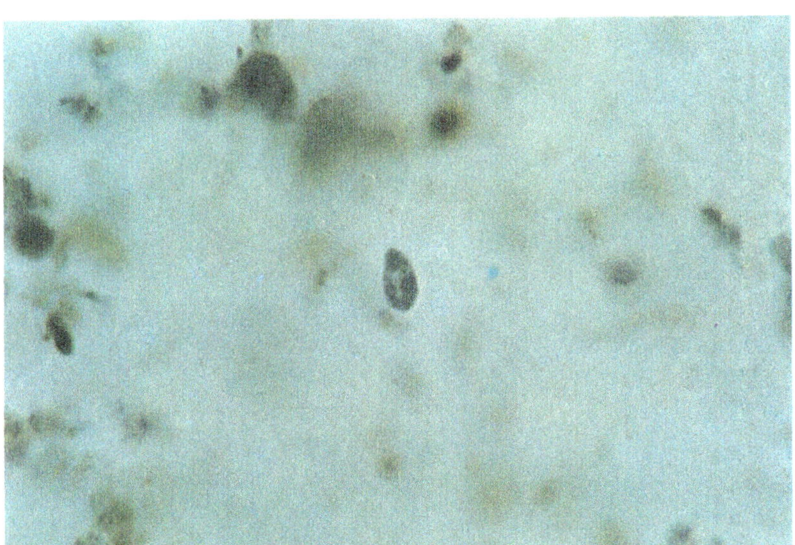

28

**29 Superficial ulceration of the large intestine by E. histolytica.** H and E. X400. Enlarged by 9.6.

**30 Moderately severe ulceration of the large intestine by E. histolytica.** The muscularis mucosa is intact. H and E. X400. Enlarged by 9.6.

**31 Severe ulceration of the large intestine by E. histolytica.** A typical conical (flask-shaped) ulcer can be seen. The muscularis mucosa has been penetrated. H and E. X400. Enlarged by 9.6.

**32 E. histolytica lying in the submucosa.** There is generally an empty space surrounding each trophozoite. This space is probably caused by enzymic activity of the parasite. H and E. X400. Enlarged by 5.4.

**33 A micro abscess in the submucosa.** Containing a large number of *E. histolytica* trophozoites mostly at the periphery. H and E. X400. Enlarged by 5.4.

**34 E. histolytica lying in a crypt of large intestine.** PAS stain. This stain is good for screening as the trophozoites take a reddish colour and can be easily distinguished from host tissues. X400. Enlarged by 5.4.

**35 An amoebic liver abscess being aspirated.** Note the reddish brown colour of the pus ('Anchovy-sauce'). This colour is due to the breakdown of liver cells. Enlarged by 5.4.

**36 Fluorescent antibody test.** With serum from a case of amoebic abscess. Note the presence of fluorescence around each trophozoite. The antigen consists of *E. histolytica* growing axenically (without bacteria) in Diamond's medium. X400. Enlarged by 5.4.

29

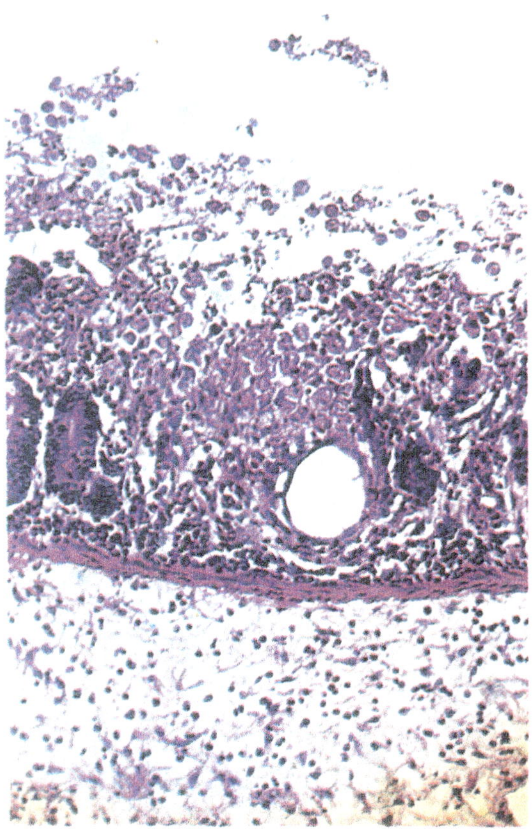

30

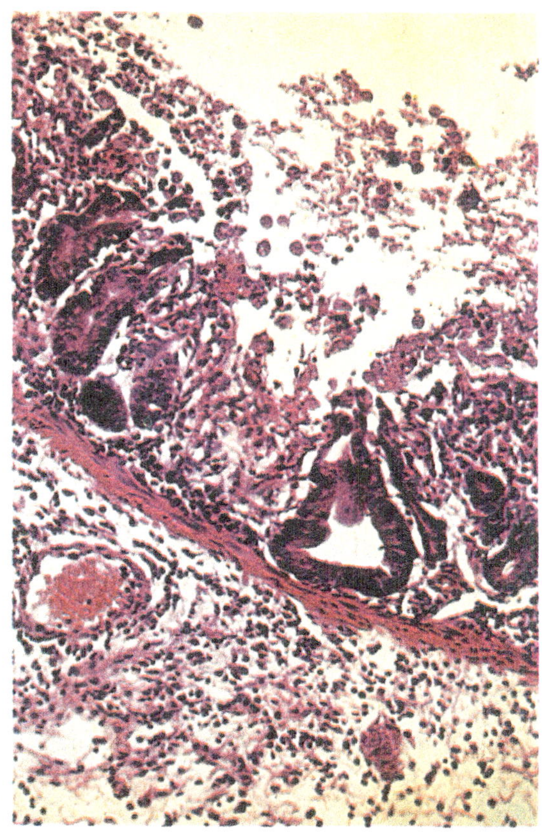

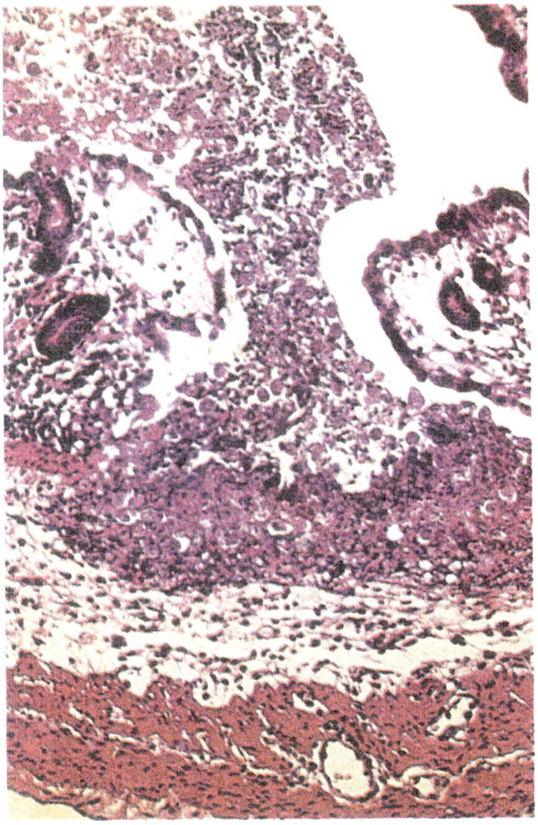

31

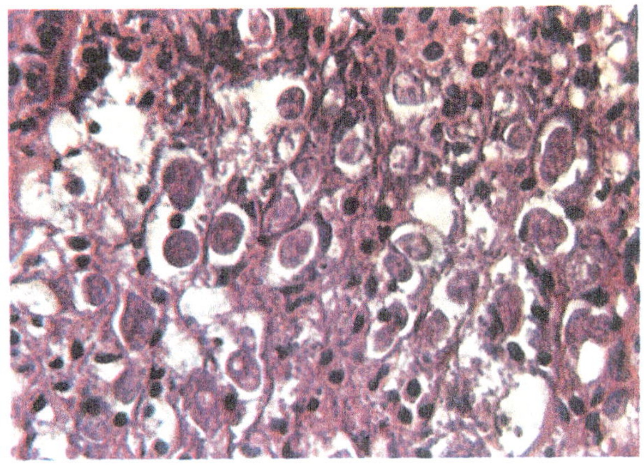

32

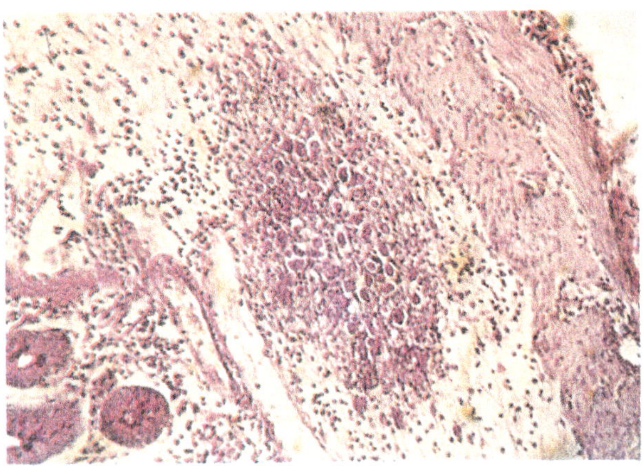

33

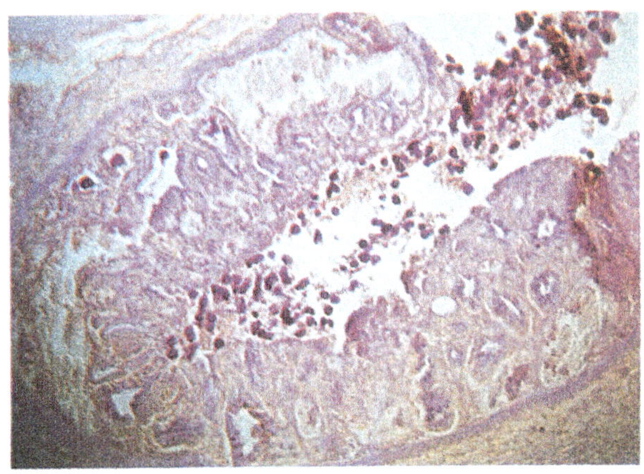

34

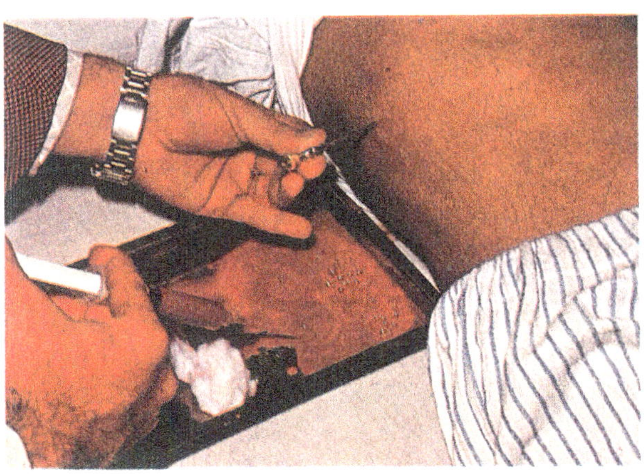

35

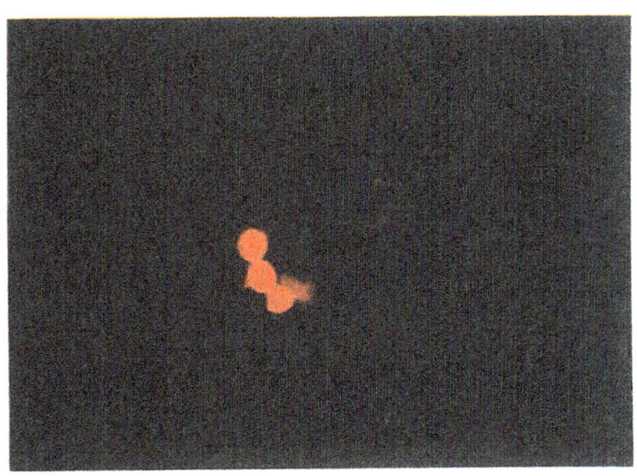

36

# Free-living amoebae

## Naegleria and Acanthamoeba (Hartmanella) Infections

These are free living amoebae found in soil and water. Human infection has been reported from many parts of the world and is acquired by swimming. The infection with *Naegleria* appears to be more common than *Acanthamoeba*. The species involved is *Naegleria fowleri* (Syn. *N. aerobi*) and *Acanthamoeba castellani* (Syn. *H. culbertsoni*). The amoebae invade the brain through the nasal cavity by penetrating the cribriform plate of the ethmoid bone. Clinically the condition is described as 'primary meningo-encephalitis' as the central nervous system is primarily involved, unlike infections with *Entamoeba histolytica* where brain involvement is usually secondary.

### *Life cycle of Naegleria* [37]

The main difference in the life cycle of this parasite as compared to *Acanthamoeba* is that it can develop into a flagellate stage, when in contact with water. The flagellate transforms back to the amoeboid form before encystation.

### *Clinical aspects*

The microscopic examination of the infected brain shows nests of amoebae with extensive haemorrhagic reaction mostly involving the basilar portion of cerebrum and cerebellum. In the case of *Acanthamoeba* cysts may also be found.

### *Diagnosis*

Diagnosis can be made by the microscopic examination of the CSF which may reveal the presence of the trophozoites. The CSF contains red cells and is bacteriologically sterile. The amoebae can also be cultured by inoculating CSF on to non-nutrient agar plates, previously seeded with *Escherichia coli.*

*Table I.* A comparison of *Naegleria-Acanthamoeba* morphology with *Entamoeba histolytica*

| *Naegleria-Acanthamoeba* | *Entamoeba* |
|---|---|
| 1. Nucleolus large and distinct | 1. Nucleolus small and indistinct |
| 2. Contractile vacuoles present | 2. Contractile vacuoles absent |
| 3. Single nucleus in cyst | 3. Four nuclei in a mature cyst |
| 4. No glycogen and chromatoid bodies in cyst | 4. Glycogen and chromatoid bodies in cyst |
| 5. Mitochondria present | 5. Mitochondria absent |

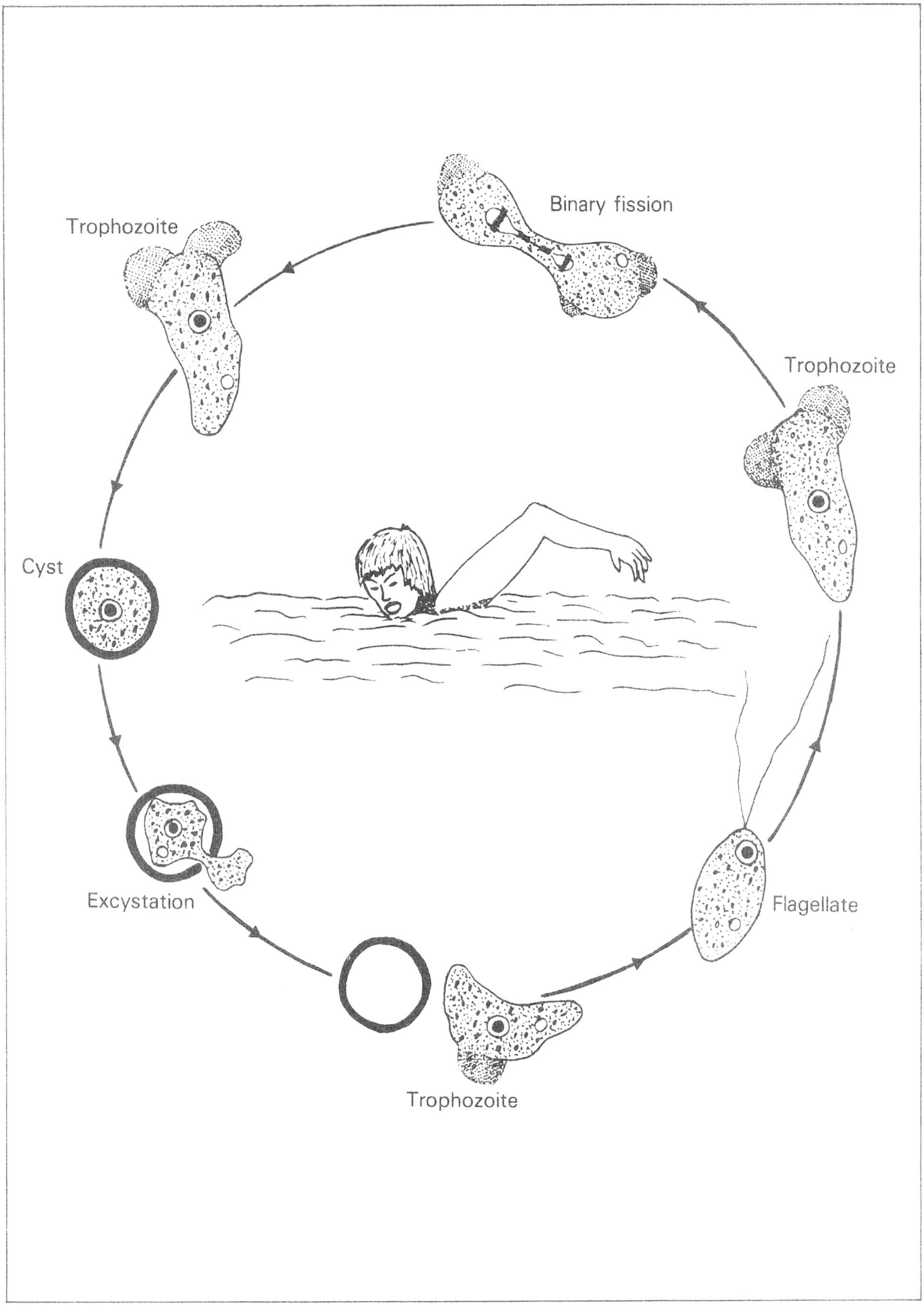
Binary fission
Trophozoite
Trophozoite
Cyst
Flagellate
Excystation
Trophozoite

37

*Table II.* A comparison of *Naegleria* morphology with *Acanthamoeba*

| *Naegleria* | *Acanthamoeba* |
|---|---|
| 1. Trophozoite displays broad pseudopods | 1. Trophozoite displays filamentous pseudopods (acanthopodia) |
| 2. Actively motile | 2. Sluggishly motile |
| 3. Form flagellate stage | 3. Does not form flagellate stage |
| 4. Thin walled cysts | 4. Double walled cysts |
| 5. Cyst wall has no pores | 5. Cyst wall may have pores or osteioles |
| 6. Does not encyst in tissues | 6. May encyst in tissues |

## *Life cycle of Acanthamoeba* [38]

The amoebae multiply in soil and water and humans are only accidentally infected. The thick walled cysts are able to withstand dessication and may be dispersed in the environment by wind.

## Reference

Carter, R.F.: Primary amoebic meningo-encephalitis. An appraisal of present knowledge. Transactions of the Royal Society of Tropical Medicine and Hygiene 66: 193-213 (1972).

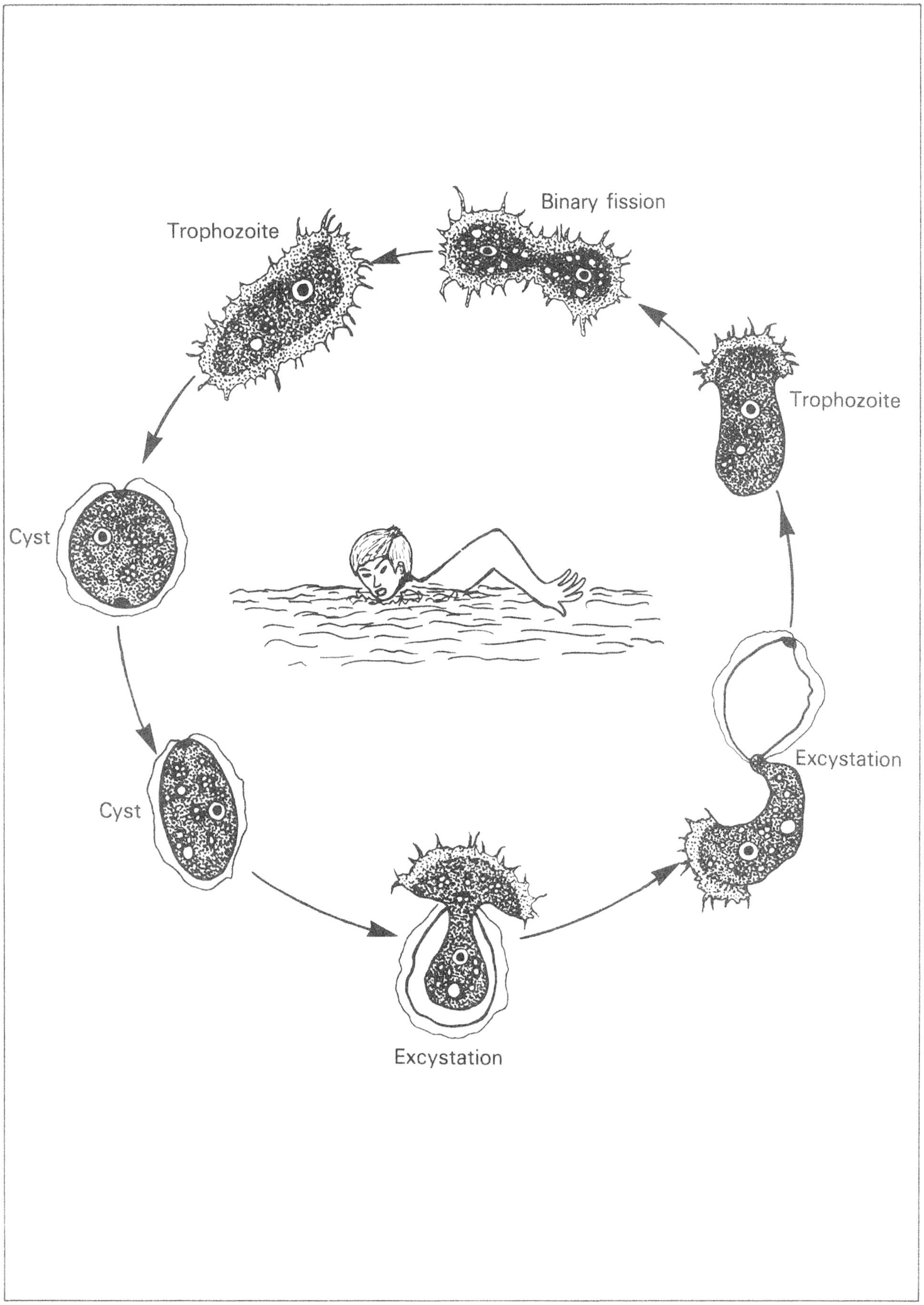
Binary fission
Trophozoite
Trophozoite
Cyst
Cyst
Excystation
Excystation

38

**39 Naegleria fowleri.**
**a) Trophozoite** — Moves actively with broad pseudopods. During movement the division between ectoplasm and endoplasm is sharp. The nucleolus is large and can be seen in the centre. Dark field. X1000.
**b) Cyst** — Has thin wall. A single nucleus with a large nucleolus can be seen in the centre of some cysts. Phase contrast. X1000.

**40 Acanthamoeba castallani.**
**a) Trophozoite** — Moves sluggishly with filamentous pseudopods. The nucleolus is large and can be seen in the centre. Phase contrast. X1000.
**b) Cyst** — Is characterised by the presence of double walls. It has a single nucleus which is not visible in this preparation. Phase contrast. X1000.

a

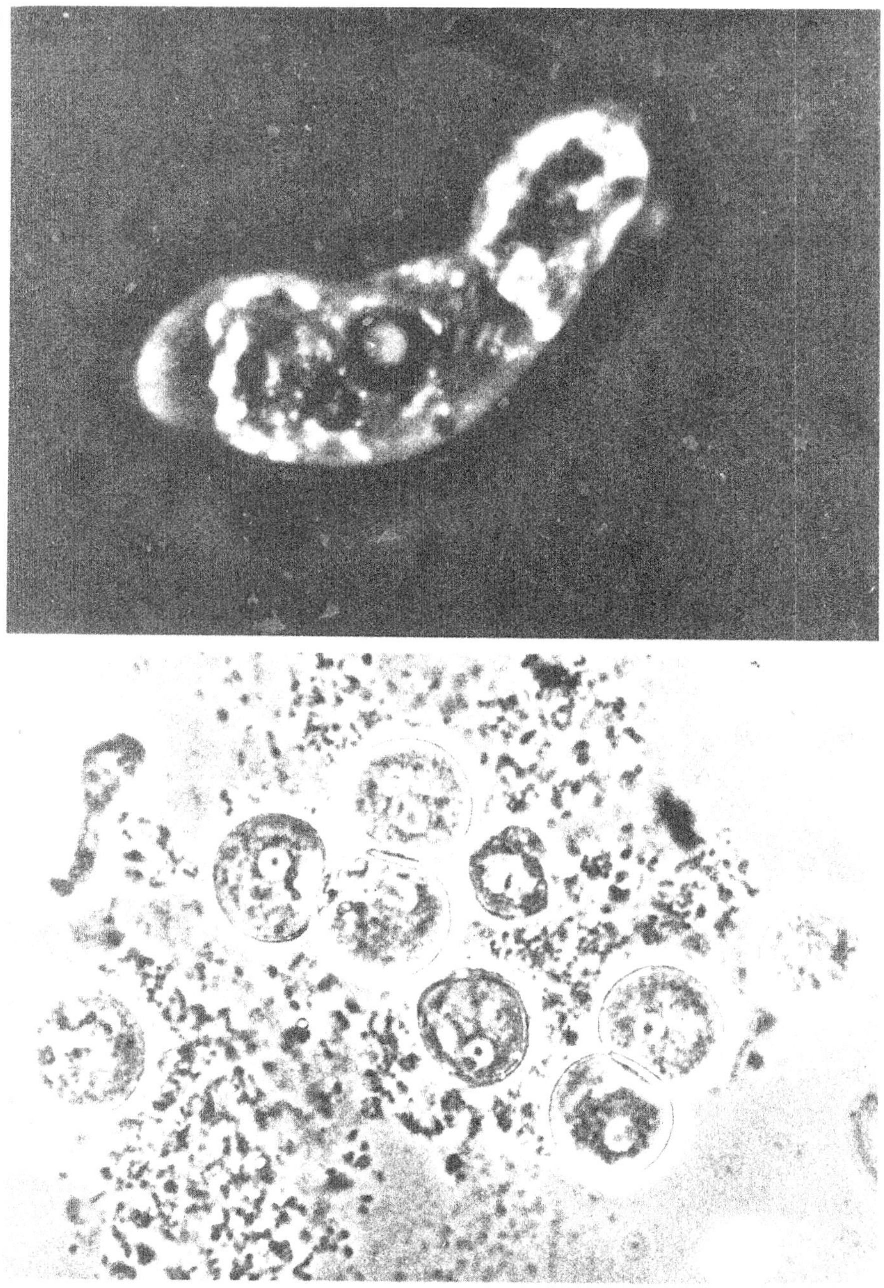

b

39

a

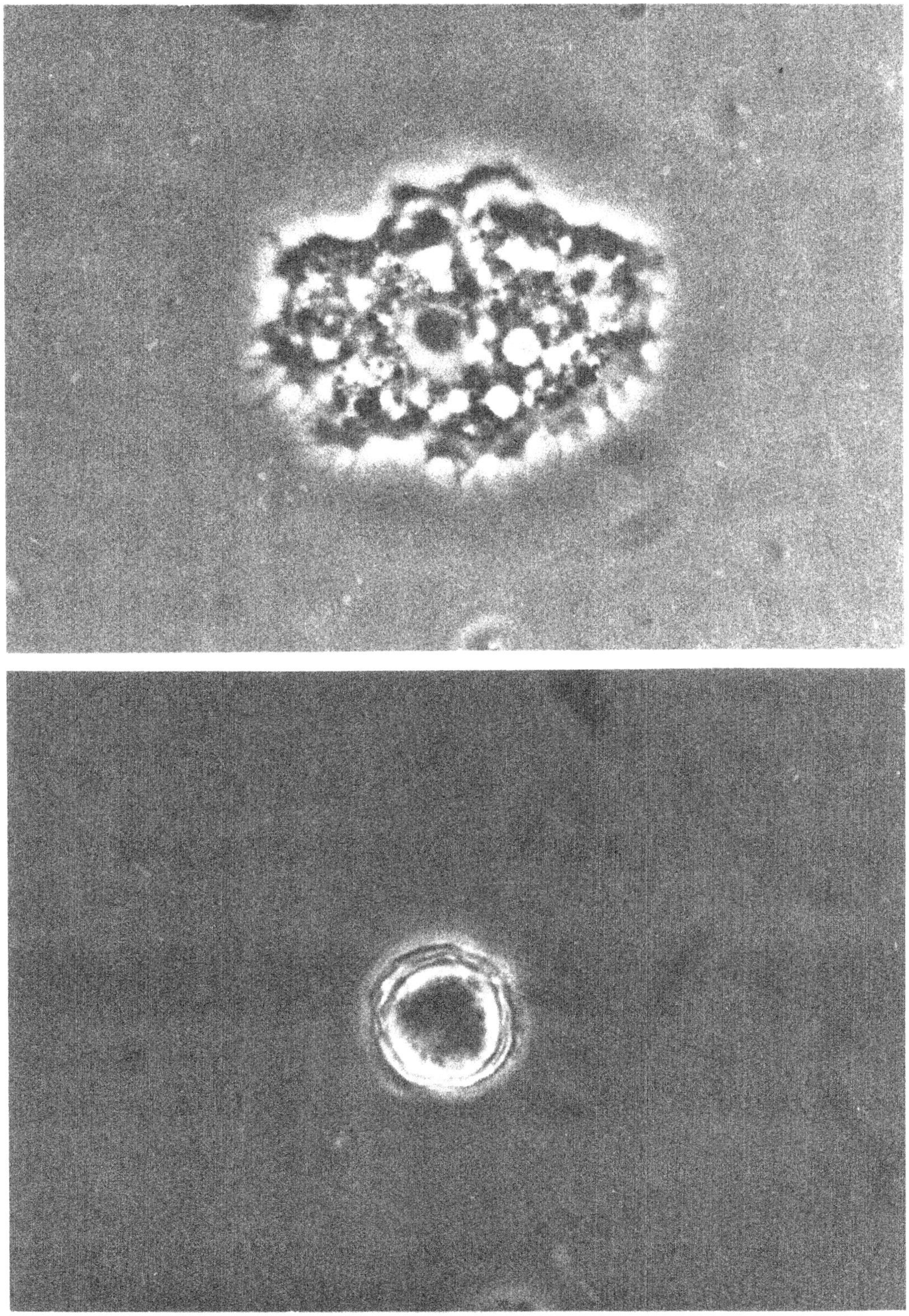

b

40

**41 Naegleria trophozoites in brain.** The nucleoli appear as darkly stained dots in the cytoplasm of the trophozoites. H and E. X800. Enlarged by 5.4.

**42 Naegleria infection** is generally associated with haemorrhage in the brain. In the centre are trophozoites surrounded with such haemorrhagic areas. H and E. X400. Enlarged by 5.4.

**43 Naegleria trophozoites in brain** showing ingested red cells. H and E. X1000. Enlarged by 5.4.

**44 Acanthamoeba trophozoites stained with acridine orange.** The large nucleolus appears as a greenish structure surrounded by a halo. The trophozoite in the centre has formed 2 nuclei, prior to cytoplasmic division. X800. Enlarged by 9.6.

**45 Acanthamoeba cyst** lying in monkey kidney tissue culture. This amoeba encysts in animal tissues as well. X800. Enlarged by 5.4.

41
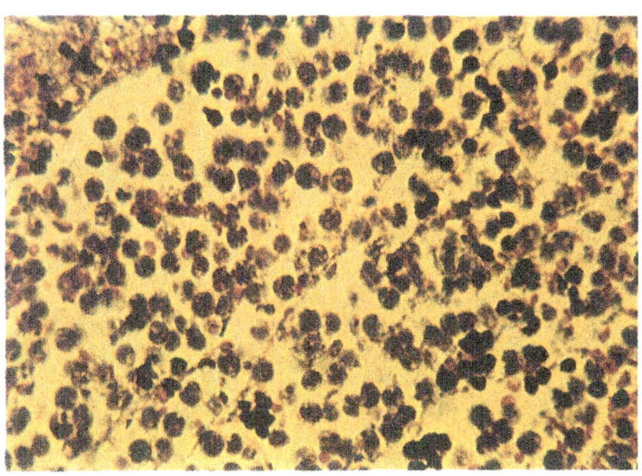

42
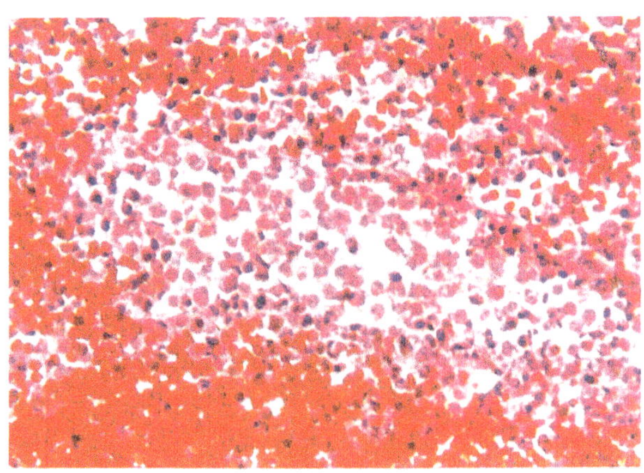

43 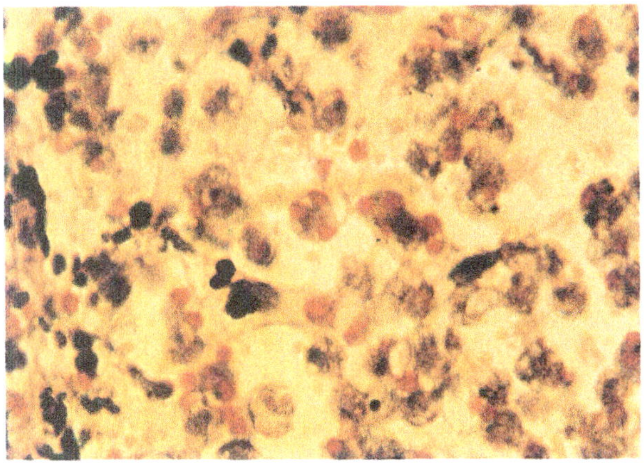

44 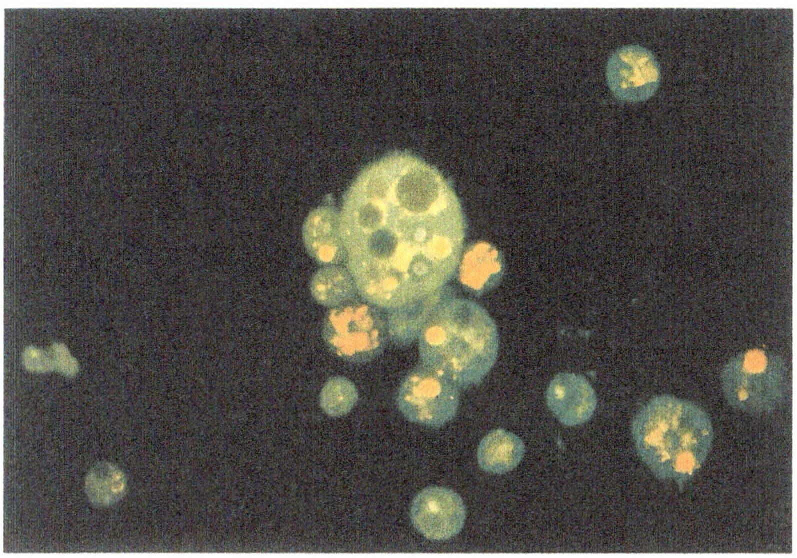

45 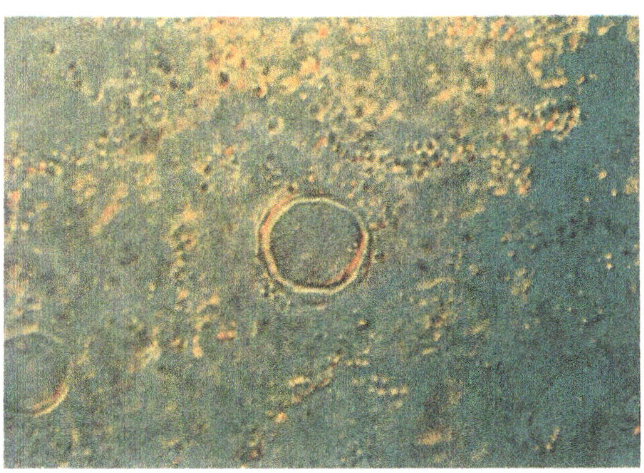

# Section 3
# Intestinal and Genital Flagellates

## Terms Used in Relation to Flagellates

**Kinetoplast** — Is an oval or rod shaped body seen in haemoflagellates. It stains with nuclear dyes and contains DNA. It is regarded as a modified part of the mitochondrium.

**Flagellum** — Is an elongated hair-like organelle used for locomotion. At ultrastructure level it reveals one pair of central tubules and nine pairs of peripheral tubules.

**Pleomorphic** — When a number of morphological types occur in 1 life cycle.

**Monomorphic** — When only a single morphological type occurs in 1 life cycle.

**Metacyclic trypanosome** — Infective forms of trypanosomes which develop in the vector.

**Xenodiagnosis** — A method of diagnosis in which a vector is fed on a suspected case and later examined for the presence of the parasite.

**Haemocoele** — The body cavity of arthropods.

**Peritrophic membrane** — A membrane which is secreted from the anterior end of the midgut in some blood-feeding arthropods. This membrane encloses the blood meal.

**Stercorian trypanosome** — Infective forms which develop in the faeces of the insect vector and enter the vertebrate host by the contamination of the bite area. This is also known as the posterior station development.

**Salivarian trypanosome** — Infective forms which develop in the mouth parts or salivary glands and enter the vertebrate host by inoculation during the act of biting. This is also known as the anterior station development.

**Undulating membrane** — Is a membranous structure which connects the flagellum to the body of the parasite. It is thrown into folds as the parasite moves giving it an undulating appearance.

**Costa** — A cytoplasmic thickening seen at the base of the undulating membrane in some flagellates.

**Axostyle** — A central supporting rod seen in some flagellates.

**Amastigote** — Also known as the 'leishmanial' stage. It is round or oval in shape without any free flagellum.

**Promastigote** — Also known as the 'leptomonad' stage. It is elongated with the kinetoplast anterior and distal to the nucleus. The flagellum emerges from the anterior end. There is no undulating membrane.

**Epimastigote** — Also known as the 'crithidial' stage. It is elongated with the kinetoplast anterior and close to the nucleus. There is a short undulating membrane.

**Trypomastigote** — Also known as the 'trypanosome' stage. It is elongated with the kinetoplast posterior and distal to the nucleus. There is a long undulating membrane.

**Axoneme** — A delicate filament extending from the region of the kinetoplast to the cell membrane. It represents the cytoplasmic part of the flagellum.

# The Intestinal and Genital Flagellates

The following species are found in humans:
*Enteromonas hominis*
*Retortamonas intestinalis*
*Chilomastix mesnili*
*Giardia lamblia*
*Trichomonas vaginalis*
*Trichomonas tenax*
*Trichomonas* (Pentatrichomonas) *hominis*

*Enteromonas hominis* and *Retortamonas hominis* are non-pathogenic and infrequently seen. The remaining parasites are frequently seen and out of these only *Giardia lamblia* and *Trichomonas vaginalis* are pathogenic.

*Chilomastix mesnili* — Is a non-pathogenic organism dwelling in the large intestine. It has 2 stages in its life cycle. The trophozoite has a long cytostome which begins from the broader end of the parasite and extends to the middle of the body. There are 3 anterior flagellae and 1 cytosomal flagellum. The cysts containing a single nucleus are more often seen than the trophozoites. They have a characteristic lemon shaped appearance.

## Giardia lamblia

*Giardia lamblia* — Is located in the small intestine and forms a trophozoite and a cystic stage. The trophozoite is symmetrical in shape with 2 oval nuclei and 4 pairs of flagellae. During movement the parasite tends to roll on itself displaying what is known as the 'falling leaf' movement. Cysts of *Giardia lamblia* have a thick wall and contain 4 small nuclei. The cytoplasm is generally shrunk away from the cyst wall leaving a space between the 2. Examination of the ultrastructure reveals a striated disc on its ventral surface. The peripheral part of the striated disc appears to be the grasping organ which lies over the villi of the small intestine (Friend, 1966). Both the trophozoites and cysts reveal a row of vacuoles at the periphery, when observed in the electron microscope.

*Table I.* Comparative morphology of intestinal and genital flagellates

| | *Enteromonas hominis* | *Retortamonas intestinalis* | *Chilomastix mesnili* | *Giardia lamblia* | *Trichomonas vaginalis* | *Trichomonas tenax* | *Trichomonas hominis* |
|---|---|---|---|---|---|---|---|
| *Trophozoite* | | | | | | | |
| Shape | Ovoid | Ovoid | Pyriform | Pyriform | Pyriform or Ovoid | Pyriform | Pyriform |
| Mean length (μm) | 8 | 6 | 14 | 14 | 18 | 12 | 12 |
| Flagellae | 4 | 2 | 4 | 8 | 4 | 4 | 5 |
| Undulating membrane | None | None | None | None | Present | Present | Present |
| Axostyle | None | None | None | None | Present | Present | Present |
| Cytostome | None | None | Present | None | Present | Present | Present |
| Nuclei | 1 | 1 | 1 | 2 | 1 | 1 | 1 |
| *Cyst* | | | | | | | |
| Shape | Oval | Oval | Lemon shaped | Ellipsoid | — | — | — |
| Mean length (μm) | 7 | 5 | 10 | 11 | — | — | — |
| Nuclei | 1 to 4 | 1 | 1 | 4 | — | — | — |

**46** a) **Chilomastix mesnili**. Morphology after iron-haemotoxylin staining.
b) **Giardia lamblia.** Morphology after iron-haemotoxylin staining.

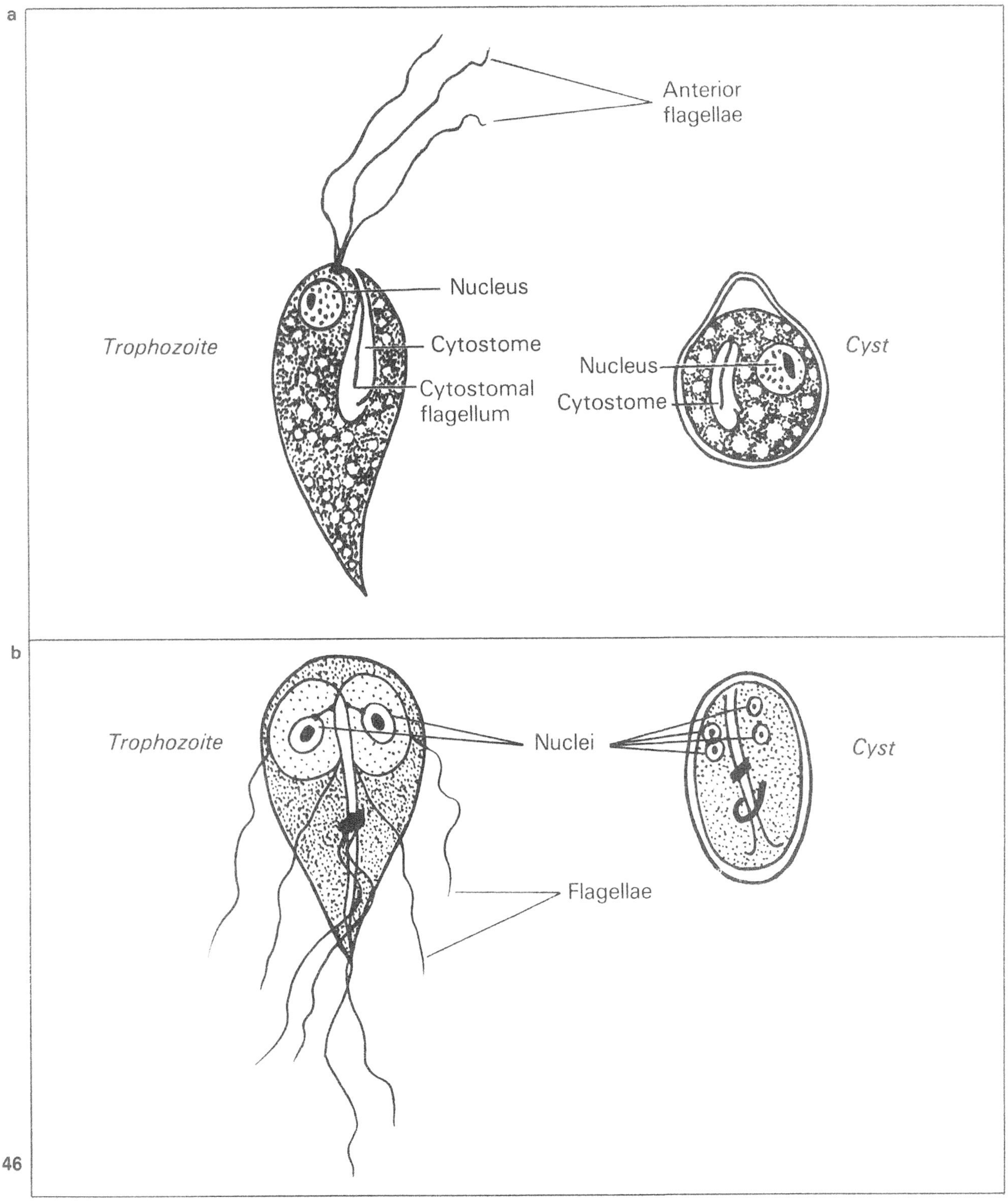


46

### *Clinical aspects*

The parasite may produce a wide range of gastrointestinal symptoms especially in children. These include vomiting, flatulence and diarrhoea. The parasite has been implicated in the intestinal malabsorption syndrome and cholecystitis.

### *Diagnosis*

Diagnosis is made by the examination of stools for trophozoites or cysts or by duodenal aspiration. Stool examination is less reliable than duodenal aspiration.

## Trichomonas sp and Trichomoniasis

*Trichomonas* (Pentatrichomonas) *hominis* in the large intestine and *Trichomonas tenax* in the mouth are of no medical significance. *Trichomonas vaginalis,* on the other hand, produces urogenital pathology and is the most widely prevalent venereal infection. The parasite has no cystic stage and infection is normally transmitted by an infected male who acts as a carrier.

The trophozoite is actively motile and moves by a gliding motion which is quite different from *Giardia.* The body may be ovoid or pyriform. From the anterior end 4 flagellae arise. A fifth flagellum turns back and is attached to the body by an undulating membrane. At the posterior end the axostyle is seen to project out of the body. It is believed that the axostyle is used for attachment to host tissues and may be responsible for irritation produced by the parasite.

*T. vaginalis* has a single nucleus and divides by binary fission. It ingests food particles through a cytostome found at the anterior end.

### *Clinical aspects*

In the female, trichomoniasis can give rise to vaginitis resulting in a frothy and creamy white discharge. The vulva and cervix may be inflamed. The majority of males remain asymptomatic, but in some, urethritis and prostatitis may occur.

### *Diagnosis*

Diagnosis is easily made by observing the parasite in the vaginal secretion or in the case of the male, in the semen (Hoffman et al., 1961). The parasite can also be grown in artificial media.

### *Life cycle of Trichomonas vaginalis* [47]

In the majority of cases the parasite is transmitted during sexual intercourse, the male acting as a carrier. Cysts are not formed during the life cycle.

### References

Friend, D.S.: The fine structure of *Giardia muris.* Journal of Cell Biology 29: 317 (1966).

Hoffman, B.; Kilczewski, W. and Malyszko, E.: Studies on trichomoniasis. British Journal of Venereal Disease 37: 172 (1961).

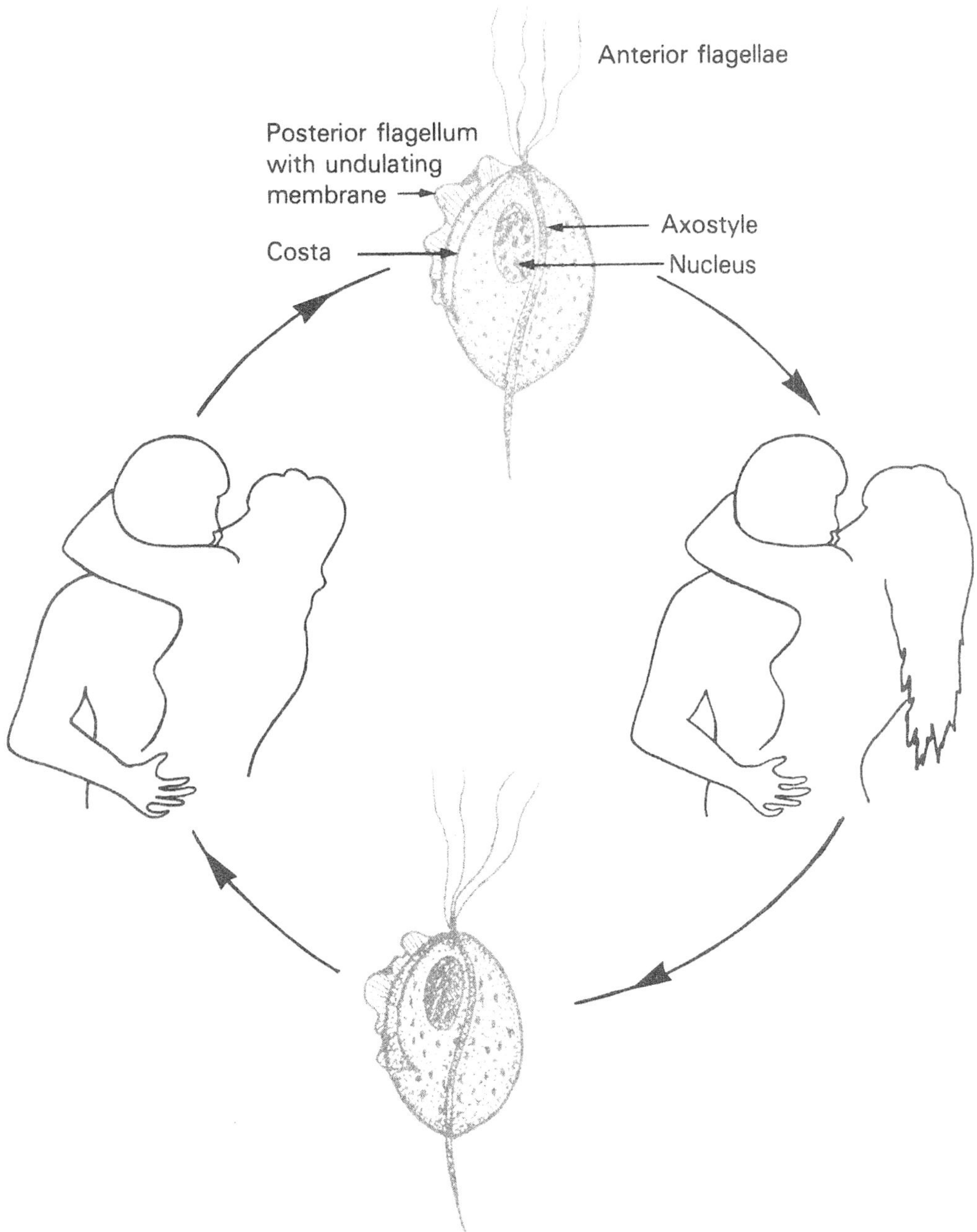
Anterior flagellae
Posterior flagellum with undulating membrane
Axostyle
Costa
Nucleus

47

**48 Cyst of Giardia lamblia.** The cyst wall consists of a thick outer layer and an inner layer made up of a double membrane. The cytoplasm is shrunk away from the cyst wall. X60,000. Electron micrograph.

**49 Trophozoite of Giardia muris lying near the intestinal tissue.** The parasite generally detaches from the mucous membrane during fixation. X40,000. Electron micrograph.

**50 Diagram showing the method of attachment of Giardia to the mucous membrane.** The central striated disc keeps the parasite in an arched position. The peripheral part is used for attachment.

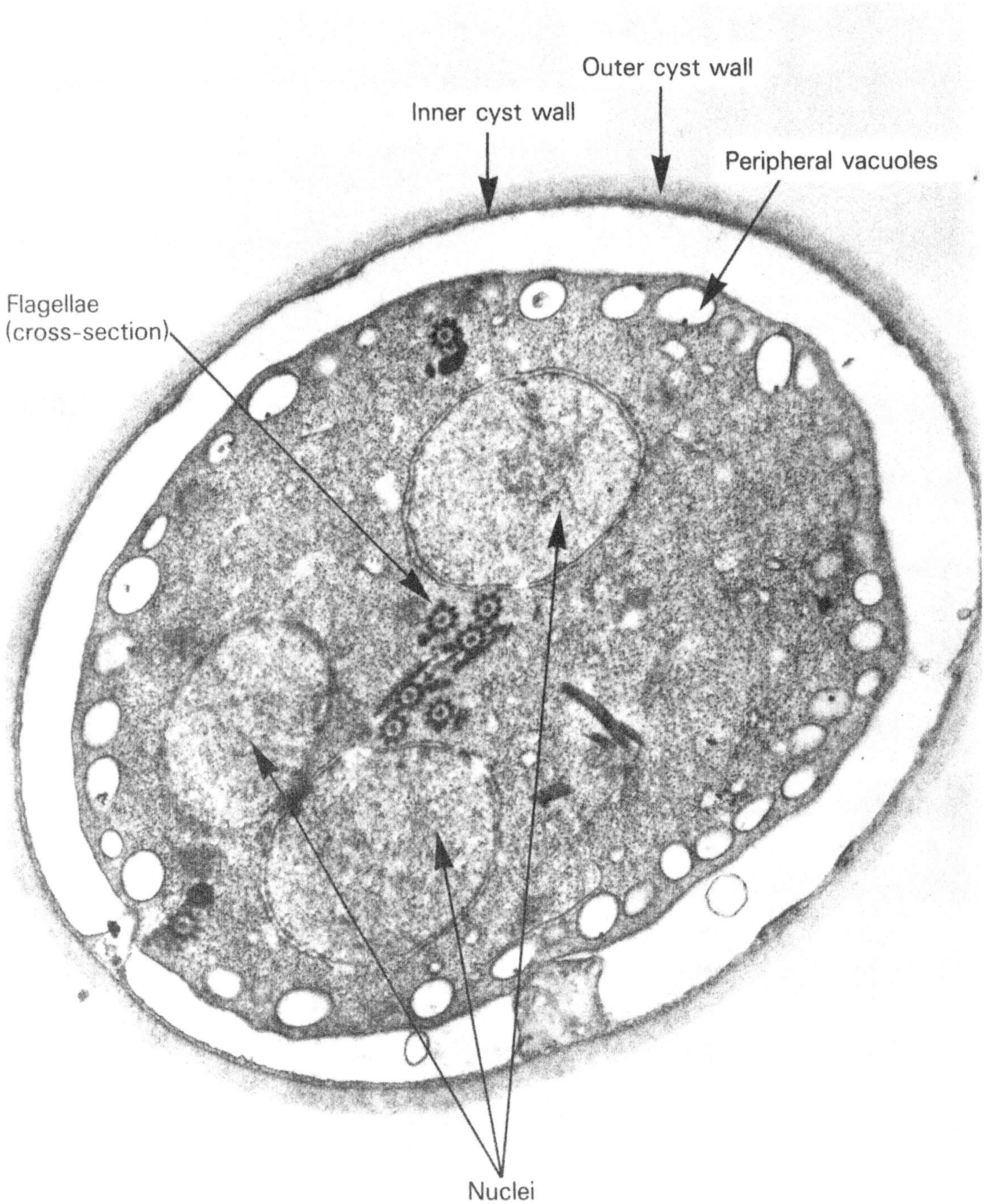


48

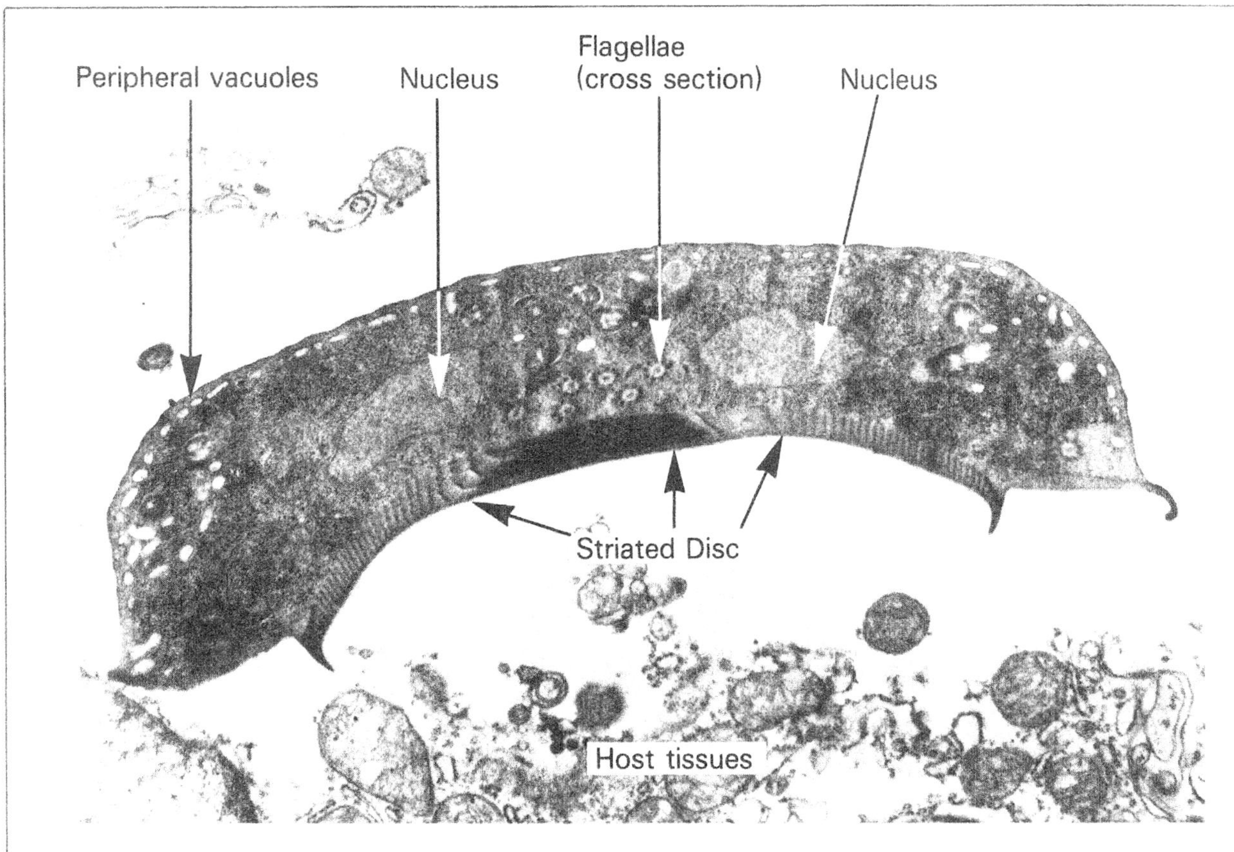
Peripheral vacuoles
Nucleus
Flagellae
(cross section)
Nucleus
Striated Disc
Host tissues

49

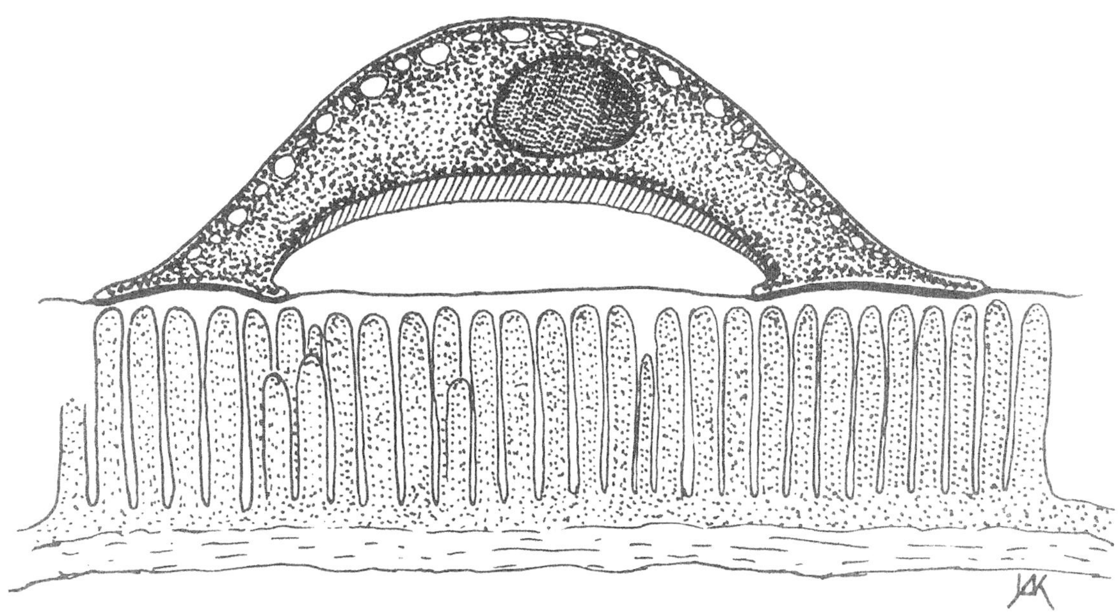
JAK

50

**51 Trichomonas vaginalis.** The axostyle projects from the posterior end. The crystalline structures in the cytoplasm are starch grains, ingested by the parasite from the medium. Interference contrast. X800. Enlarged by 23.4.

**52 Chilomastix mesnili cysts.** The cysts are lemon shaped. In 1 of them the nucleus is clearly visible. Iodine. X800. Enlarged by 9.6.

**53 Giardia trophozoite.** Ventral surface. The central disc (corresponding to the striated disc in the electron microscope) and the flagellae can be seen. Dark field. X1000. Enlarged by 5.4.

**54 Giardia trophozoite.** Two nuclei and the flagellae can be seen. In 1 of the nucleus, 2 nucleoli are seen indicating the beginning of binary fission. Iron-haematoxylin. X800. Enlarged by 5.4.

**55 Giardia cysts.** The cyst wall appears as a thin structure with the cytoplasm retracted from it. Nuclei are partially visible. Iron-haematoxylin. X1000. Enlarged by 5.4.

**56 Longitudinal section of villi.** Showing *Giardia* trophozoites lying in the crypts. Iron-haematoxylin. X800. Enlarged by 5.4.

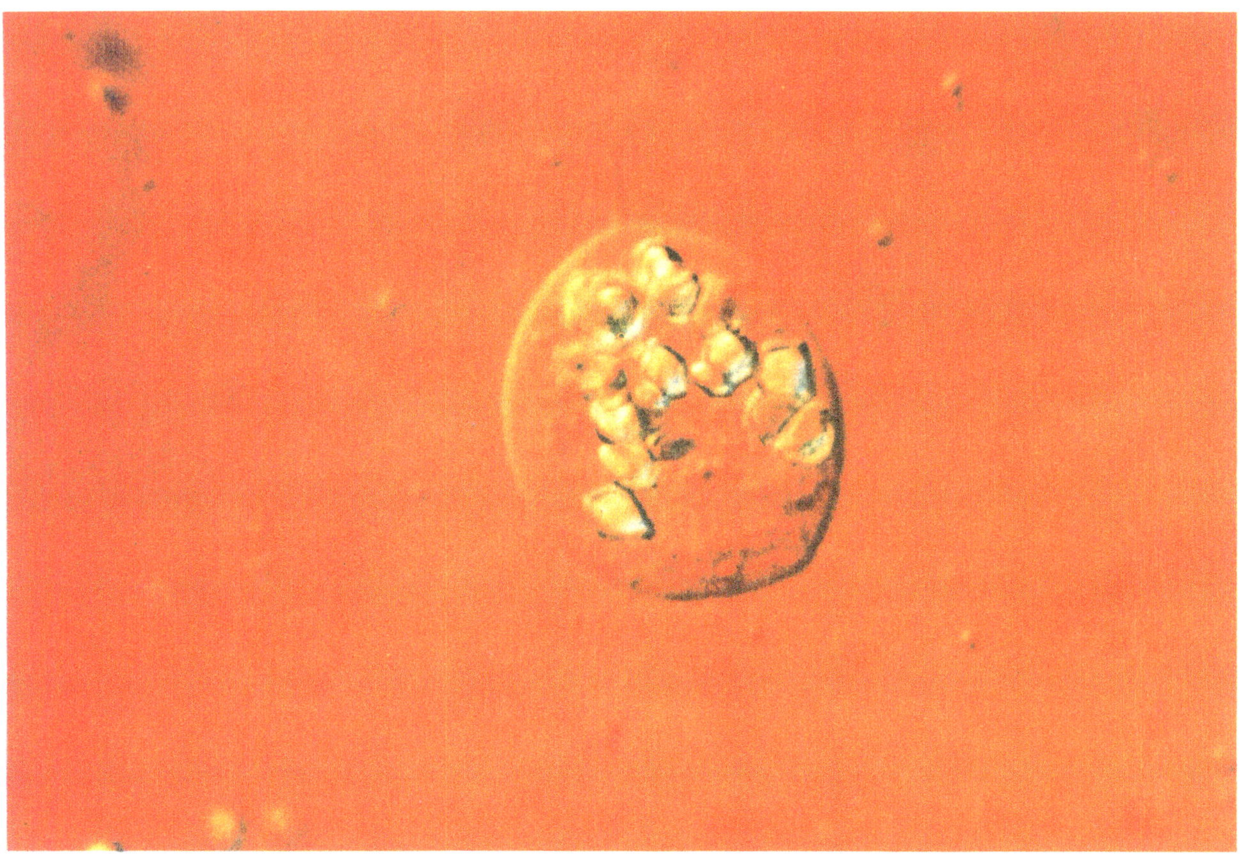

51

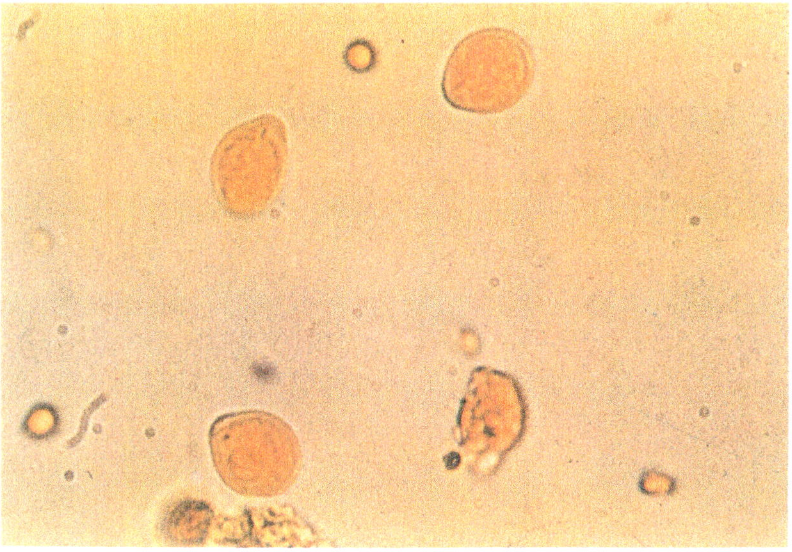
52

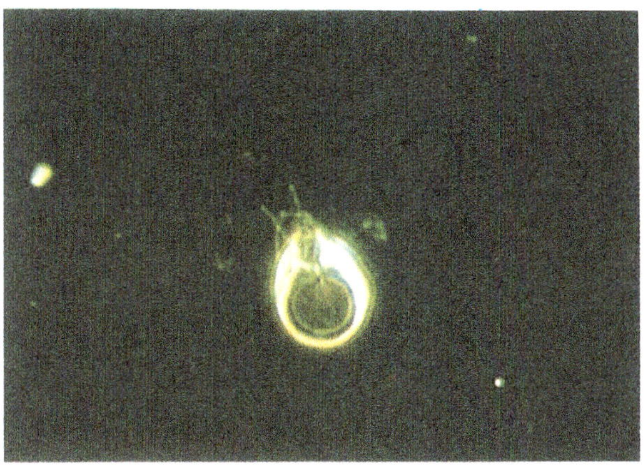
53

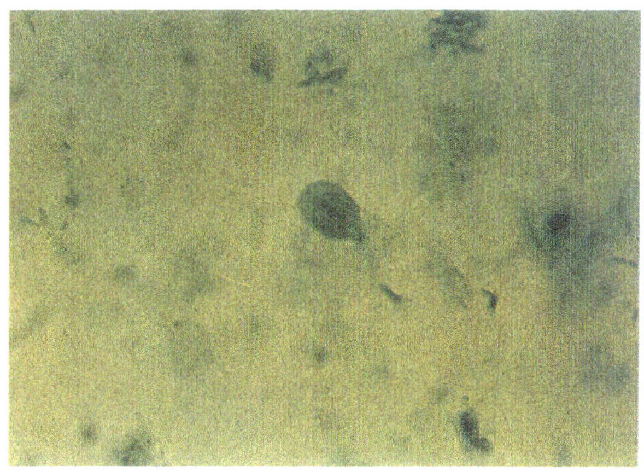
54

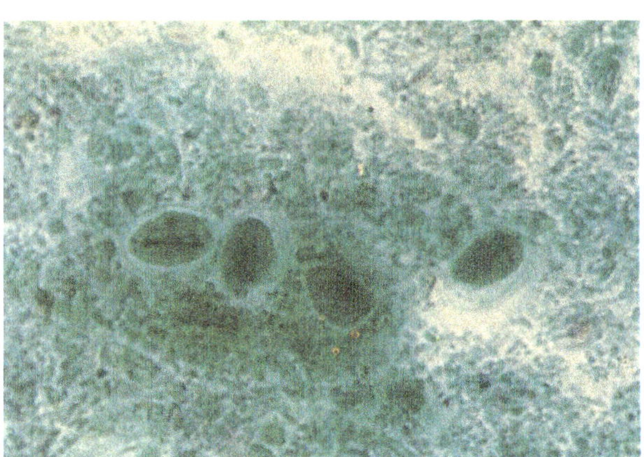
55

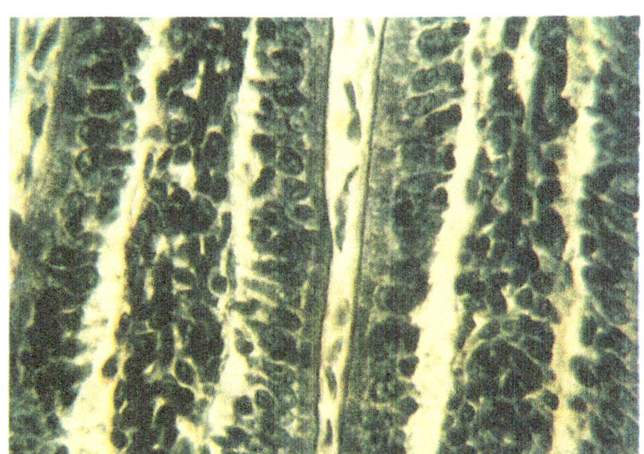
56

# Section 4 Haemoflagellates

Two genera are of medical importance:-
1) *Leishmania*
2)*Trypanosoma*

## Leishmania

Three main species are involved.

*L. donovani* causes visceral leishmaniasis. This is the most virulent form in the group. The epidemiology is complex and varies in different parts of the world (see page 47).

*L. braziliensis* causes muco-cutaneous leishmaniasis. This is of intermediate virulence. A number of subtypes are now recognized. These are:

a) Muco-cutaneous subtypes: Espundia subtype *(L. braziliensis braziliensis),* Guiana subtype *(L. b. guyanensis)*
b) Zoonotic, nodular subtypes: Mexican subtype *(L. b. mexicana),* Peruvian subtype *(L. b. peruviana)*

*L. tropica* causes cutaneous leishmaniasis. This is the least virulent form. Two types of lesion are encountered:

a) Wet-type lesion *(L. tropica major)*
b) Dry-type lesion *(L. tropica minor)*

*Table I.* Morphological forms in haemoflagellates. Their distribution in the genus *Leishmania* and *Trypanosoma* is as follows:

| | *Trypo-mastigote* | *Epimas-tigote* | *Promas-tigote* | *Amas-tigote* |
|---|---|---|---|---|
| *Leishmania* spp | | | | |
| Animal Host | — | — | — | + |
| Vector | — | — | + | — |
| *Trypanosoma* spp (African) | | | | |
| Animal Host | + | — | — | — |
| Vector | + | + | — | — |
| *Trypanosoma cruzi* | | | | |
| Animal Host | + | + | — | + |
| Vector | + | + | — | — |

Morphologically, the amastigote stages of various species of *Leishmania* are indistinguishable. Multiplication occurs by binary fission and starts by the division of the kinetoplast followed by the nucleus. After nuclear division the outer membrane gradually splits giving rise to 2 parasites. In the electron microscope a row of 130-200 microfibrils can be detected under the outer membrane. The flagellar vacuole is distinct and contains electron dense material which is a possible source of soluble antigens.

### *Life-cycle* [57]

The vectors are sandflies belonging to the genus *Phlebotomus.* Only the female sandflies are involved in the transmission. The promastigote forms of *Leishmania* are injected into the tissues by the sandfly. These enter the various phagocytic cells where they lose their flagellum and become amastigotes. The amastigotes multiply by binary fission inside the host cells. After some time they come out of the host cell and invade new cells. The infection is transmitted to the vector by the ingestion of host cells containing the parasites. Inside the vector the amastigotes rapidly change to promastigotes.

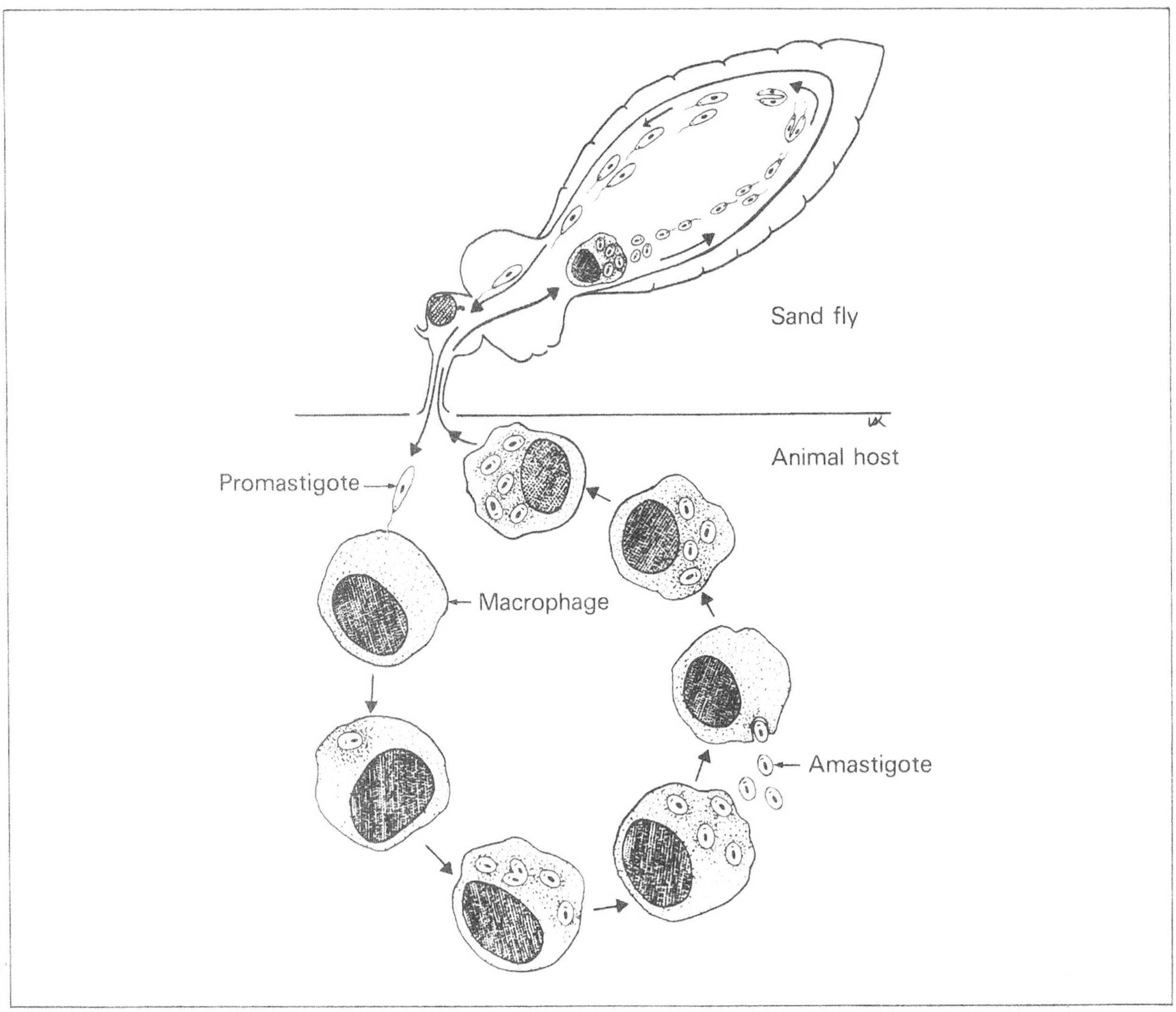


57

## Jungle Cycle of Visceral Leishmaniasis

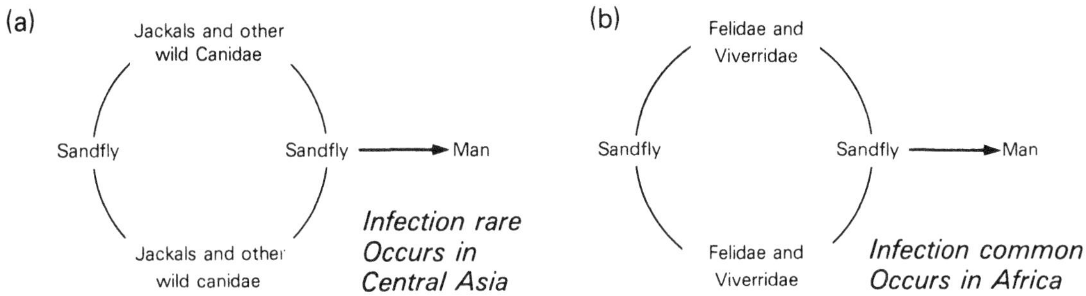


## Urban Cycle of Visceral Leishmaniasis

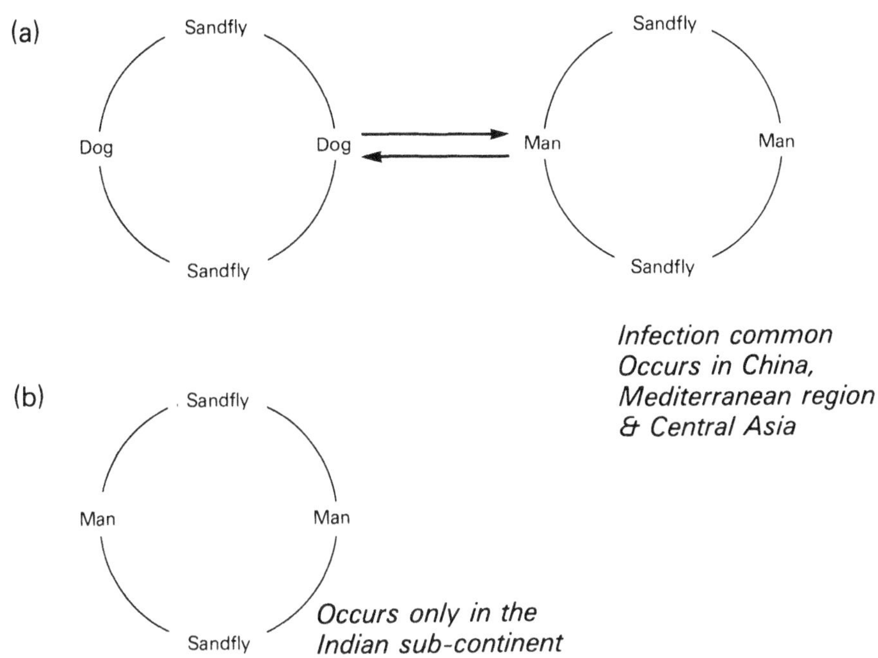


## Rural Cycle of Visceral Leishmaniasis

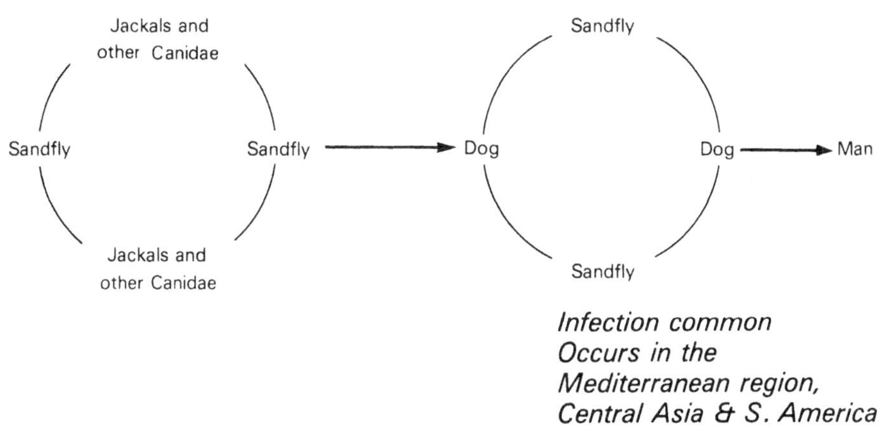

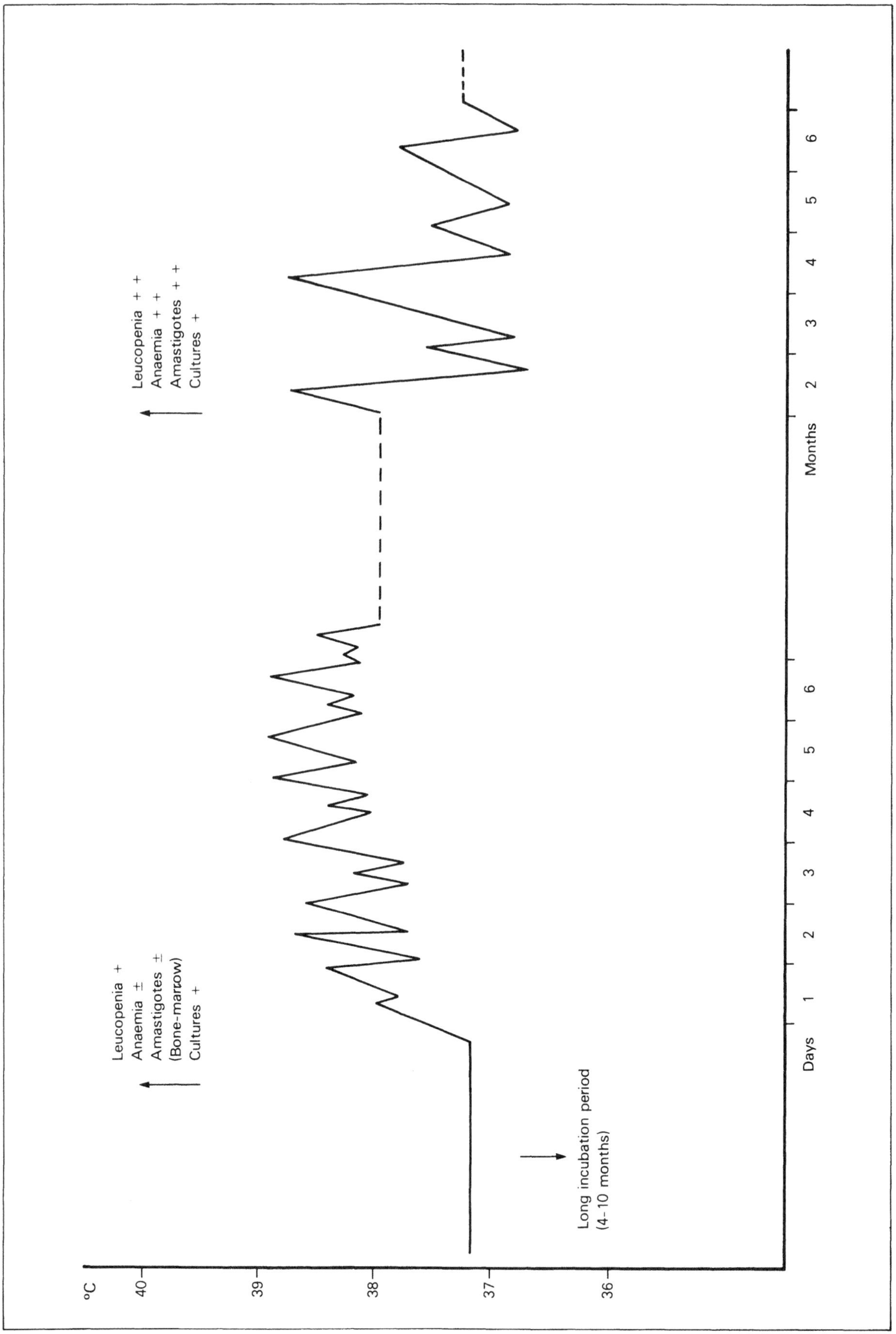


**Temperature Chart in Kala-azar.** During the first few weeks of disease patient often shows double peaks in 24 hours. As the disease becomes chronic fever pattern becomes intermittent with the peaks becoming lower than in the early stages.

## *Clinical aspects*

### Visceral Leishmaniasis

This condition is also known as Kala-azar. The incubation period is usually 4 to 10 months. The earliest symptom is generally low grade fever with malaise and sweating. In later stages fever becomes intermittent and both liver and spleen become grossly enlarged. The skin may show increased pigmentation usually over the dorsum of the hands and face. Post Kala-azar dermal leishmaniasis seen in India and Sudan generally occurs 1 or 2 years after the visceral disease.

### Muco-Cutaneous Leishmaniasis

This usually starts as a pustular swelling in the mouth or on the nostrils. The lesion may be ulcerative after many months and then extend into the naso-pharyngeal mucous membrane. Secondary infection is very common with destruction of the nasal cartilage and the facial bone.

### Cutaneous Leishmaniasis

It starts as a painless papule on exposed parts of the body, generally the face. The lesion ulcerates after a few months producing a circular or an oval ulcer with an indurated margin. The ulcer heals spontaneously after a few months leaving behind a scar. Diffuse cutaneous leishmaniasis (DCL) as seen in parts of Africa consists of nodules and a thickening of skin generally without any ulceration (Bryceson, 1970).

## *Diagnosis*

In visceral leishmaniasis (Kala-azar) blood changes occur early in the disease. These consist of a low white cell count and anaemia. These progressively increase as the disease becomes chronic. The gamma globulin levels are also raised. As a result of the rise in gamma globulins certain non-specific tests such as the formol-gel test becomes positive. This consists of placing a drop of commercial formaldehyde added to 1ml of patient's serum resulting in the formation of a coagulum. Specific antibodies to *Leishmania* can be demonstrated by the fluorescent antibody test and the indirect haemagglutination test (Bray and Lainson, 1967). Definite diagnosis is based on the demonstration of the parasite in biopsy tissues or by the positive cultures. The most useful tissue for biopsy is the bone marrow or the spleen.

In cutaneous and muco-cutaneous leishmaniasis the parasite can be isolated from the margin of the ulcers. A useful diagnostic test is the Montenegro test which consists of intradermal injection of 0.1ml of antigen prepared from cultures of promastigotes. In positive cases erythema and induration occurs in 48 hours. However, false positive reactions have been reported from Africa (Southgate and Manson-Bahr, 1967).

## References

Bray, R.S. and Lainson, R.: Studies of the immunology and serology of leishmaniasis V. The use of particles as vehicles in passive agglutination tests. Transactions of the Royal Society of Tropical Medicine and Hygiene 61: 490 (1967).

Bryceson, A.D.M.: Diffuse cutaneous leishmaniasis in Ethiopa. Transactions of the Royal Society of Tropical Medicine and Hygiene 64: 380 (1970).

Southgate, B.A. and Manson-Bahr, P.E.C.: Studies in the epidemiology of East African leishmaniasis 4. The significance of the positive leishmanin test. Journal of Tropical Medicine and Hygiene 70: 29 (1967).

**58 Leishmania tropica from a ruptured histiocyte.** Fresh specimen. Dark field. X1000.

**59 Leishmania tropica inside a polymorph.** Four parasites (1 of them dividing) are lying inside a phagocyte, in this case a polymorph. Electron micrograph. X36,000.

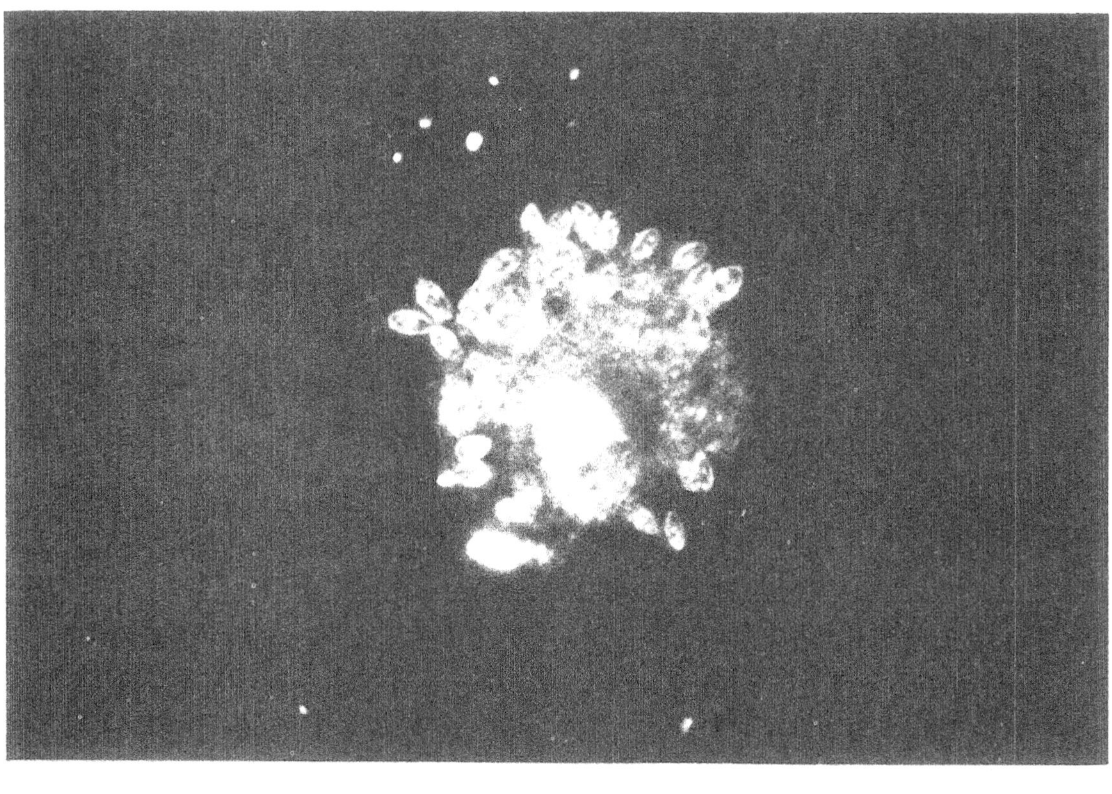

58

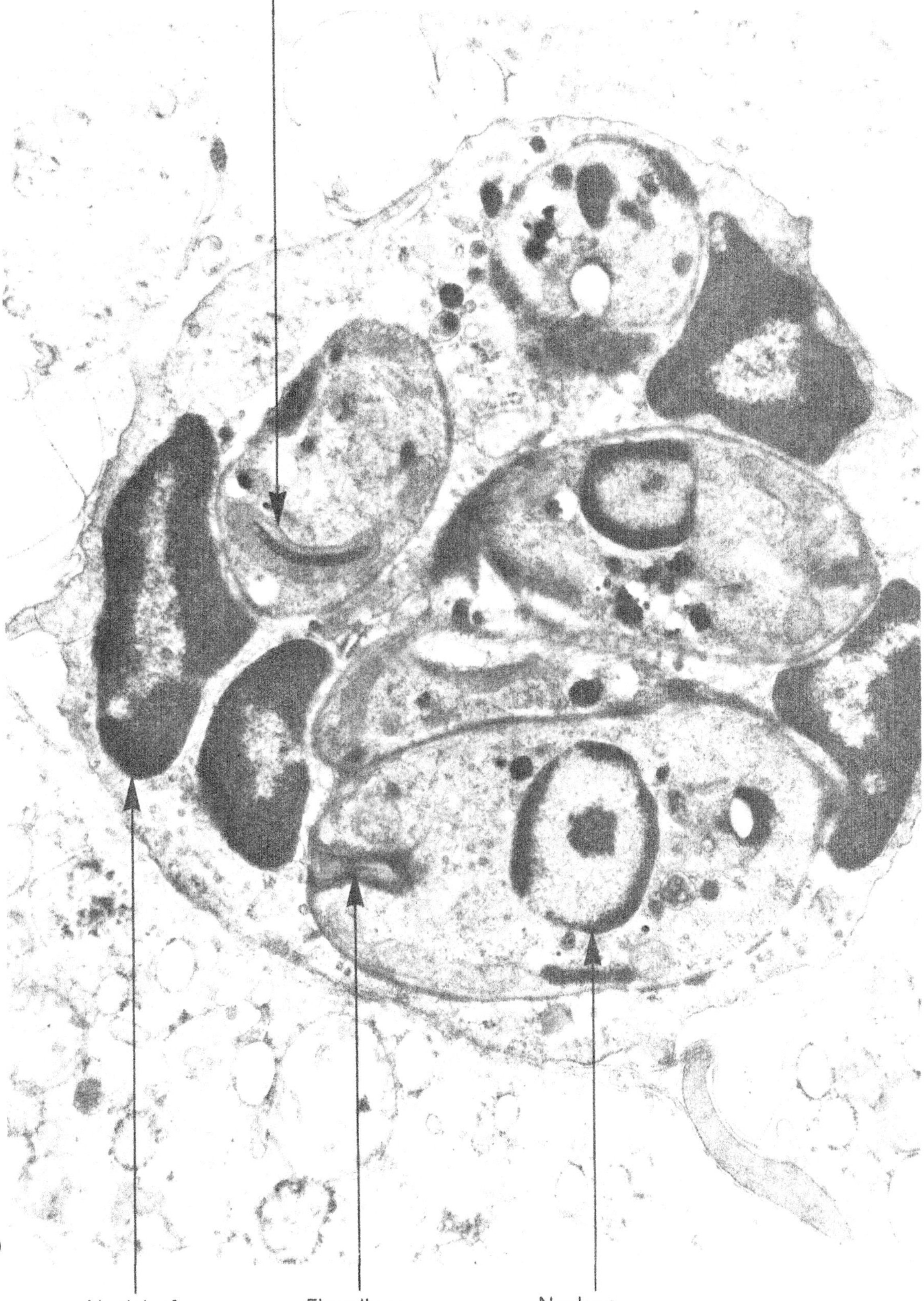


59

**60 Morphological forms in haemoflagellates. Trypomastigote** — Kinetoplast is at the extreme posterior end. **Epimastigote** — Kinetoplast is lying anterior to the nucleus. **Promastigote** — Kinetoplast is at the extreme anterior end. **Amastigote** — Free flagellum is absent and the cell has become ovoid.

**61 Leishmania undergoing binary fission.**
**a)** Shows the formation of two parasites, 1 of them shows a kinetoplast, a flagellar vacuole and flagellum clearly. The flagellar vacuole contains electron opaque material.
**b)** Shows the beginning of detachment of the 2 parasites.
**c)** Shows almost complete separation of the 2 parasites. Electron micrograph. X36,000.

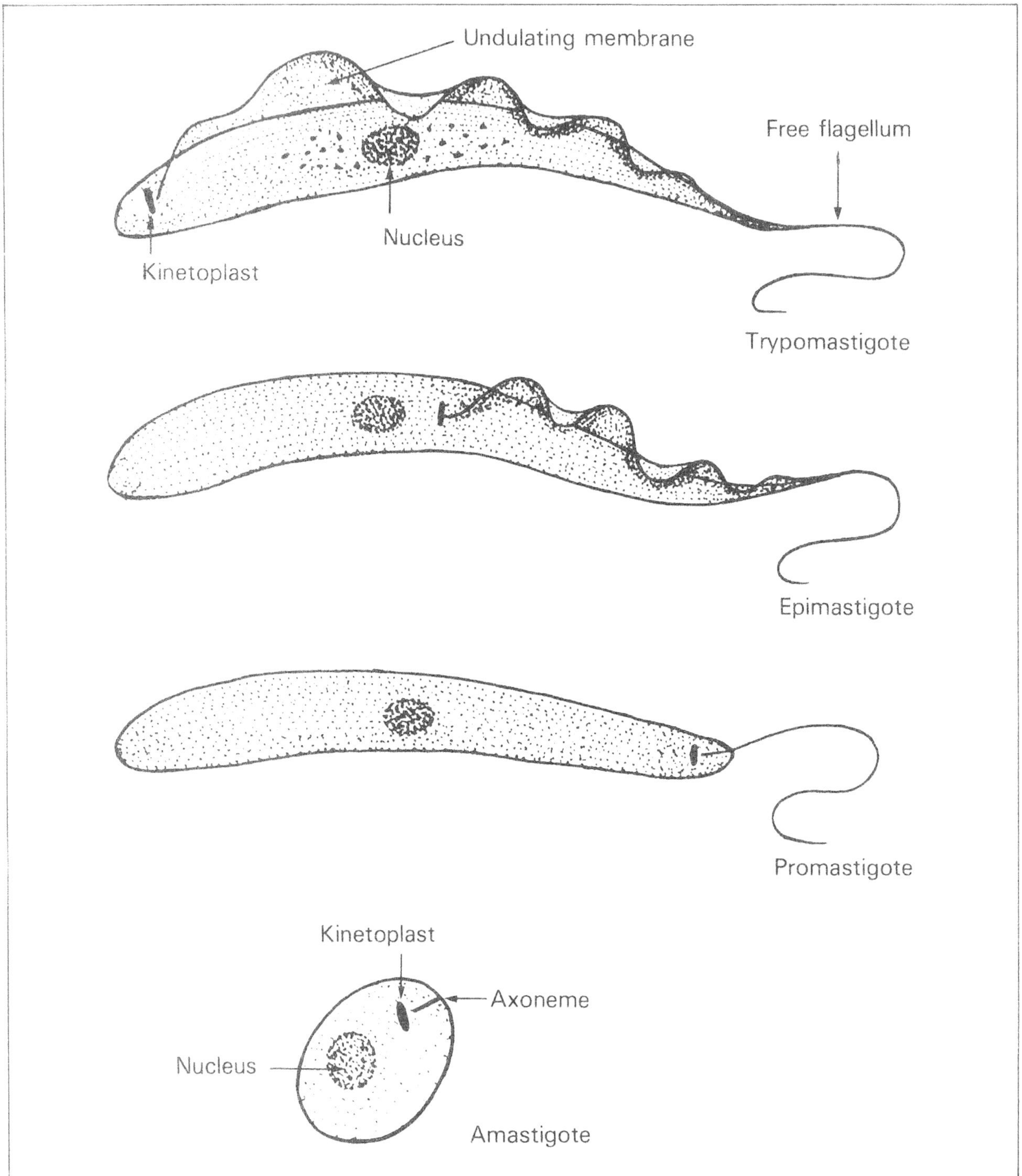


60

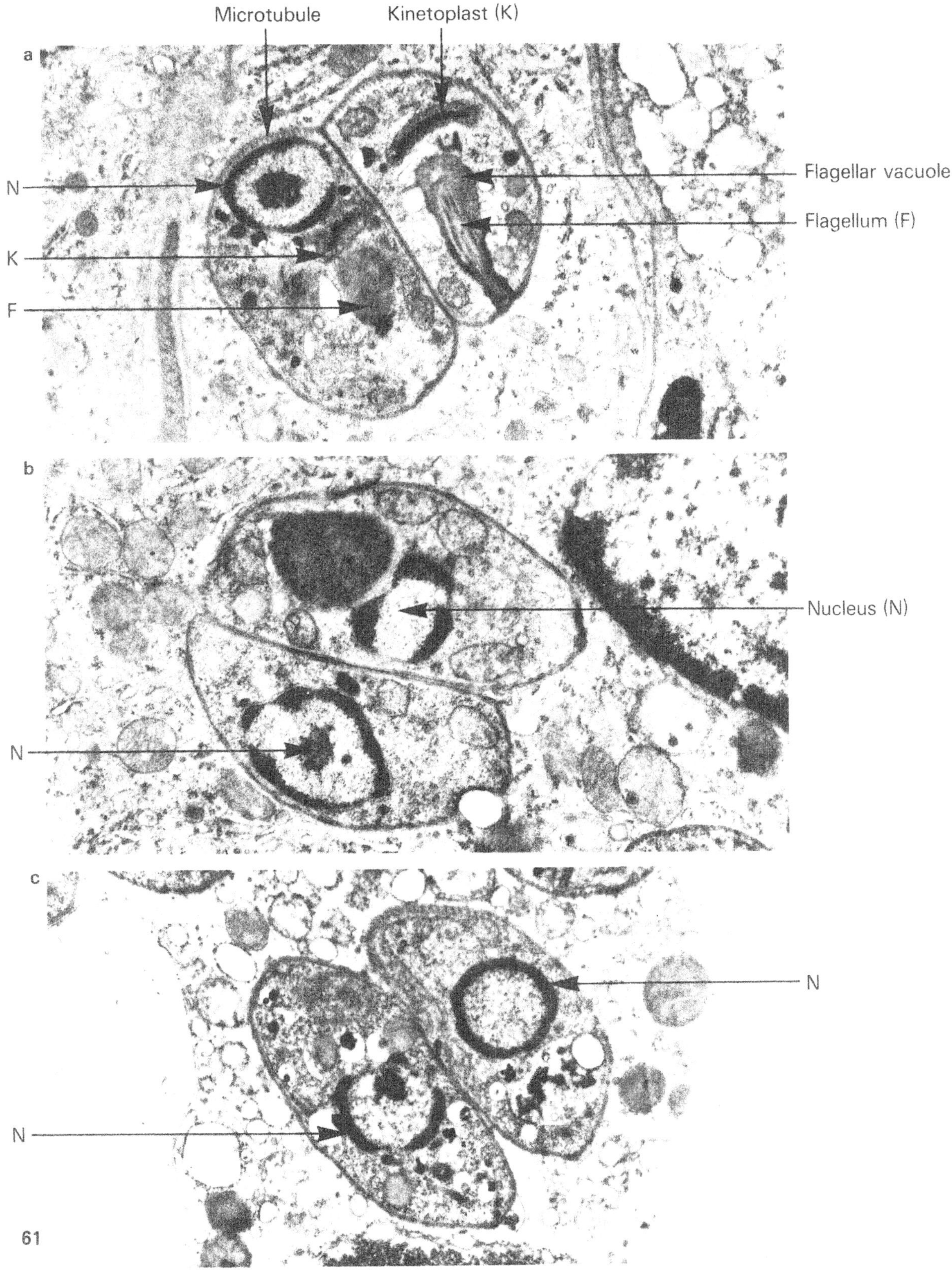
Microtubule
Kinetoplast (K)
a
N
K
F
Flagellar vacuole
Flagellum (F)
b
Nucleus (N)
N
c
N
N

61

**62 Sandfly belong to the genus** *Phlebotmus.* Note the presence of hairs on the body. The wing is surrounded by a fringe of hairs and the wing venation is parallel. Enlarged by 9.6.

**63 Leishmania (Leishman-Donovan or LD bodies).** Lying in macrophage cells from liver. Giemsa. X12000. Enlarged by 9.6.

**64 Leishmania tropica.** From ruptured histiocytes from skin. Giemsa. X1000. Enlarged by 9.6.

**65 Leishmania tropica cultures from NNN medium.** The promastigotes often form clumps and become tied together at the flagellar end. Anoptral phase. X400. Enlarged by 23.4.

62

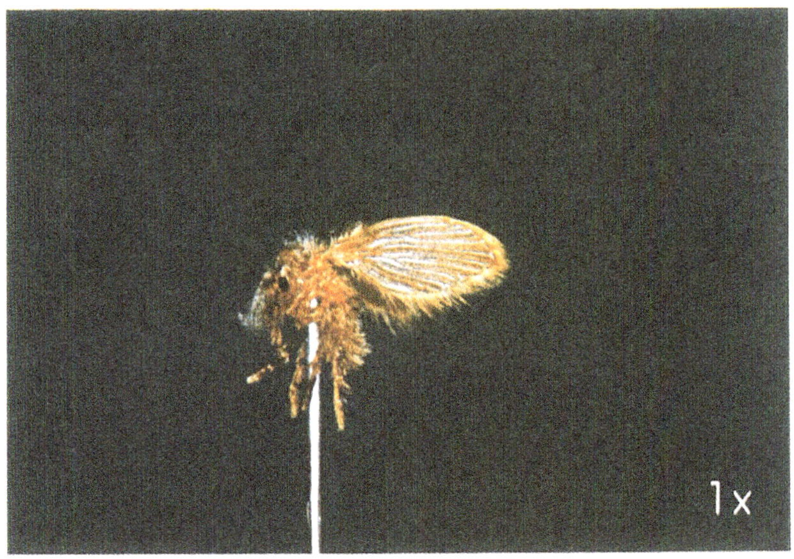


63

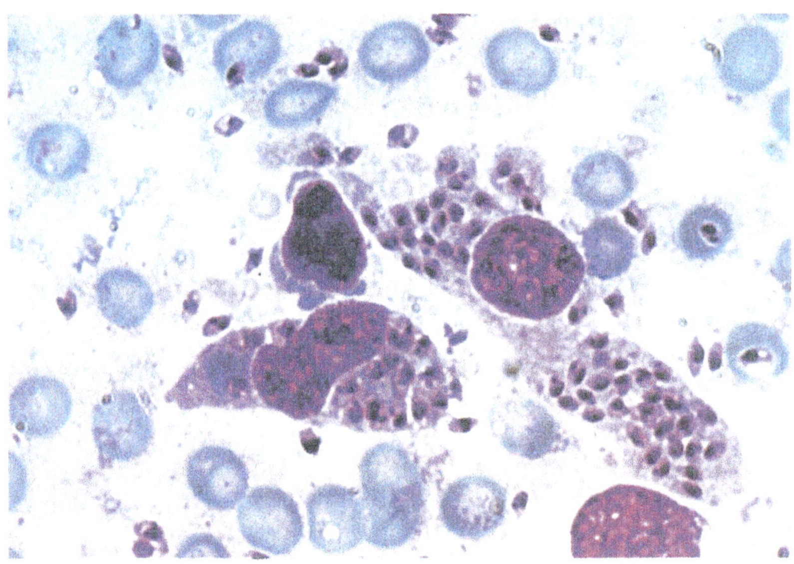

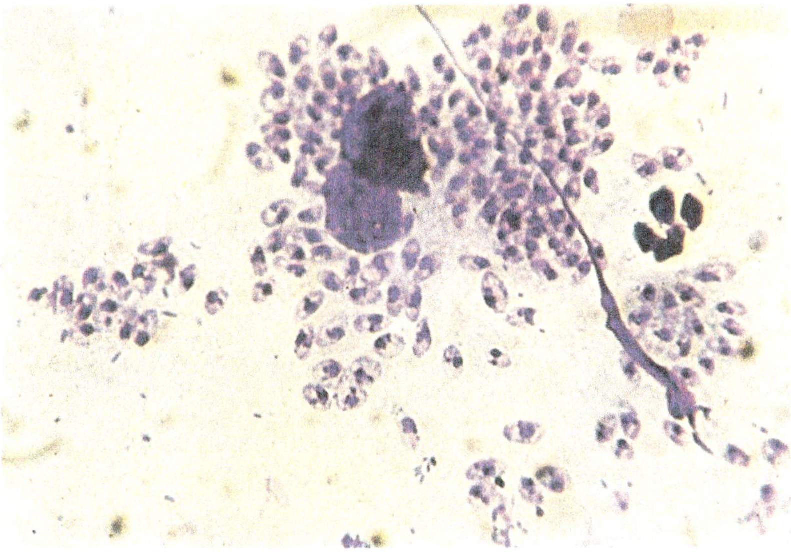

64

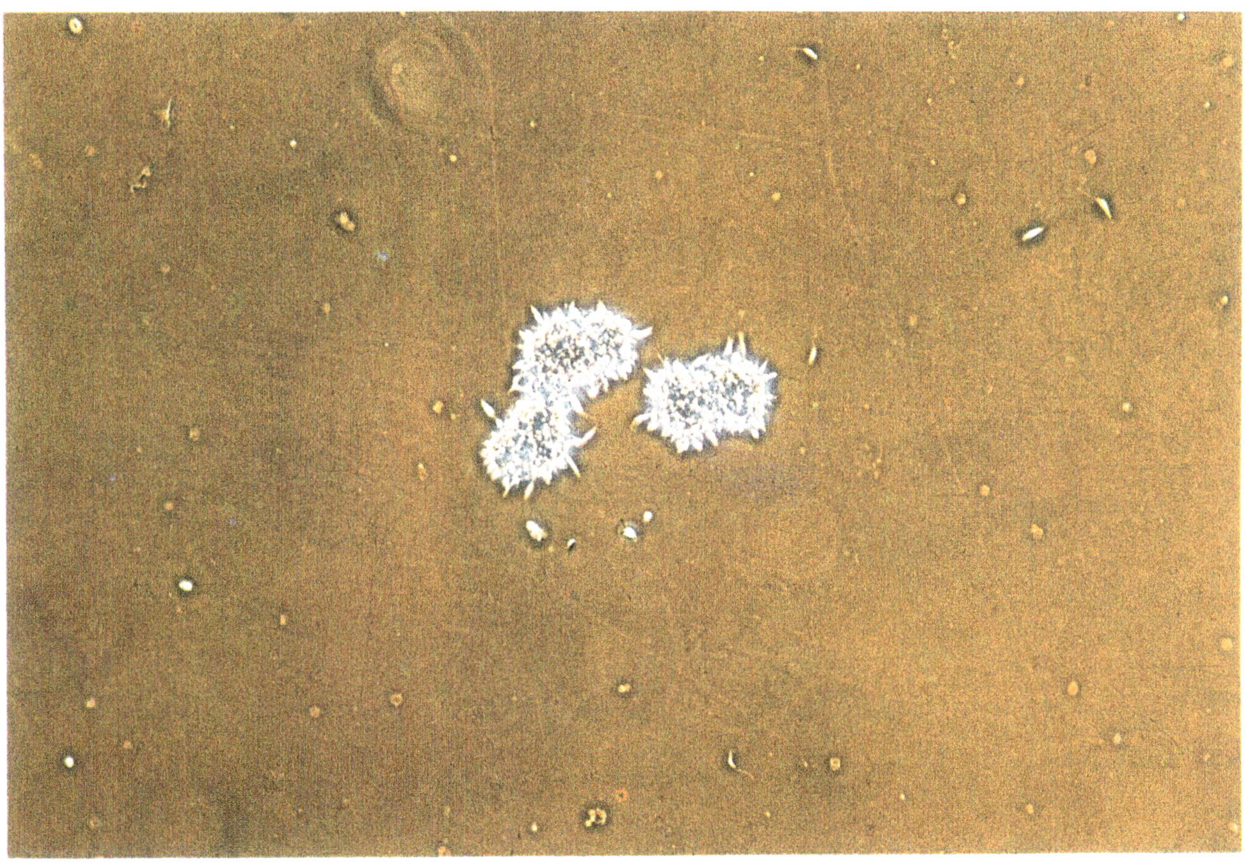

65

**66 Section of skin showing LD bodies.** In sections stained with HE the parasites appear as dots and the differentiation between the nucleus and the kinetoplast cannot be made. Histologically these lesions could be confused for Histoplasmosis. X1000. Enlarged by 5.4.

**67 Section of skin showing** *Histoplasma capsulatum*. The oval bodies of *H. capsulatum* have a PAS positive granule which takes a reddish colour. LD bodies are not PAS positive. X1000. Enlarged by 5.4.

**68 Smear from a skin of a case with DCL.** In this condition the skin is heavily infected with parasites unlike cutaneous leishmaniasis in which the infiltration is not extensive. X1000. Enlarged by 5.4.

**69 Cutaneous lesion due to L. tropica.** The ulcer has a clear cut margin with raised indurated edge. The lesion is generally located on the exposed part of the body. In this case it is on the lower leg. (Courtesy Dr C.H. Tay)

**70 'Wet type' cutaneous lesion due to L. tropica.** There is a marked inflammatory reaction around the ulcer and it has a tendency to spread. (Courtesy Dr A.D.M. Bryceson).

**71 Early DCL.** The lesions are in the form of small nodules from a primary lesion on the nose. The patient is from Ethiopia. (Courtesy Dr A.D.M. Bryceson).

**72 Late DCL.** The nodules are large and resemble lepromatous leprosy. This patient is also from Ethiopia. (Courtesy Dr A.A. Buck)

**73 Post Kala-azar dermal leishmaniasis in an Englishman.** Diffuse hypo-pigmented macular rash is distributed extensively on the body. In this case the back is shown. (Courtesy Dr A.D.M. Bryceson).

66

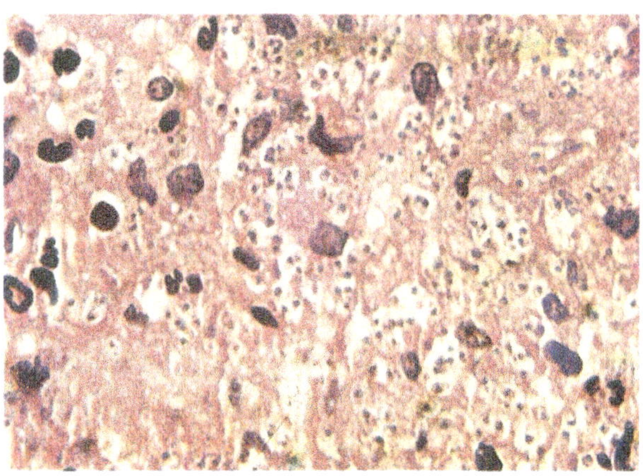

67

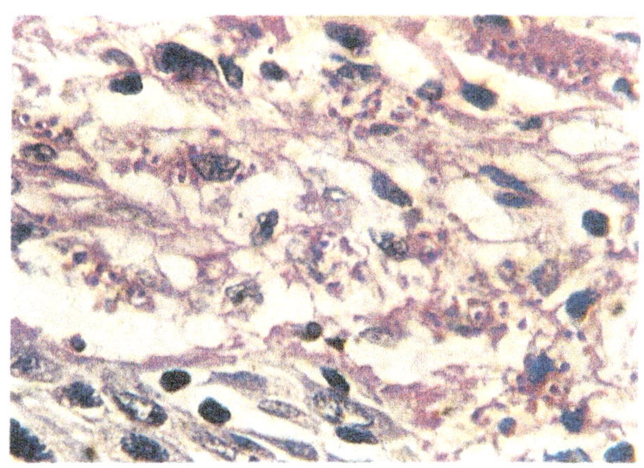

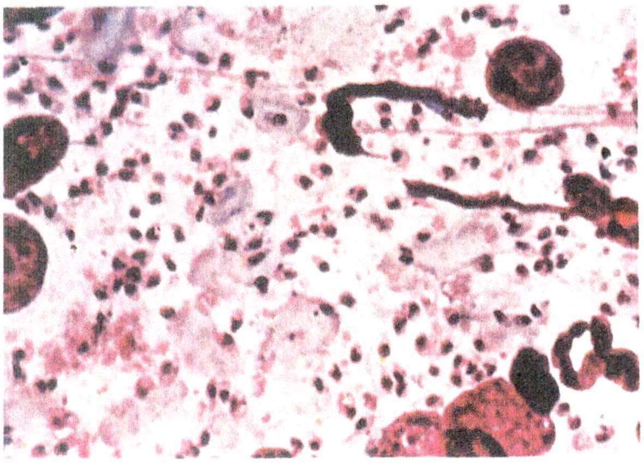

68

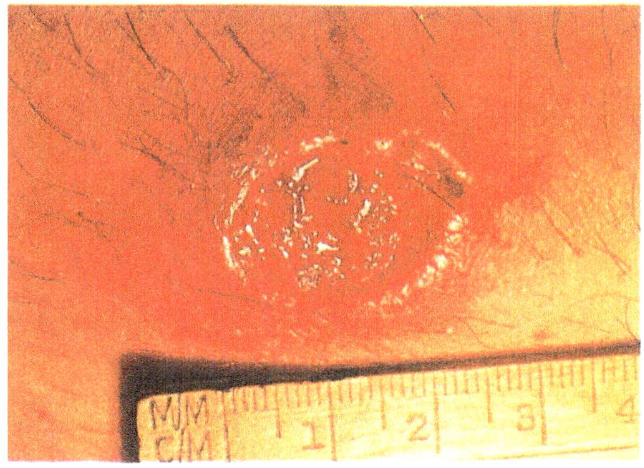

69

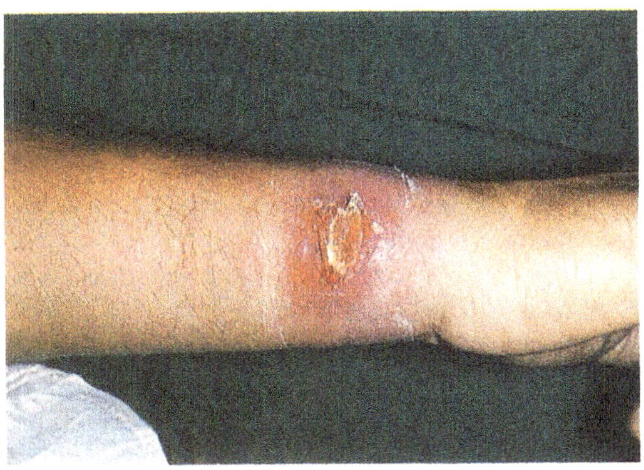

70

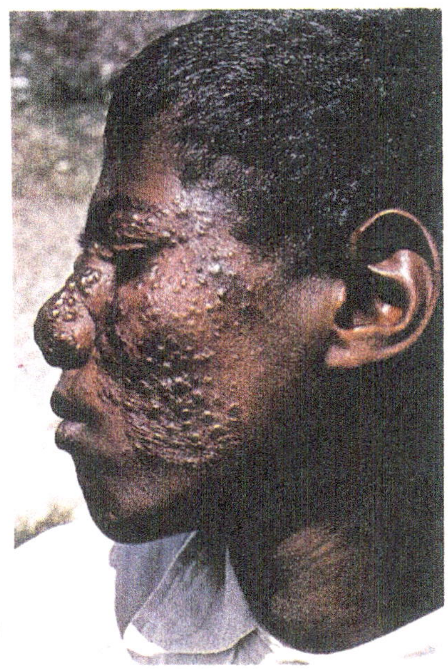

71

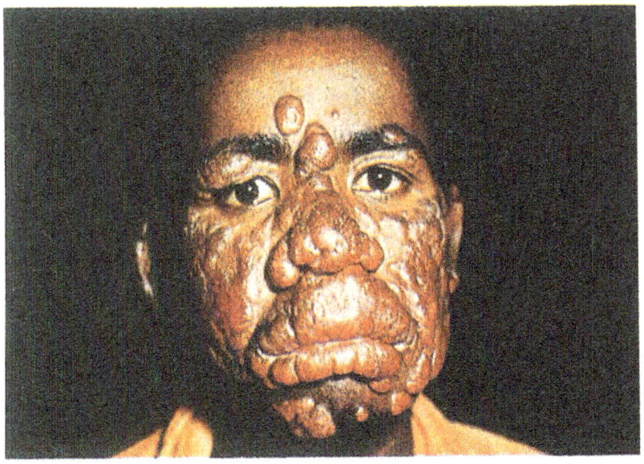

72

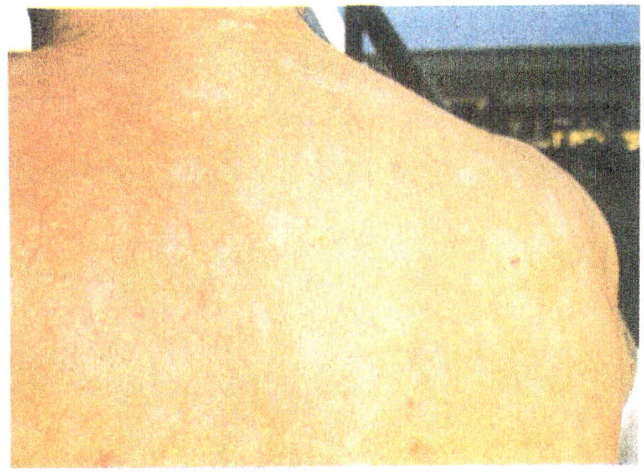

73

# Trypanosomes Infecting Man

*Trypanosoma* infecting man belong to the following species:

*T. rhodesiense*
*T. gambiense*
*T. cruzi*
*T. rangeli*

The first two produce a disease known as sleeping sickness in Africa and the third parasite is the aetiologic agent of Chaga's disease in Central and South America. The last *Trypanosoma* is non-pathogenic and also occurs in Central and South America.

## African Trypanosomes

The African trypanosomes *(T. rhodesiense* and *T. gambiense)* are transmitted by biting flies belonging to the genus *Glossina* (commonly known as the tsetse fly). The transmission occurs by a cyclical method in which the parasite undergoes a complex development in the fly before becoming infective. In the blood of the host the parasites divide by binary fission and are polymorphic, showing slender, intermediate and stumpy forms. In fresh preparations of the blood, trypanosomes can be recognised as actively moving elongated organisms. The movement of the parasite occurs in the direction of the flagellum. In stained preparations, the nucleus, kinetoplast, undulating membrane and the free flagellum can be recognized.

Ultrastructural study of the trypanosome reveals a row of microtubules under the surface membrane. The microtubules extend from one end of the parasite to the other. The function of the microtubules is not known but it is generally believed that they provide skeletal support to the trypanosome. In the bloodstream forms, a single tubular mitochondrium with poorly developed cristae is seen. The flagellum comes out of a flagellar pocket or a reservoir. The reservoir acts as a source of entry for the macromolecules and a source of exit for the secretory and excretory products. In the African form of trypanosomiasis only the trypomastigote stage is seen in the blood, but there is a possibility that amastigotes and 'sphaeromastigotes' (small spherical forms with a short free flagellum) may exist in the choroid plexus at some stage of the disease, as this has been demonstrated in rats. The 2 species of African trypanosomes are morphologically identical but show certain clinical, epidemiological and chemotherapeutic differences.

### *Life Cycle of T. rhodesiense and T. gambiense* [74]

Trypanosomes undergo morphological changes during their life cycle. In the animal host (blood stream forms) the slender, intermediate and the stumpy forms can be seen. The stumpy forms initiate the infection of *Glossina*. Inside *Glossina* trypanosomes are initially confined to an area surrounded by the peritrophic membrane and rapidly change to elongated forms (1). They next move into the space between the peritrophic membrane and the gut wall, where they become slightly shorter (2). The forward movement continues and they enter the main food channel by penetrating the peritrophic membrane. At the tip of the proboscis, they turn back into the hypopharynx to enter the salivary gland. At this stage they become epimastigotes (3). In the salivary gland further development occurs and they finally change to metacyclic forms, which are injected into the tissues with the saliva (4).

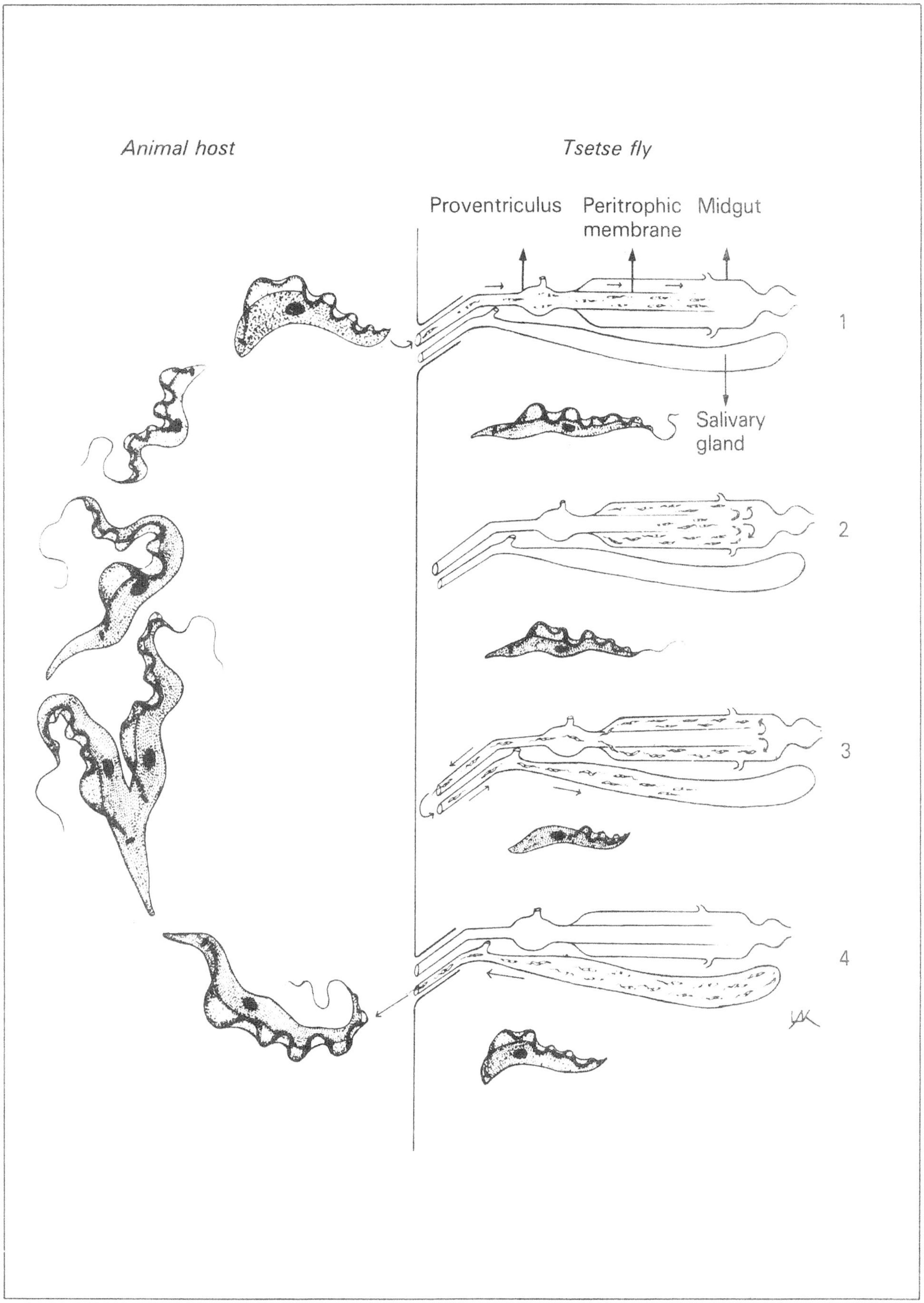
Animal host
Tsetse fly
Proventriculus
Peritrophic membrane
Midgut
1
Salivary gland
2
3
4

74

## *Clinical aspects*

A primary reaction occurs at the site of inoculation of *Trypanosoma*, consisting of a small subcutaneous nodule in the beginning which later becomes a chancre of 25 to 100mm in diameter. The primary reaction resolves in 2 to 3 weeks' time. The first systemic manifestation is in the form of fever with headache. In *T. gambiense* infection, lymph glands, mainly in the cervical and suboccipital region, become enlarged (Winterbottom's sign).

The fever becomes intermittent in later stages and is followed by the involvement of the CNS. In a typical case there is daytime sleeping, psychological changes, tremors, convulsions and finally coma. Death generally occurs by intercurrent infection. In *T. rhodesiense* infection the tempo of the disease is faster with the CNS involvement occurring within a few months.

The pathology of the CNS consists of leptomeningitis and cellular infiltration by mononuclear cells, mainly around the blood vessels. The infiltrating cells are lymphocytes, plasma cells and Mott's morula cells.

## *Diagnosis*

1) Demonstration of parasites by lymph gland puncture in *T. gambiense* infection.
2) Demonstration of parasites in blood in *T. rhodesiense* infection.
3) Concentration of parasites from blood by DEAE cellulose column separation.
4) Demonstration of antibodies using CFT and FAT which becomes positive in 90% of the cases.
5) Rise in IgM levels in blood and CSF. The rise of IgM in CSF is especially pathognomonic of trypanosomiasis.
6) Inoculation of rats with *T. rhodesiense,* which develop heavy parasitaemia.

*Table I.*

| | *T. gambiense* | *T. rhodesiense* |
|---|---|---|
| *Virulence* | Less virulent to humans and laboratory animals | More virulent to humans and laboratory animals |
| *Reservoir* | Mainly humans | Mainly animal |
| *Vector* | Mainly *G. palpalis* | Mainly *G. morsitans* |
| *Geographical distribution* | Mainly W. Africa | Mainly E. Africa |
| *Chemotherapy* | Responds to tryparsamide | No response to tryparsamide |

**75 Diagram of the ultrastructure of Trypanosoma rhodesiense in its intermediate blood stream form.** The pellicular microtubules, which are found over the whole surface of the flagellate, are shown only at the anterior end. Note the presence of a single tubular mitochondrion with poorly developed cristae, indicating low metabolic activity. In the vector, the mitochondrion becomes active and enlarges into a highly complex network. (Redrawn with permission from 'The Protozoa' by Vickerman and Cox; John Murray, London 1967).

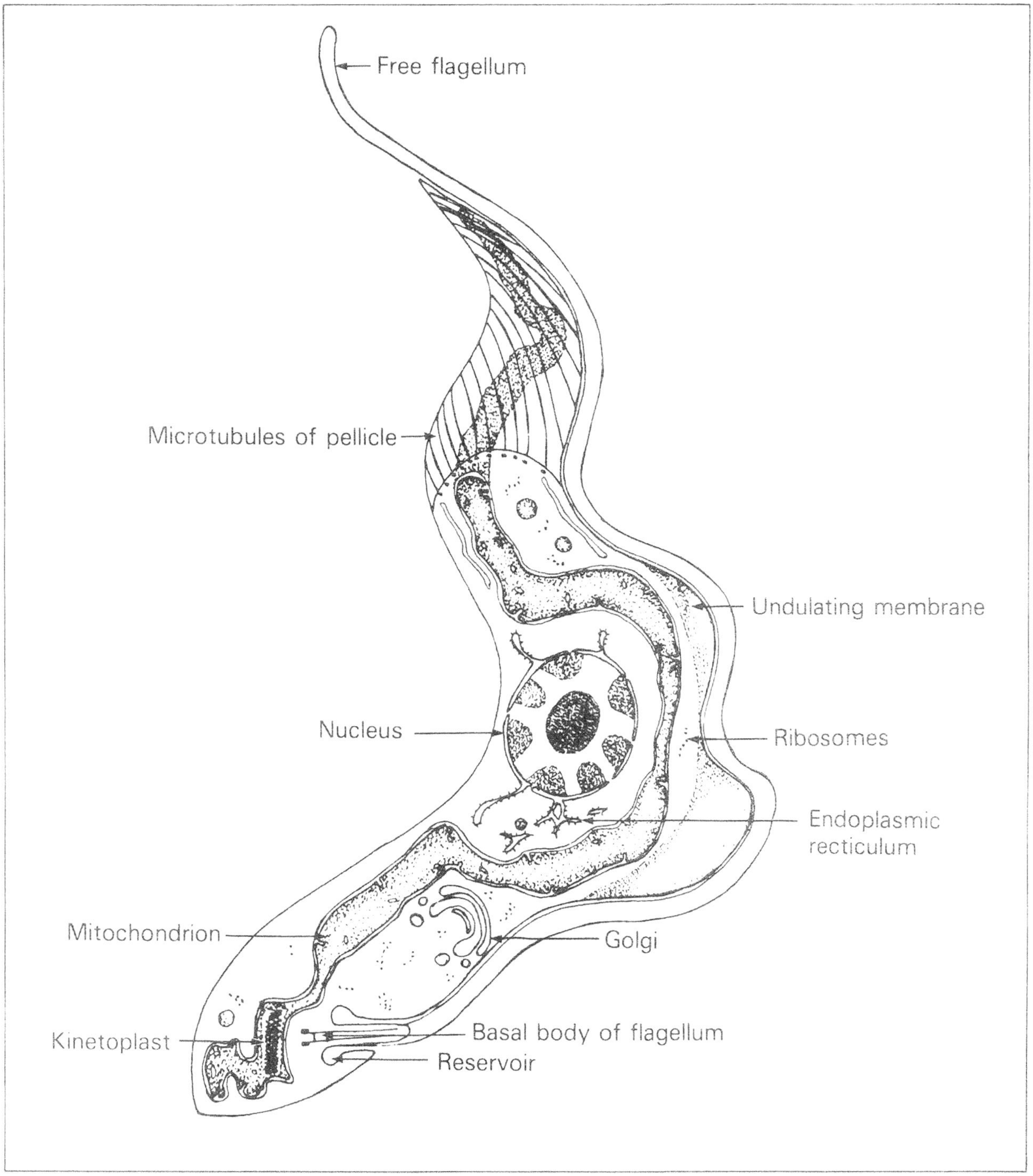


75

**76 Ultrastructural changes at the anterior end of the Trypanosome during division.**
1) The division starts by bipartition of the kinetoplast and the development of a new axoneme.
2) Separation of the kinetoplast-mitochondria complex occurs and the new axoneme pushes forward, lifting the pellicle.
3) The new flagellum is covered by the body of the pellicle. The cytoplasmic division then starts splitting the whole parasite into two.

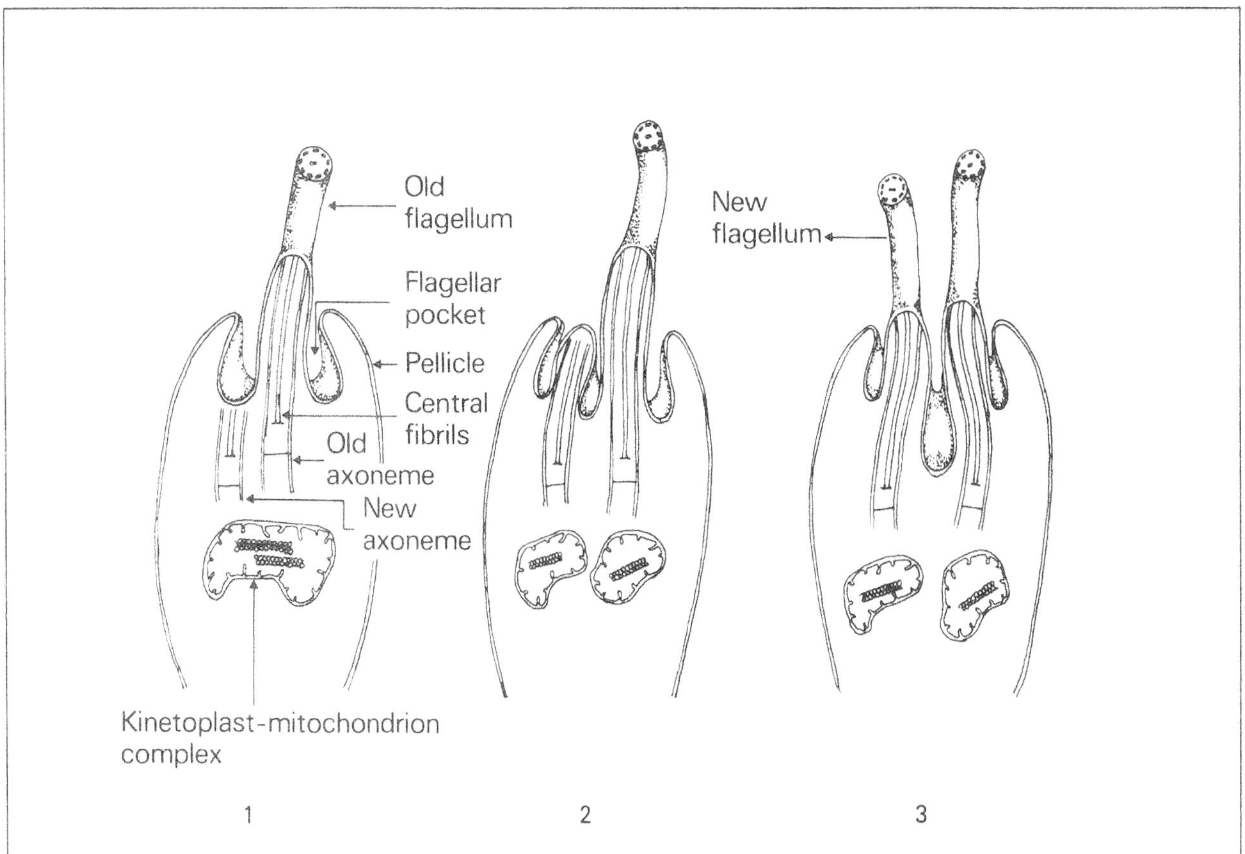


76

**77 Adult Glossina in a resting position.** The wings partially or completely overlap each other and extend beyond the body. It can be easily distinguished from other biting flies by a prominent forwardly projecting proboscis. The flies vary in length from 6-15mm.

**78 This mounted specimen shows the position of the mouthparts before feeding.** The palps are kept horizontal to the body and the fascicle (biting parts) is pushed vertically downwards.

**79 The wing venation of Glossina is characteristic in that it shows an enclosed area between the 4th and 5th veins.** This enclosed area is also known as the 'hatchet' cell.

**80 Mouthparts of Glossina.** On the sides are non-segmented hairy palps. In the centre is the labium which appears as a long gutter containing labrum and hypopharynx. Interference contrast. X100. Enlarged by 5.4.

77

78

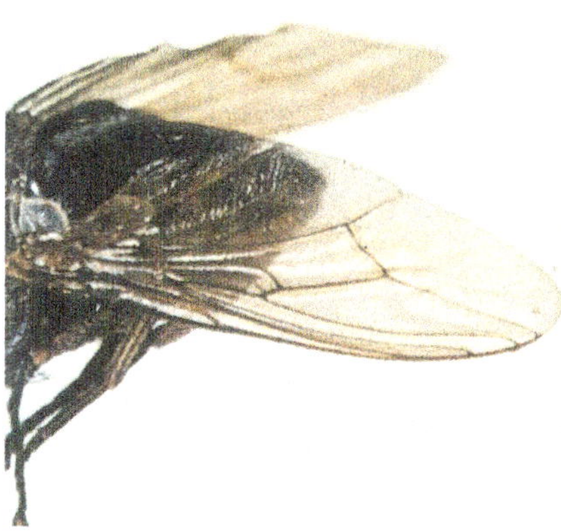

79

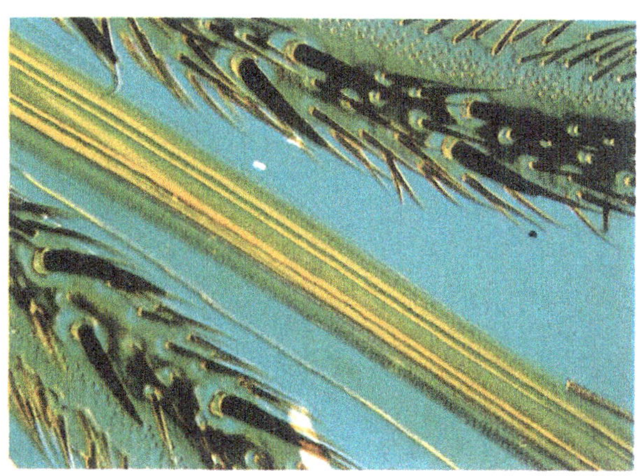

80

**81 Glossina grips the skin of the animal host with its claws before piercing the skin.** The claws consist of a pair of curved chitinous structures projecting from the distal end of the legs. Interference contrast. X100. Enlarged by 9.6.

**82 The puparium of Glossina is barrel shaped varying between 5 to 8mm in length.** It is dark brown or black in colour and has 2 distinct lobes (polypneustic lobes) at one end. These lobes are used for respiration.

81 

82 

**83 T. rhodesiense in rat blood.** These are morphologically indistinguishable from *T. gambiense*. However, *T. gambiense* are normally not infective to rats. Note the elongated bodies with minute kinetoplast. The darkly staining structure in the centre is the nucleus. Giemsa. X1000. Enlarged by 5.4.

**84 T. rhodesiense in rat blood.** Stained with Acridine orange. Only the parasites and the white cells fluoresce making it a suitable method for screening blood smears. The reddish colour of the parasite is due to the presence of RNA. X400. Enlarged by 5.4.

**85 Section of brain from a T. rhodesiense case.** There is marked perivascular cuffing and infiltration with round cells. Scattered amongst the round cells are large cells with eosinophilic or greyish cytoplasm. These are the 'morula cells of Mott'. X100. Enlarged by 5.4.

**86 Section of brain from a T. rhodesiense case.** Showing 5 or 6 morula cells lying with a large number of round cells. It is believed that the morula cells are involved in IgM production. X1000. Enlarged by 5.4.

83

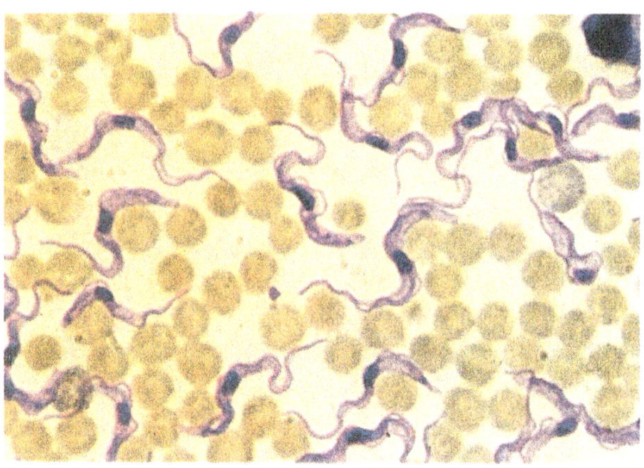

84

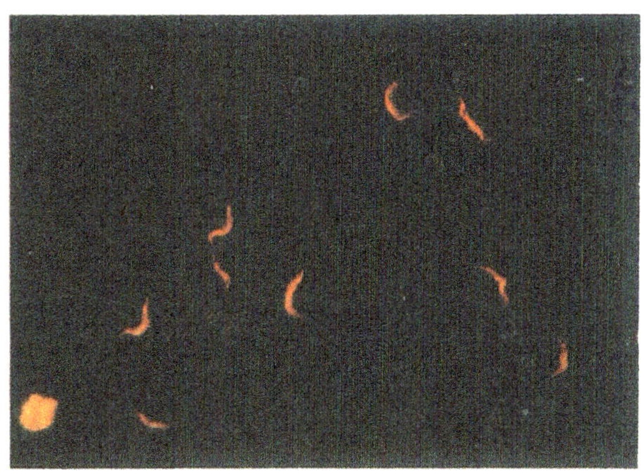

85

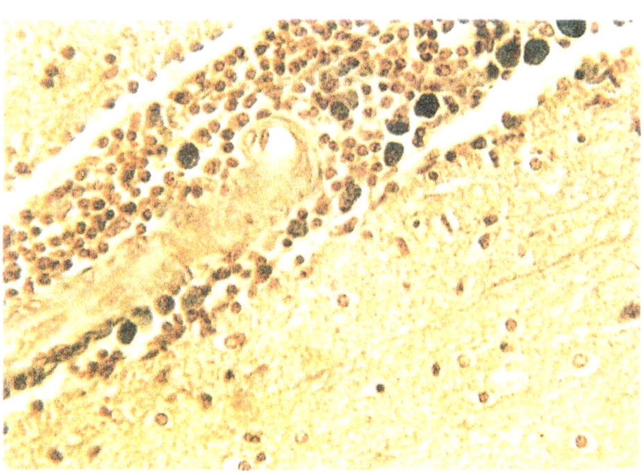

86

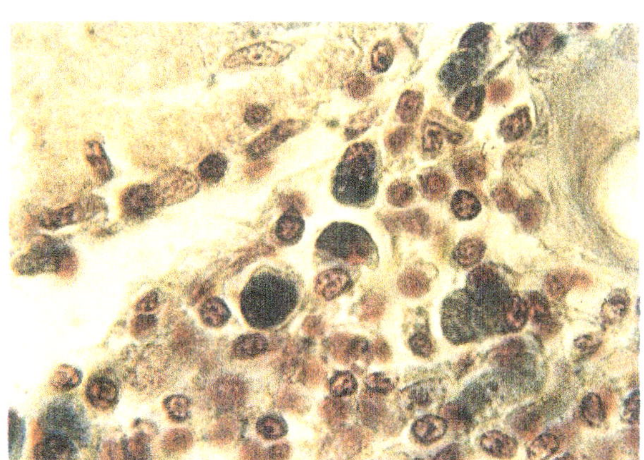

## American Trypanosomiasis

Human infection with *T. cruzi* is common in South and Central America. The vector is the blood sucking reduviid bugs (Genus *Triatoma, Panstrongylus* and *Rhodnius).* After the ingestion of the blood stages of trypanosomes by the vector, the metacyclic forms appear in the hindgut in 8 to 10 days' time. Infection occurs when the faeces of the infected bug are rubbed into the wound caused by the bite. *T. cruzi* has a large animal reservoir and is found in armadillos, opossums, cats, dogs and pigs in the endemic regions.

### *Life Cycle of Trypanosoma cruzi* [87]

The metacyclic forms are passed in the faeces, enter the blood stream and penetrate the cardiac muscle. Inside the cardiac muscle they change to amastigote, epimastigote and the trypanosomal forms. The trypanosomal forms may re-invade the cardiac muscle. In the insect vector the trypanosomal forms change to epimastogotes before becoming metacyclic.

The parasites produce focal lymphangitis and oedema (Chagoma) at the site of entry, after which they enter the blood stream and find their way into cells of mesenchymal origin, mainly the cardiac muscle. In the muscle cells they get transformed to amastigote stage and multiply to form a pseudocyst. The amastigote stages then get transformed to epimastigote and eventually the trypanosomal forms. The trypanosomal forms emerge from the pseudocyst and enter the blood stream. In the blood, the trypanosomal forms do not divide and could be differentiated from *T. gambiense* and *T. rhodesiense* by a large kinetoplast and their 'C' or a 'S' shaped appearance.

### *Clinical aspects*

The primary lesion in the form of chagoma is generally found on the face near the eyelids. This produces swelling of the eye and temporal region with conjunctivitis (Romana's sign).

The most constant feature of chronic disease is cardiomyopathy. In mild cases this may lead to extra systole and slight tachycardia. In severe cases partial or complete heart block may develop, eventually leading to cardiac failure.

In a certain number of cases megaoesophagus (dilation of oesophagus) and megacolon (dilation of colon) may occur. The dilation of oesophagus may lead to dysphagia and other intestinal complications.

### *Diagnosis*

1) Isolation of *T. cruzi* from the blood:-
   a) Trypanosomes may be found by direct examination but this is rare and concentration methods such as centrifugation may be required.
   b) Culture on NNN medium is very useful.
2) Xenodiagnosis — At least 6 clean uninfected laboratory bred reduviid bugs are allowed to feed on the suspected patient and 2 weeks later the hindgut is examined for epimastigotes.
3) Serological methods — An indirect haemagglutination test and a FAT are available and become positive in the majority of infected individuals.

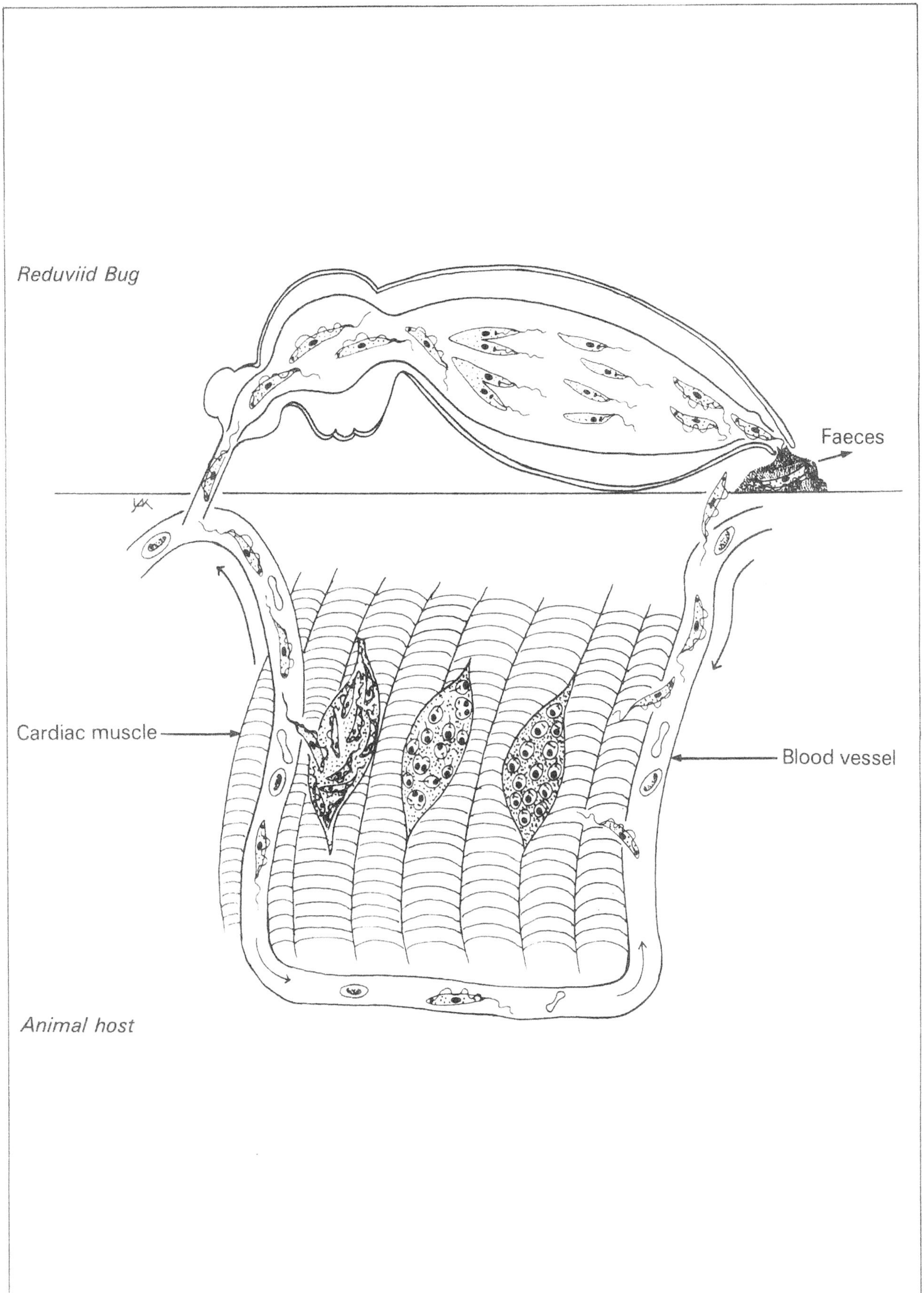
Reduviid Bug
Faeces
Cardiac muscle
Blood vessel
Animal host

87

**88 Reduviid bug, in this case T. rubrofasciata, is seen feeding.** Note the head is elongated and the dorsum of the thorax is covered by a shield like structure (pronotum).

**89 A close up view of the proboscis and the head regions during the act of feeding.** The long proboscis makes an angle before entering the tissues. The 4 segmented antennae arise from the side of the head.

**90 The bug generally defaecates on the skin of the host after or during the taking of a blood meal.** The faeces are black in colour and are teeming with trypanosomes in an infected animal.

**91 Triatoma usually live in cracks and crevices in walls and under the wooden planks.** They emerge during the night time to take a blood meal. In this case a large number of larvae are seen feeding on an anaesthetized mouse.

**92 Dividing epimastogote stages from the gut of an infected Triatoma.** Note the presence of large kinetoplasts lying anterior to the nuclei. Giemsa. X1000. Enlarged by 9.6.

88

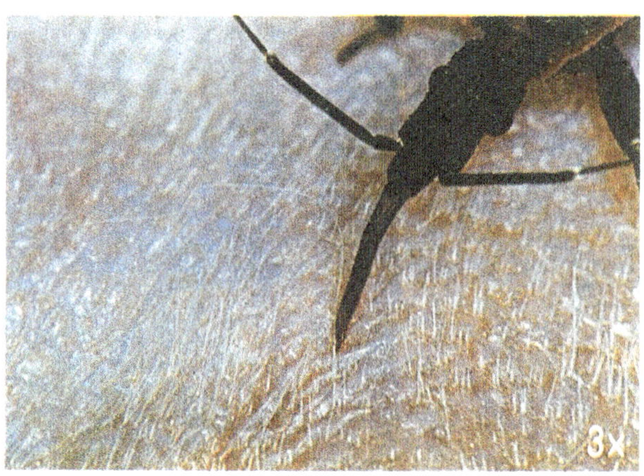


89

90

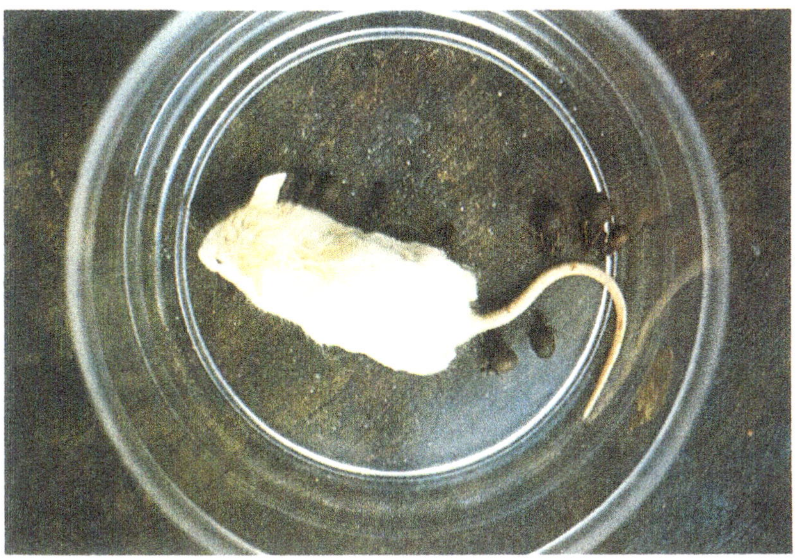

91

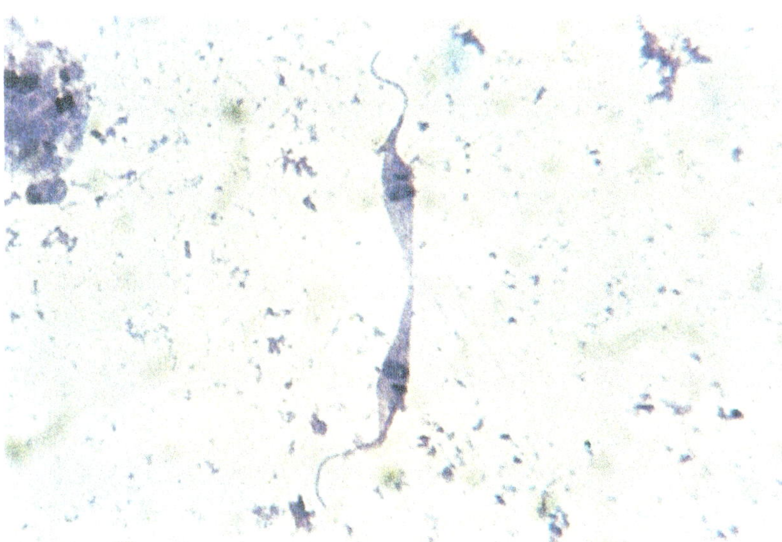

92

**93 Trypanosomal stages in the blood stream.** In this case the posterior ends are pointing inwards giving the parasites a 'C' shaped appearance. This is the shape which parasites generally assume. The kinetoplasts are distinct. Giemsa. X1000. Enlaraged by 9.6.

**94 Trypanosomal stages in the blood stream.** In this case the posterior ends are pointing outwards giving the parasites a 'S' shaped appearance. Giemsa. X1000. Enlarged by 9.6.

**95 Amastigotes lying in a pseudocyst in human cardiac muscle.** H and E. X1000. Enlarged by 9.6.

**96 Trypanosoma rangeli. Epimastogote forms from cultures.** These trypanosomes are also conveyed by reduviid bugs, but can be easily differentiated from *T. cruzi* being longer and thinner with a small kinetoplast. They are non-pathogenic to man. Giemsa. X1000. Enlarged by 9.6.

93

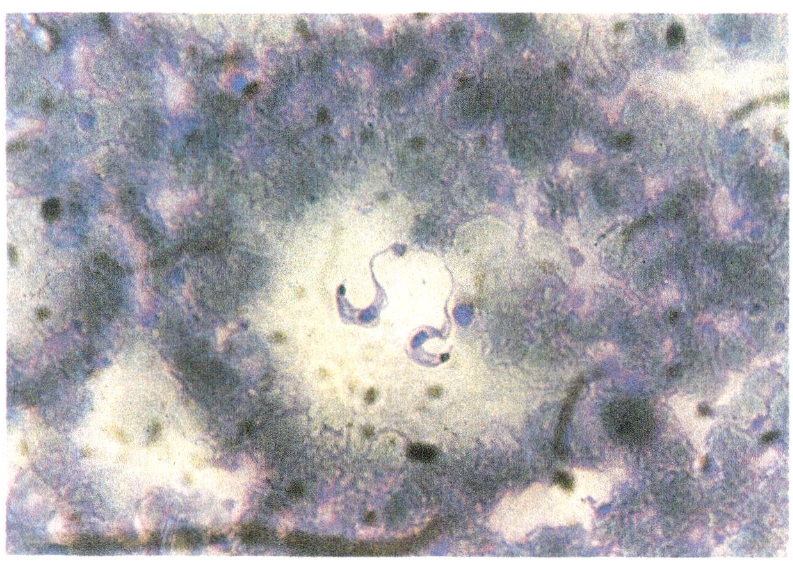

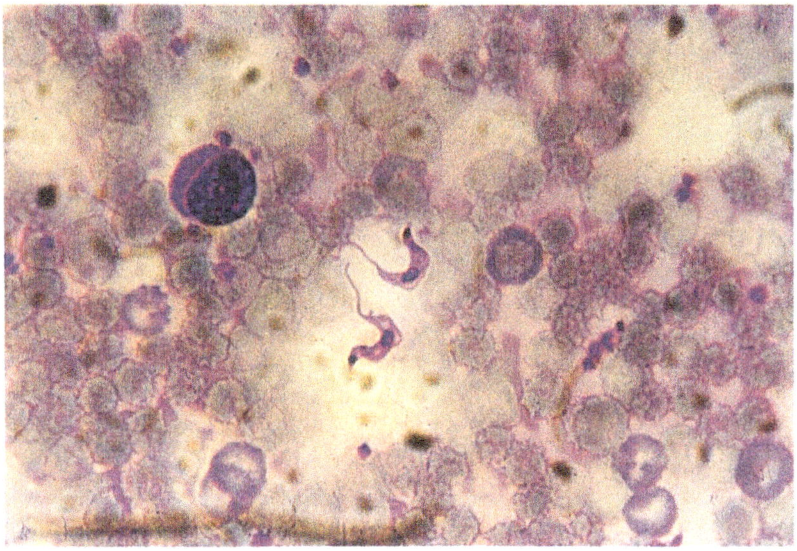
94

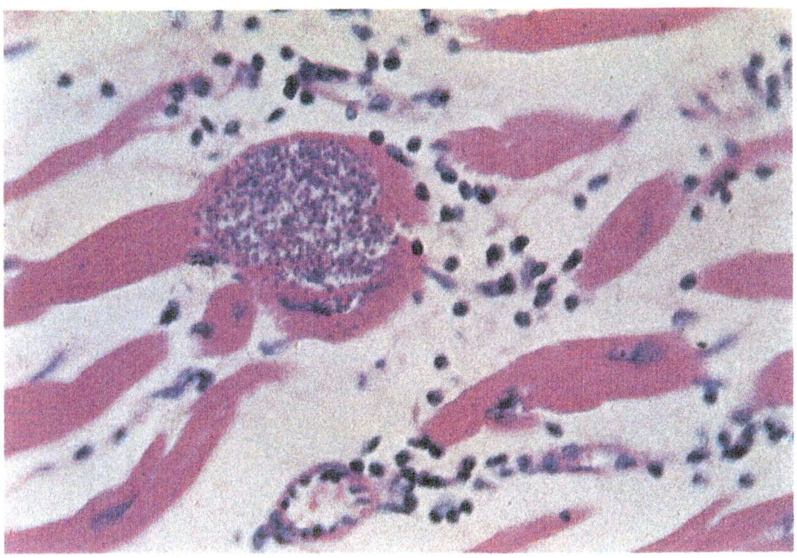
95

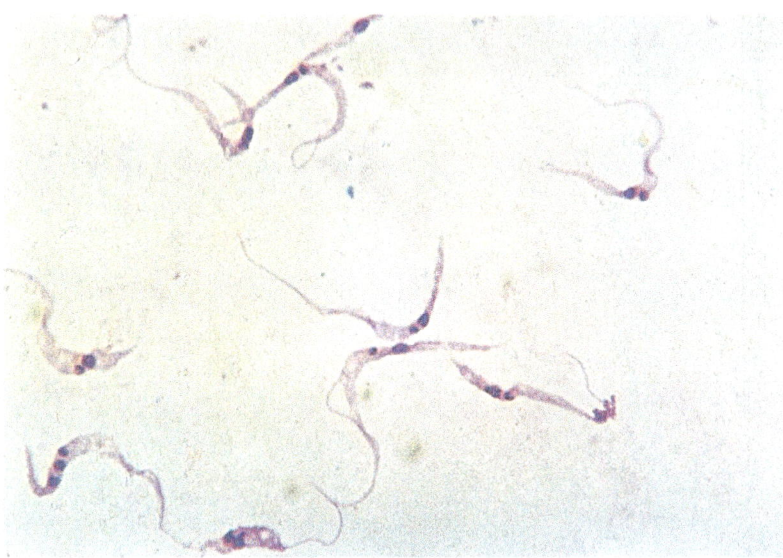
96

# Section 5
# Balantidium coli

The parasite has 2 stages in its life cycle. The trophozoite stage and the cystic stage. Both these stages have a large macronucleus which is often reniform in shape. A smaller micronucleus may be observed in the concavity of the macronucleus. The trophozoite is 30µm to 300µm in size and has 2 contractile vacuoles. At the anterior end it has a funnel shaped depression known as the peristome which leads to the mouth or the cystotome. The cilia, which are the organs of locomotion are embedded in the pellicle in longitudinal rows. These rows (kineties) can be observed in phase contrast, interference contrast or by silver staining.

The cyst of *Balantidium* is spherical or ovoid having a diameter of 40 to 60µm. The cyst wall is thick and transparent and the parasite is visible inside the cyst.

In addition to binary fission *Balantidium* undergoes conjugation in which 2 parasites approach each other and become attached at the anterior end. Conjugation generally continues for a few minutes after which the parasites detach and move away.

## *Life cycle of Balantidium coli* [97]

Human infection occurs from close association with pigs. Once the infection is established in the human, direct human to human transmission can also occur. The infective stage of the parasite is the cyst. The excystation occurs in the small intestine and the parasites multiply in the large intestine by binary fission. They may be passed out in the faeces as cysts or trophozoites.

## *Clinical aspects*

The pathology and the clinical picture of Balantidiasis is similar to Amoebiasis excepting that extra-intestinal involvement does not occur. In the infected intestine trophozoites are often found in the submucosa and are easily recognized by their characteristic macronucleus.

## *Diagnosis*

Diagnosis is made by the demonstration of *B. coli* in the faeces. In acute dysentery actively motile trophozoites are seen.

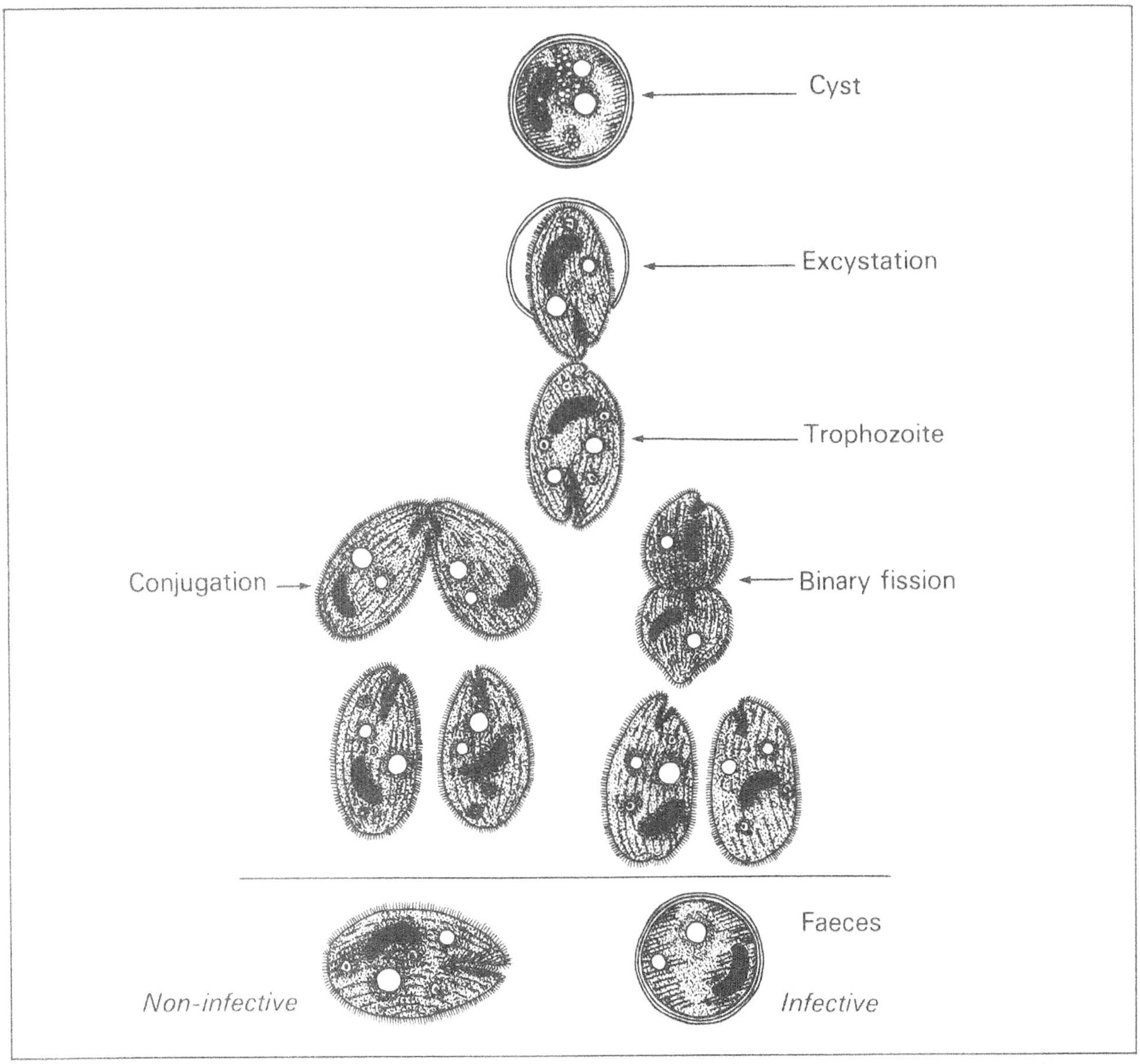


97

**98 Balantidium undergoing conjugation.** The parasites remain motile during this process. It does not lead to increase in numbers. Phase contrast. X1000.

**99 Balantidium undergoing binary fission.** Cytostome can be seen in 1 of the dividing cells. The 2 contractile vacuoles have separated. Interference contrast. X1000.

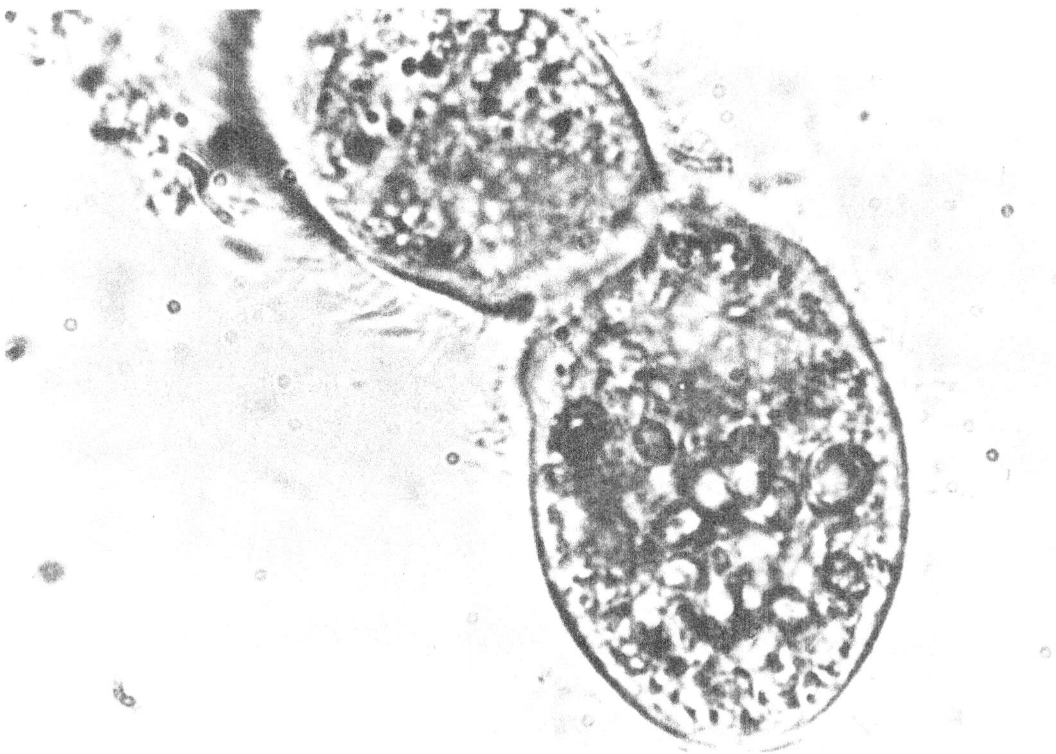

98

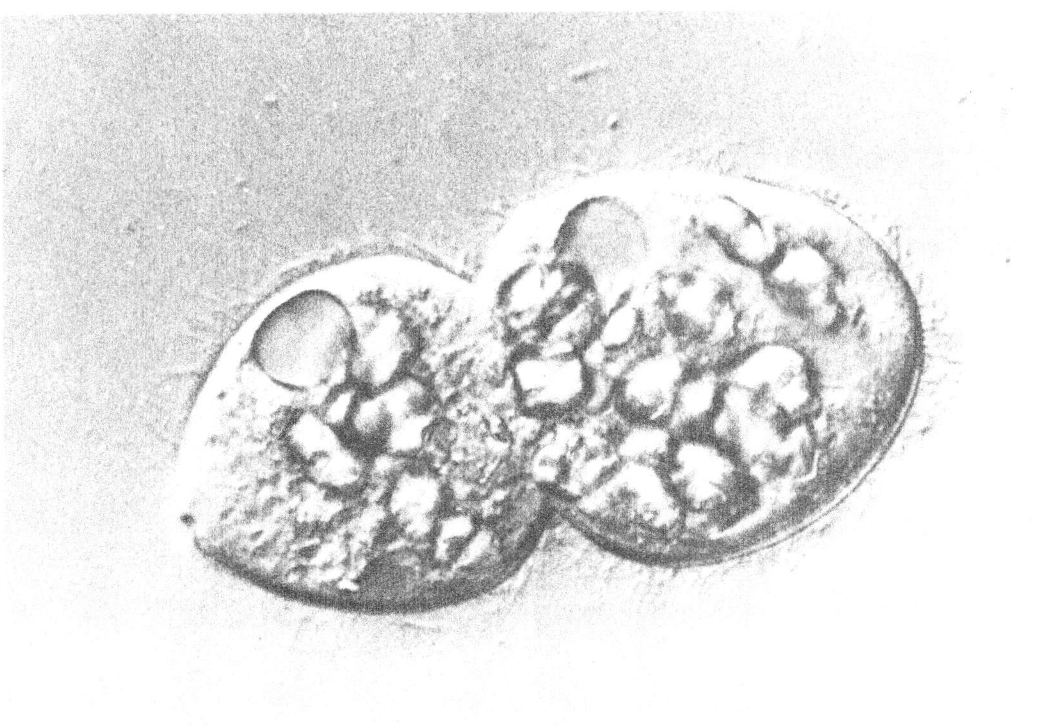

99

**100 Cyst of Balantidium.** It has a thick transparent cyst wall. The contents are visible. Phase contrast. X1000.

**101 Trophozoite of Balantidium.** The 2 contractile vacuoles can be seen at the posterior end as spherical bodies. Phase contrast. X1000.

100 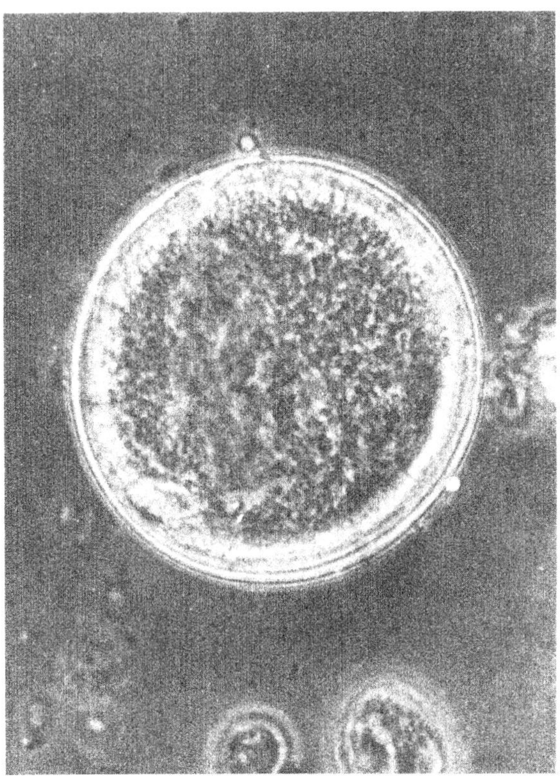

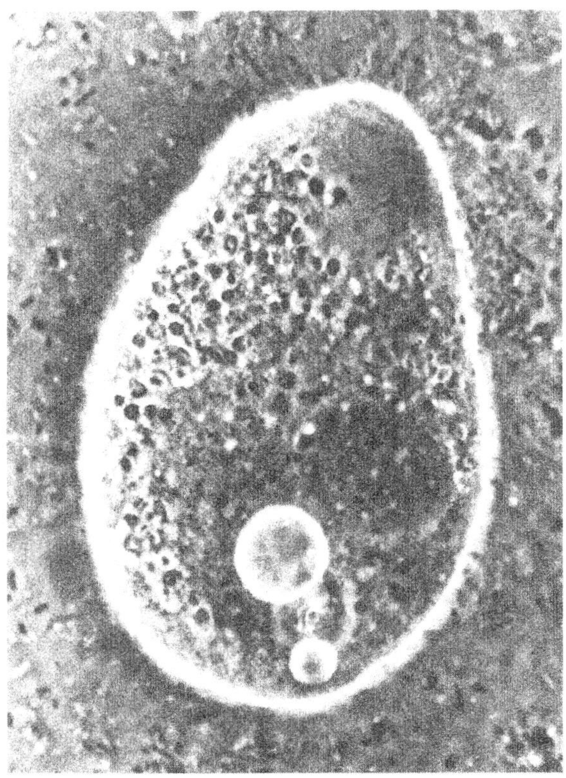 101

**102 Longitudinal section of macronucleus.** It is made up of a network of irregular electron dense structures interspaced with spherical bodies. X64,000. Electron micrograph.

**103 Cross-section of macro and micronucleus.** Starch grains have been ingested from the culture medium. X38,000. Electron micrograph.

**104 Anterior end of a trophozoite.** Showing the cytostome and the adjacent cilia. X46,000. Electron micrograph.

PG = polysaccharide granules. MN = macronucleus. MI = micronucleus. S = starch. CS = cytostome. C = cilia.

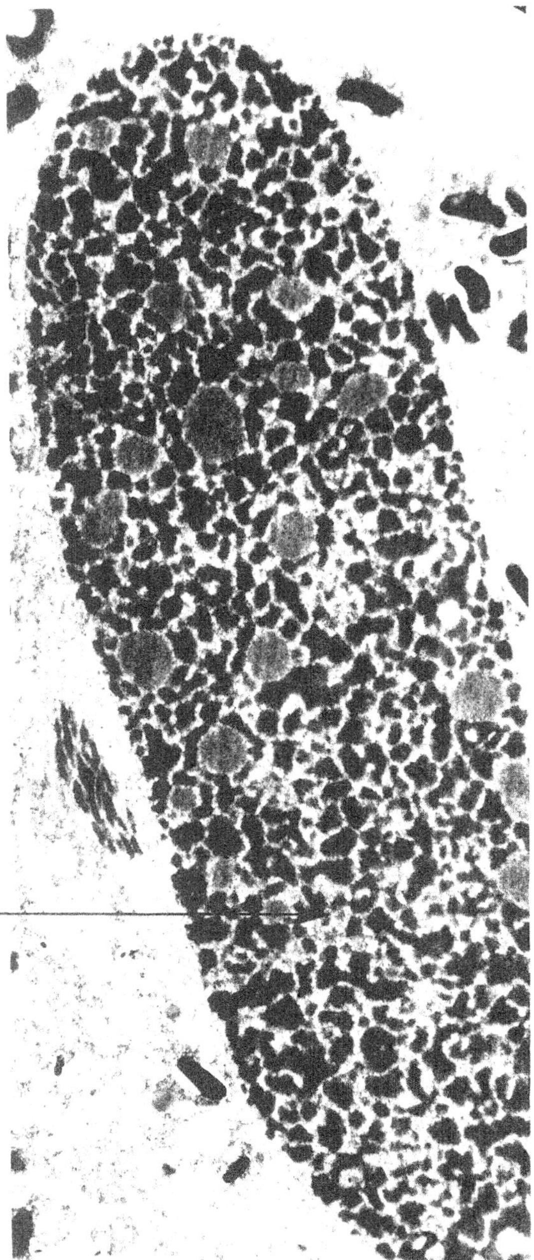


102

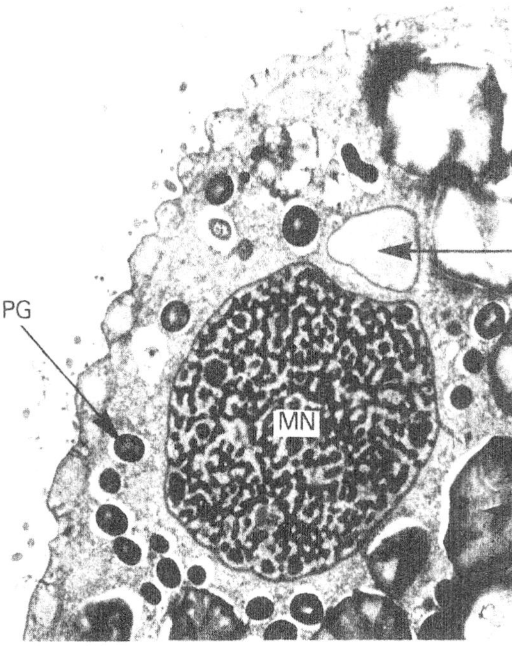


103

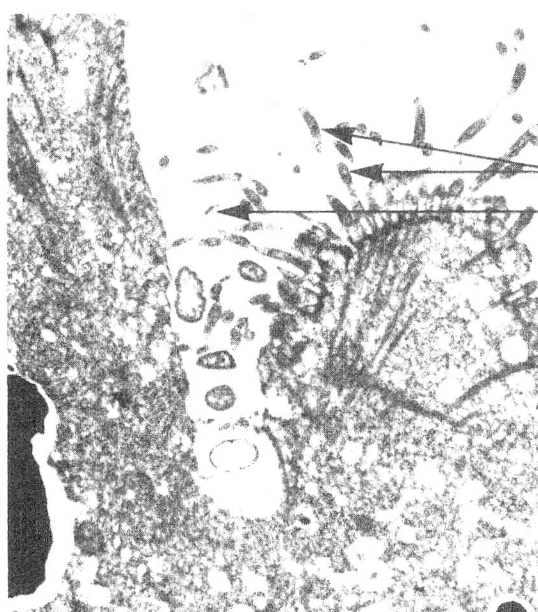


104

**105 Trophozoite Balantidium.** Showing the cytostome and the ciliary pattern. Interference contrast. X800. Enlarged by 6.5.

**106 Balantidium coli cyst.** Cyst of *Balantidium* showing the cyst wall and the 2 contractile vacuoles which are faintly visible. Interference contrast. X800. Enlarged by 5.4.

**107 Trophozoite of Balantidium.** Silver staining to show the kineties. X800. Enlarged by 9.6.

**108 Section of large intestine infected with Balantidium.** The parasites are easily diagnosed by the presence of darkly staining macronuclei. H and E. X100. Enlarged by 6.5.

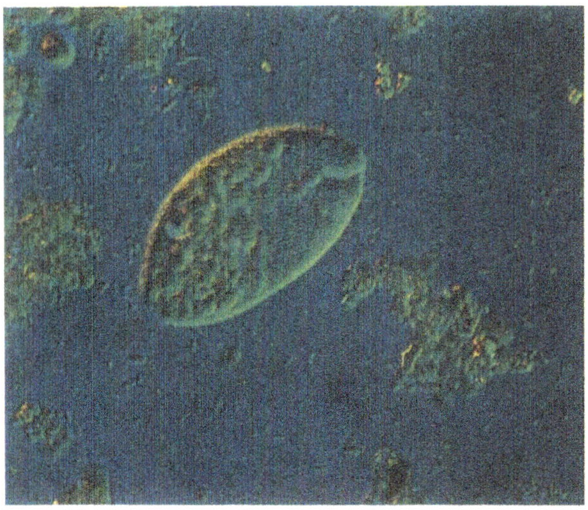

105

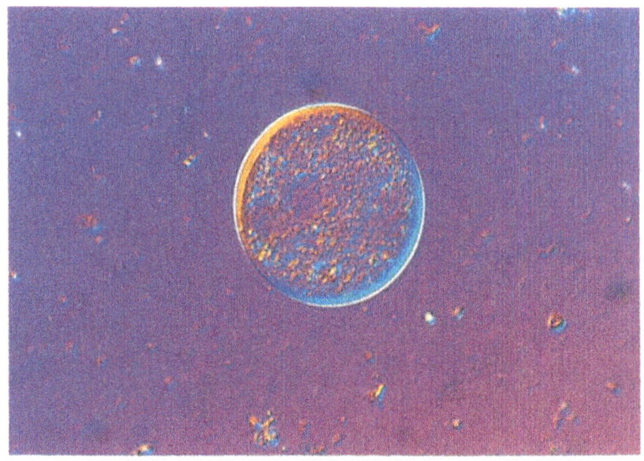

106

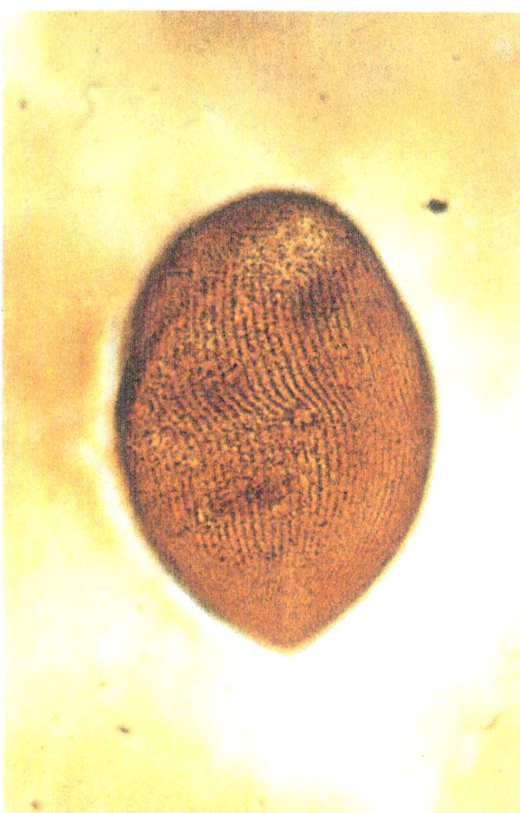

107

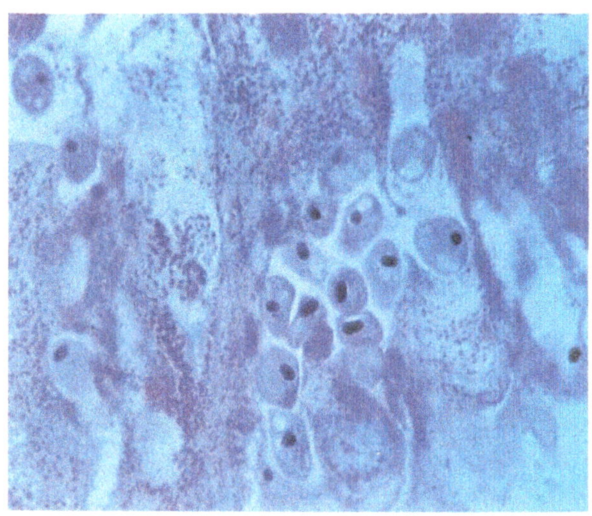

108

# Section 6
# Malaria Parasites

## Terms Used in Relation to Malaria

**Sporozoite** — The infective stage passed in the saliva of the mosquito and formed inside an oocyst by the process of sporogony.

**Schizont** — The stage undergoing asexual division by multiple fission or segmentation. These may be found in the liver cells (pre-erythrocytic schizonts) or in the erythrocytes (erythrocytic schizonts).

**Merozoite** — Is the product of division by schizogony.

**Secondary exo-erythrocytic stage** — This was the term used to denote schizonts developing in the liver as a result of invasion by merozoites from pre-erythrocytic schizont. It is now believed that re-infection of liver cells does not occur by pre-erythrocytic merozoites.

**Trophozoite** — Is the stage of the asexual form with an undivided nucleus, seen in the erythrocyte.

**Microgametocyte** — The male gametocyte which produces a number of microgametes.

**Exflagellation** — The process by which microgametes are formed from a microgametocyte.

**Macrogametocyte** — The female gametocyte which produces a single macrogamete.

**Zygote** — The fertilized ovum.

**Ookinete** — The motile stage of the zygote preceding the oocyst stage.

**Oocyst** — The zygote after the formation of the cyst wall.

**Sporogony** — The sexual phase in the life cycle of certain protozoa. In *Plasmodia* this takes place in the mosquito vector.

**Romanowsky stain** — Mixtures of methylene blue, eosin and methylene azure. Giemsa, Leishman and Wrights are examples of Romanowsky stain.

**Schuffner's dots** — Pinkish small and round stippling seen on *P. vivax* and *P. ovale* infected red cells, in Romanowsky stained films. They appear earlier and in greater numbers in *P. ovale* than in *P. vivax.*

**Maurer's dots** — Coarse and irregular stippling seen on *P. falciparum* infected red cells in Romanowsky stained films.

**Ziemann's dots** — Pinkish small dots sometimes seen in *P. malariae* infected red cells in Romanowsky stained films.

**Incubation period** — The interval between the entry of sporozoites and the first clinical manifestation.

**Pre-patent period** — The minimum time between the entry of sporozoites and the first appearance of parasites in red cells.

**Latency** — The duration between the primary attack of malaria and the relapse. There are no parasites in the circulation during this period.

**Paroxysms** — Bouts of fever due to the liberation of merozoites during the erythrocytic schizogony.

**Recrudescence** — Renewed manifestation of infection due to the survival of erythrocytic forms.

**Relapse** (Recurrence) — Renewed manifestation of infection due to the invasion of blood by merozoites from the late pre-erythrocytic stage (previously known as secondary exo-erythrocytic stage).

**Re-infection** — Renewed manifestation of infection not due to the original infection but resulting from subsequent fresh infection.

The malaria parasites belong to the genus *Plasmodium.* The important species affecting man are:

*P. malariae* — the cause of quartan malaria
*P. vivax* — the cause of 'benign tertian' malaria
*P. ovale* — the cause of tertian malaria
*P. (Laverania) falciparum* — the cause of 'malignant tertian' malaria

As *P. falciparum* has certain characteristics differing from other *Plasmodia* it has been classified into subgenus *Laverania.* In addition to the above species, humans may sometimes be infected by monkey *Plasmodia.* This is, however, of little clinical or epidemiological significance as it occurs very infrequently.

## *Life-cycle of P. vivax*

The sporozoites after being inoculated by the infected *Anopheles* remain in the blood for about half an hour. From the blood they enter the parenchyma cells of the liver where they divide to form the pre-erythrocytic schizonts. It was previously thought that in the case of *P. vivax,* merozoites liberated from pre-erythrocytic schizonts re-invade fresh liver cells, producing secondary exoerythrocytic stages and that the entry of merozoites into blood from secondary exo-erythrocytic stages was the cause of relapses. It now appears that relapses are due to differential rates of development of pre-erythrocytic schizonts resulting in 'early' and 'late' forms. The time of maturation of pre-erythrocytic schizonts is determined by the genetic make up of the individual sporozoites which initiate their formation (Coatney, 1976).

The merozoites discharged from the pre-erythrocytic schizonts enter the blood and parasitise red cells. The entry into the red cells occurs by invagination of the host cell membrane. Inside the vacuole, thus formed, the merozoite is transformed to a trophozoite. The trophozoite digests haemoglobin to form the malarial pigment (haemozoin). On maturation the trophozoite undergoes schizogony resulting in the formation of daughter merozoites. After completing a few schizogonic cycles, merozoites develop into sexually differentiated cells, the male and female gametocyte. The gametocytes continue their development in the mosquito vector.

In the midgut of the mosquito the gametocytes become male and female gametes. The male gametes (microgametes) arise by a process of exflagellation. The union of a microgamete and a macrogamete leads to the formation of a zygote. Within 4 to 6 hours of its formation the zygote is transformed into a motile organism, the ookinete. The ookinete penetrates the wall of the gut and becomes transformed into a circular body, the oocyst. Inside the oocyst, sporozoites develop from germinal cells, known as sporoblasts. The sporozoites emerge from the oocysts and migrate to the cells of the salivary glands and enter the host tissues during the act of feeding.

## *Life cycle of P. falciparum* [109]

The sporozoites discharged from the salivary gland of the mosquito develop in the liver cells and discharge merozoites into the blood stream. Some merozoites from the blood transform into gametocytes which, when taken by the mosquito, initiate sexual development in the midgut. In *P. falciparum* infection relapses do not occur. It is, therefore, assumed that

109

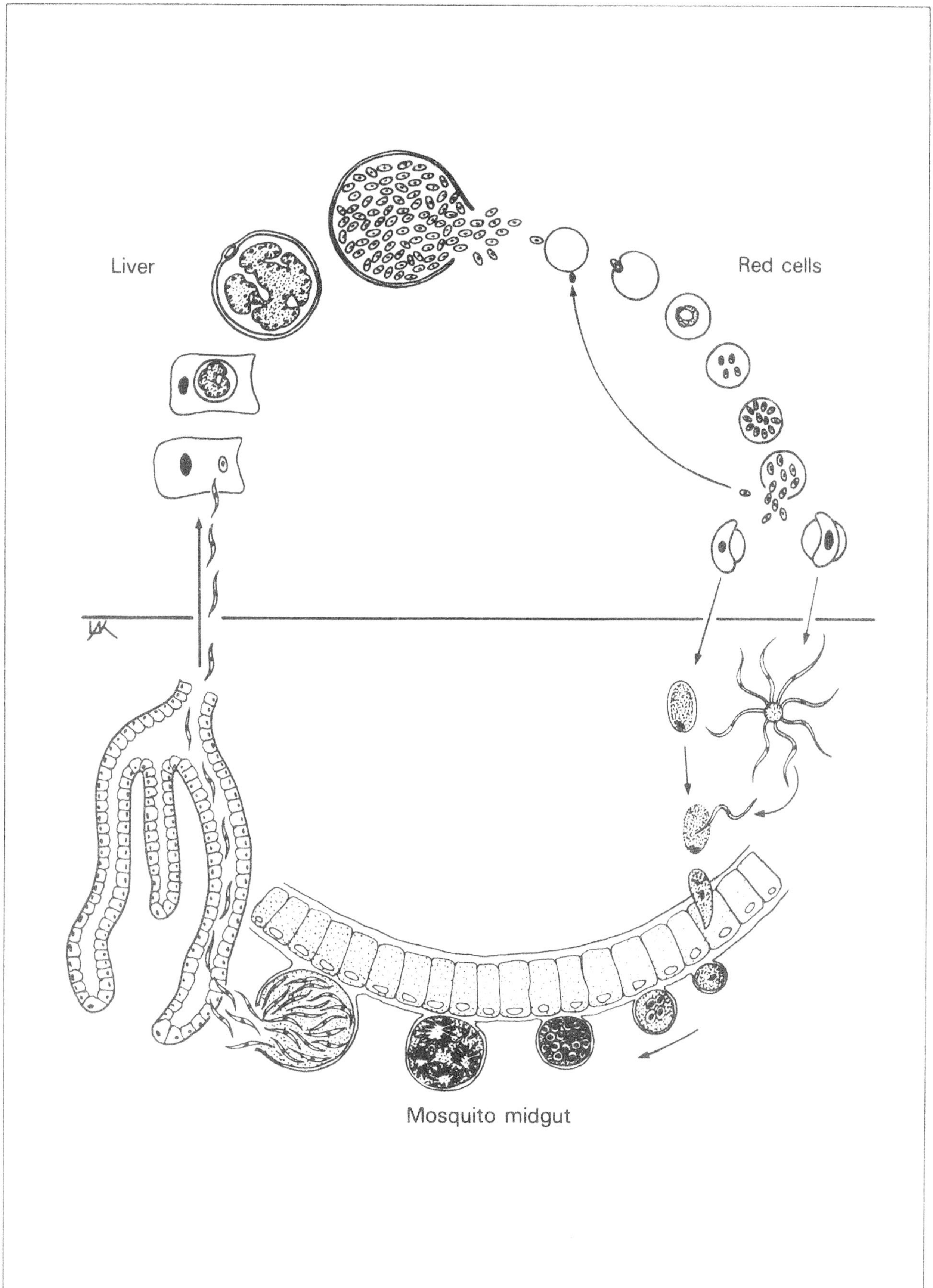

*Table I.* Differential diagnosis of malaria parasites

| | *P. falciparum* | *P. malariae* | *P. vivax* | *P. ovale* |
|---|---|---|---|---|
| Young trophozoite | Fine ring<br>Multiple infection<br>Accolé forms<br>1-2 small chromatin dots | Thick ring<br>1 chromatin dot | Thick ring<br>often irregular<br>1 chromatin dot | Thick ring<br>1 chromatin dot |
| Mature trophozoite | Ring enlarged<br>slightly irregular | Round with central chromatin and band forms<br>Pigment distinct | Irregular<br>Amoeboid | Round<br>Compact |
| Schizont | Rarely seen in peripheral blood<br>18-32 merozoites | 8-10 merozoites often arranged as a rosette<br>Pigment generally in centre | 12-18 merozoites arranged irregularly | 8-14 merozoites arranged irregularly |
| Gametocyte | Crescentic<br>Male reddish with diffuse chromatin<br>Female bluish with compact chromatin | Oval or rounded<br>Male diffuse chromatin | Oval or rounded<br>Male diffuse chromatin | Oval or rounded<br>Male diffuse chromatin |
| Size of red cell | Unchanged | Unchanged or smaller | Enlarged | Enlarged |
| Shape of red cell | Sometimes irregular and crenated | Unchanged | Unchanged | Often irregular with jagged edges |
| Stippling | Sometimes present (Maurer's dots) | Rarely present (Zieman's dots) | Often present (Schuffner's dots) | Always present (Schuffner's dots) |

the sporozoites develop uniformly producing pre-erythrocytic schizonts at the same time and these schizonts once formed, discharge all the merozoites simultaneously; i.e. they do not remain dormant as in *P. vivax.*

## *Clinical aspects*

The main clinical features in a typical case of malaria are paroxysms of fever which appear at regular intervals. Each paroxysm shows a succession of 3 stages. These are the cold stage, lasting for 30 minutes to 1 hour; the hot stage lasting for 1 to 4 hours; the sweating stage lasting for 1 to 2 hours. The paroxysm occurs due to the sudden liberation of merozoites into the blood stream.

## *Diagnosis*

The laboratory diagnosis of malaria is based on the microscopic examination of a blood film. Thick and thin blood films are used for this purpose. The thick film is useful for the detection of light infections but species diagnosis is easier with the thin film. This is because the red cell morphology is maintained in the thin film, while in thick film the red cells are completely lysed.

## **Reference**

Coatney, G.R.: Relapse in malaria — an enigma. Journal of Parasitology 62: 3 (1976).

**110 Fever Pattern and Schizogony in Malaria.**
In malaria fever occurs due to the liberation of merozoites. However, if the blood is examined during the peak of temperature early trophozoites are seen. In the case of *P. falciparum* schizogony occurs in the blood vessels of the internal organs of the body. The schizonts are, therefore, rarely seen in the peripheral blood.

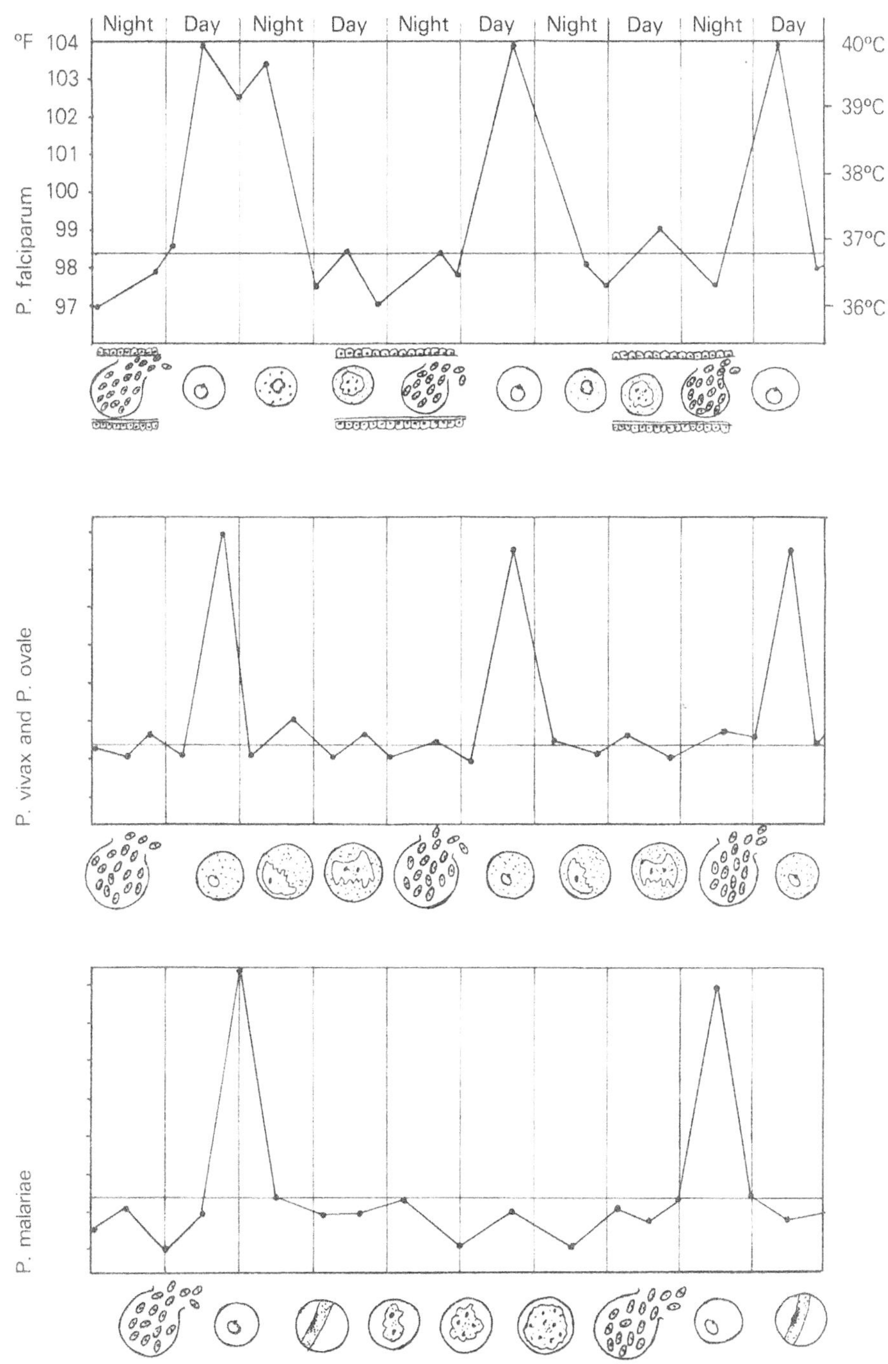


110

**111 Mechanism of relapse in P. vivav** Sporozoites differ genetically producing pre-erythrocytic stages at different time intervals and the pre-erythrocytic schizonts once formed may remain dormant. This results in the occurrence of early and late schizonts. The late schizonts give rise to relapses.

**112 Course of infection in P. vivax** The microscopic threshold is always less than the fever threshold; i.e. parasites become visible in blood before the symptoms of infection develop. As immunity increases the fever threshold rises, with the result that in endemic areas it is not uncommon to find individuals with asymptomatic parasitaemia.

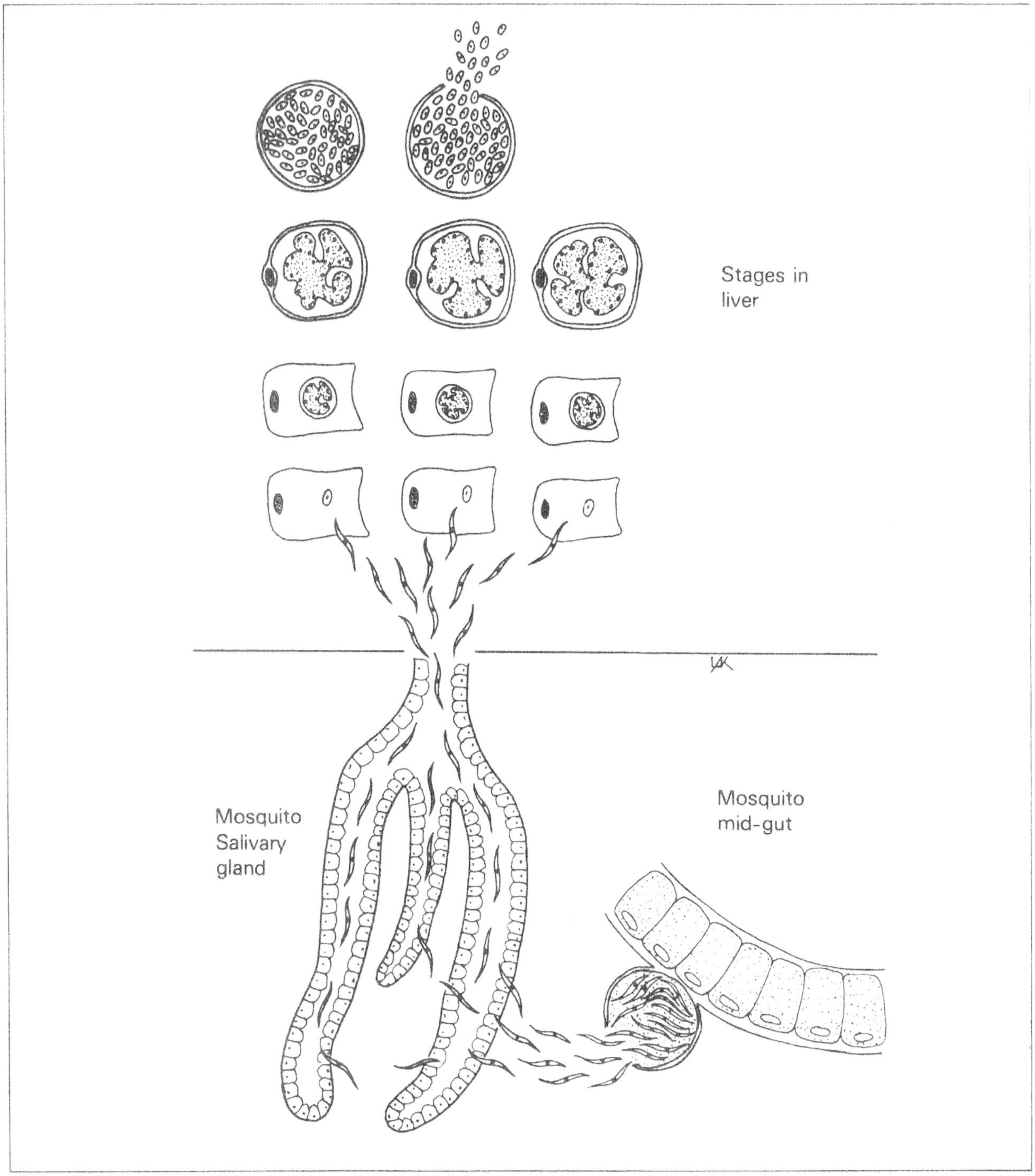


111

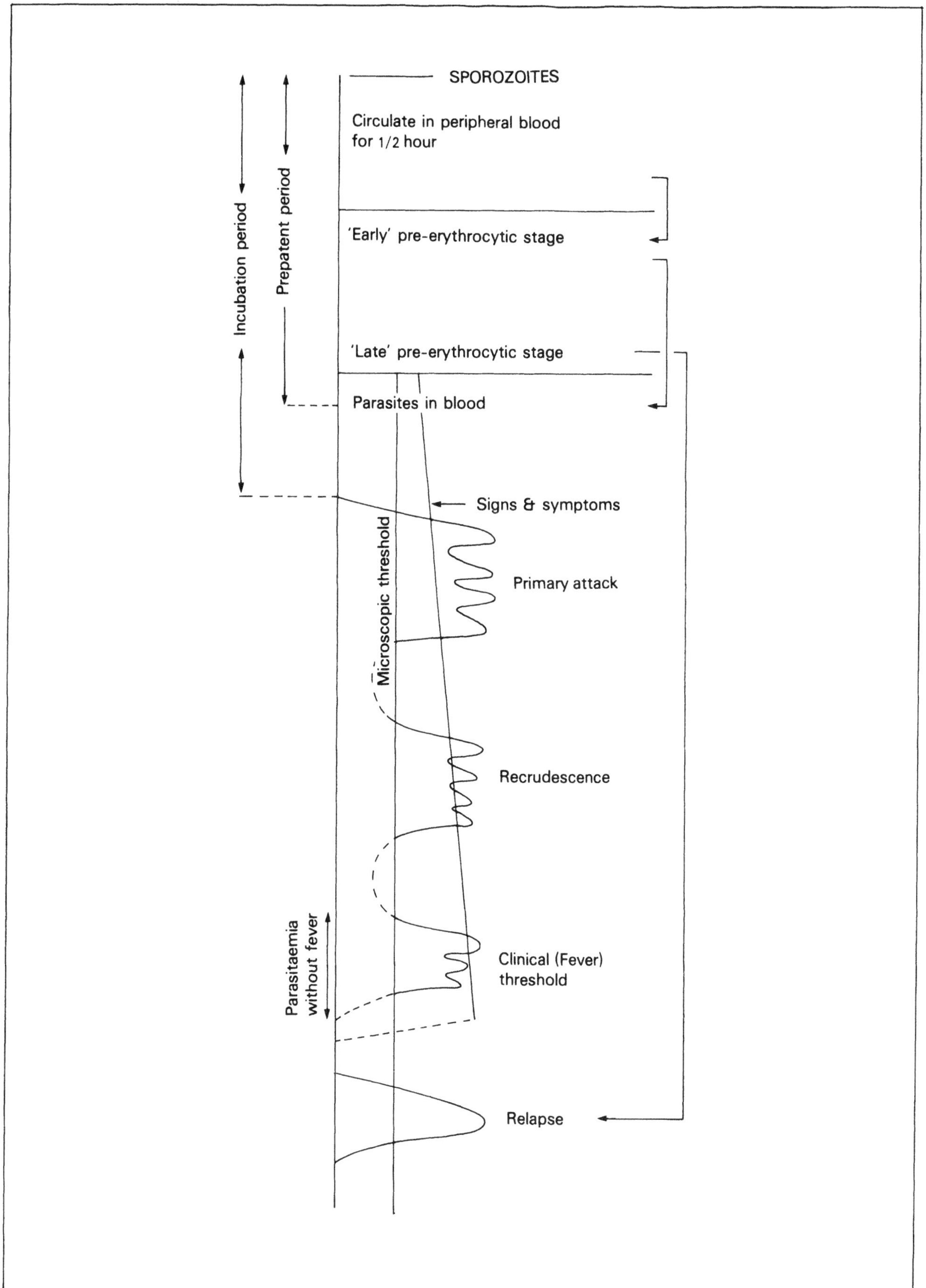
SPOROZOITES
Circulate in peripheral blood for 1/2 hour
Incubation period
Prepatent period
'Early' pre-erythrocytic stage
'Late' pre-erythrocytic stage
Parasites in blood
Signs & symptoms
Microscopic threshold
Primary attack
Recrudescence
Parasitaemia without fever
Clinical (Fever) threshold
Relapse

112

**113 Morphology of Malaria Parasite**

*P. falciparum*
1) Early trophozoite (Accolé form).
2) Early trophozoite double infection.
3) Early trophozoite double chromatin with a few Maurer's dots.
4) Late trophozoite with Maurer's dots and crenated red cell.
5) Mature schizont with merozoites and clumped pigment.
6) Macrogametocyte with bluish cytoplasm and compact chromatin.
7) Microgametocyte with pinkish cytoplasm and dispersed chromatin.

*P. vivax*
1) Early trophozoite (ring form) with Schuffner's dots.
2) Late trophozoite with Schuffner's dots and enlarged red cell.
3) Late trophozoite with amoeboid cytoplasm.
4) Late trophozoite with amoeboid cytoplasm.
5) Mature schizont with merozoites and clumped pigment.
6) Microgametocyte with irregular nucleus.
7) Macrogametocyte with compact nucleus.

*P. malariae*
1) Early trophozoite (ring form).
2) Early trophozoite with central chromatin.
3) Early trophozoite elongated form.
4) Late trophozoite band form with distinct pigment.
5) Mature schizont with merozoite forming a rosette.
6) Microgametocyte with irregular nucleus.
7) Macrogametocyte with compact nucleus.

*P. ovale*
1) Early trophozoite (ring form) with Schuffner's dots.
2) Developing schizont with enlarged red cell and Schuffner's dots.
3) Developing schizont in a red cell with jagged edges.
4) Developing schizont in an irregular red cell.
5) Mature schizont with merozoites arranged irregularly.
6) Microgametocyte with irregular nucleus.
7) Macrogametocyte with compact nucleus.

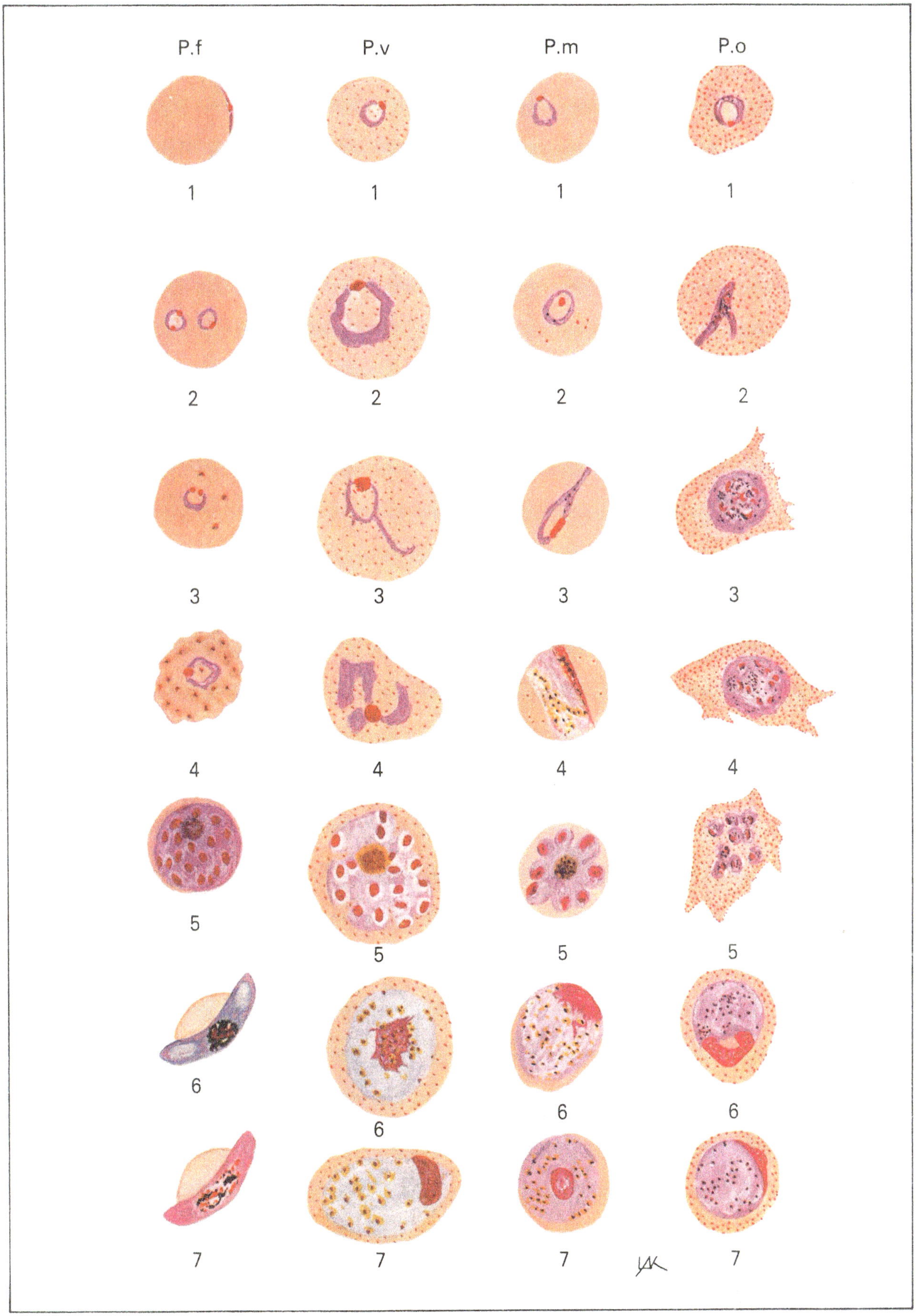


113

**114 Malarial parasite** inside a red cell, with ingested haemoglobin in its cytoplasm. X18,000. P = parasite. Hb = haemoglobin.

**115 Malarial parasite** with electron dense clumps of malarial pigment (haemozoin). The pigment is enclosed in membrane bound vacuoles. X25,000. MP = malarial pigment. Electron micrograph.

**116 Macrophage in liver** showing ingested malarial pigment. The liver macrophages are the cells mainly involved in the removal of this pigment from the circulation. X25,000. M = macrophage. MP = malarial pigment. Electron micrograph.

**117 Section of an oocyst** of *P. falciparum* showing large number of sporozoites cut at different levels. X4,500. OW = oocyst wall. Electron micrograph.

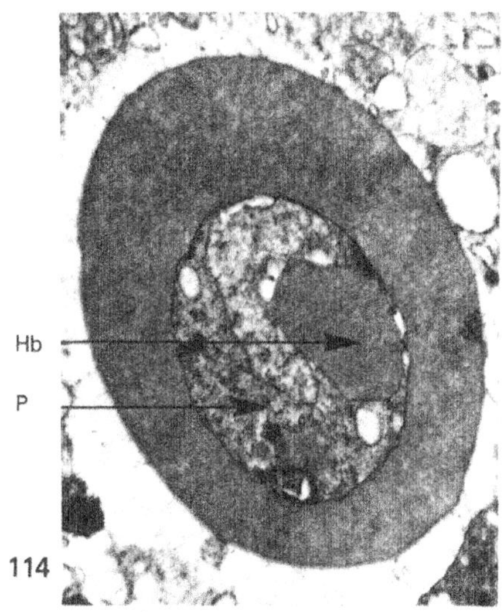


114

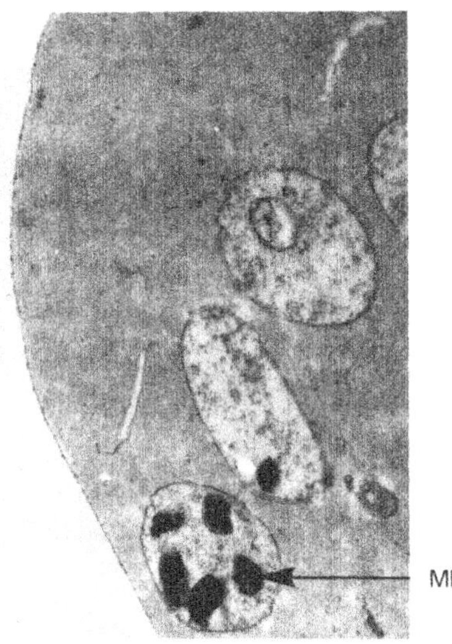


115

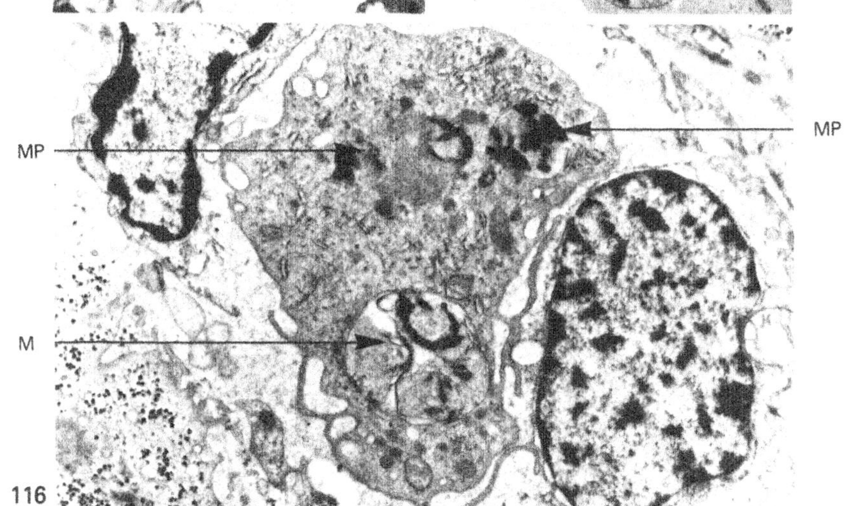


116

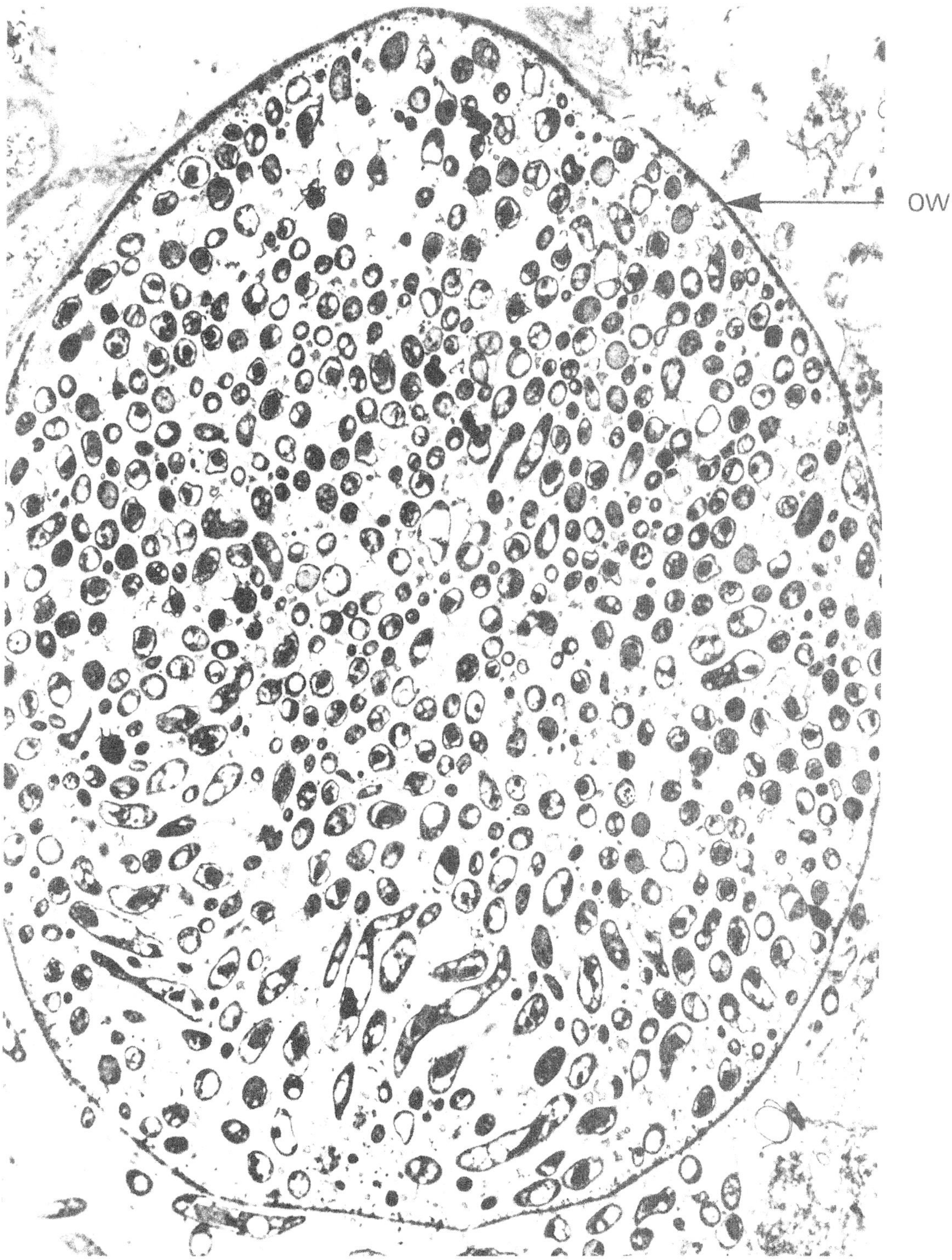


117

**118 Anopheline mosquito resting on a net**. The resting position of the adult *Anopheles* is typical in that the head, thorax and abdomen are kept in almost a straight line. The long axis of the body forms an angle of about 45° to the surface.

**119 Culicine mosquito taking a blood meal**. Unlike Anopheline, these mosquitoes sit with a hunch. During the act of feeding the labium curls backwards allowing the biting fascicle to penetrate the tissues. Both the labium and the biting fascicle can be seen in this photograph of *Aedes aegypti*.

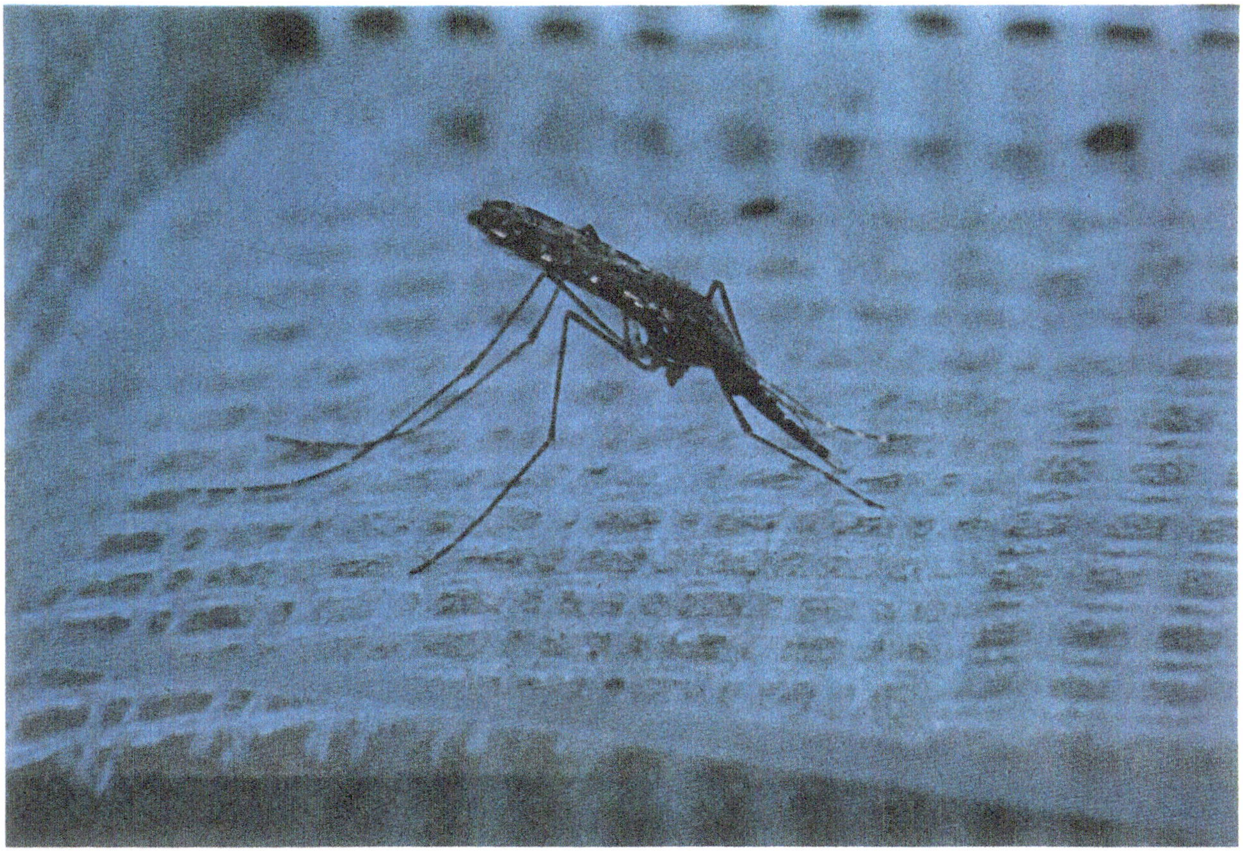

118

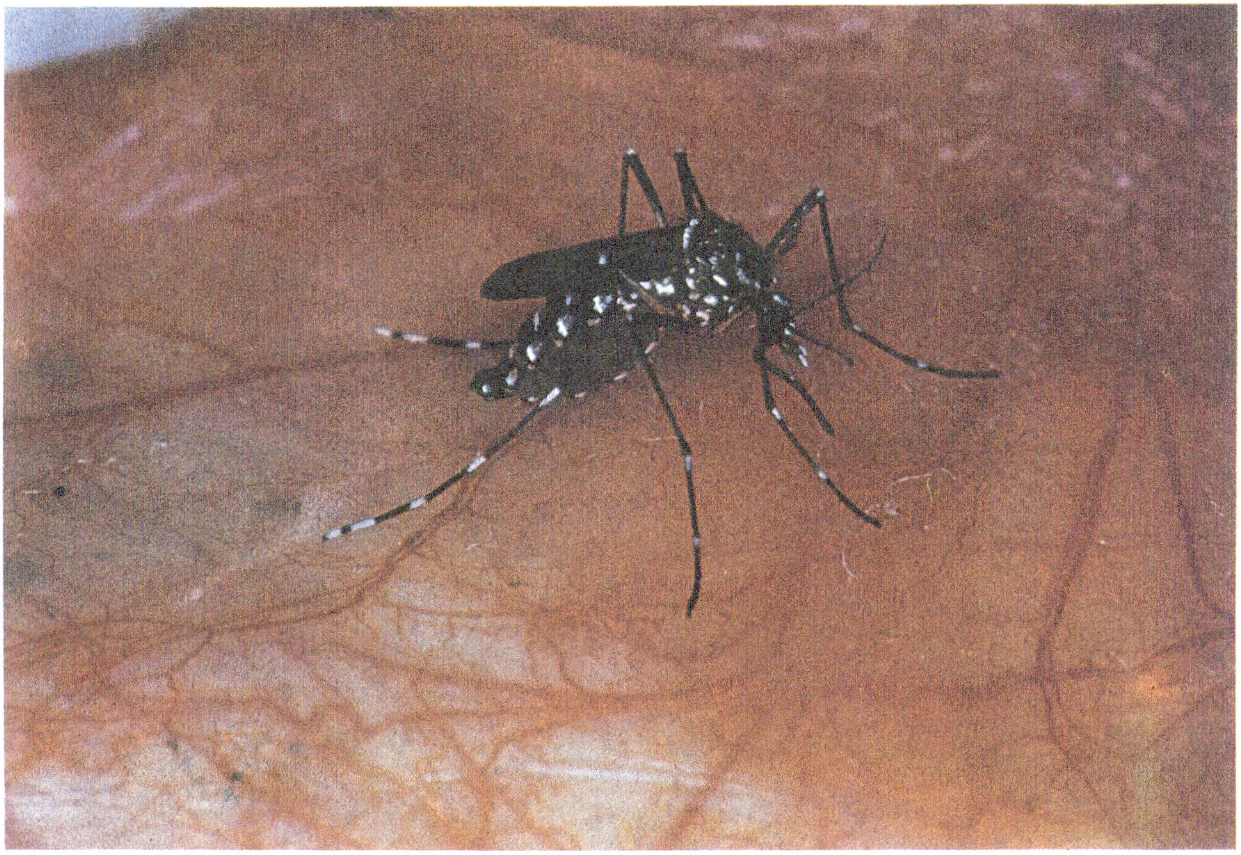

119

**120 Anopheles eggs, with a larva emerging.** The eggs have distinct lateral floats which easily differentiates them from culicine eggs.

**121 Anopheles egg, with a larva emerging.** The anterior end of the larva emerges first.

**122 A newly emerged Anopheles larva (1st instar larva).** At the centre of the head a pointed structure can be seen. This is the egg breaker and is used for cutting the egg shell.

**123 Ookinete, from the midgut of an infected mosquito.** Giemsa. X800. Enlarged by 5.4.

**124 Oocysts of** *P. falciparum* in midgut of an infected mosquito. Oocysts appear as circular bodies. Fresh preparation. X100. Enlarged by 5.4.

**125 Sporozoites from an infected salivary gland.** These are generally clumped and often lie inside secretory masses of saliva. In this case *Plasmodium gallinaceum.* Fresh preparation. Interference contrast. X400. Enlarged by 23.4.

**126 Pre-erythrocytic schizont in liver.** These mature in 6-14 days' time liberating merozoites into the blood stream. Giemsa-colophonium. X400. Enlarged by 5.4.

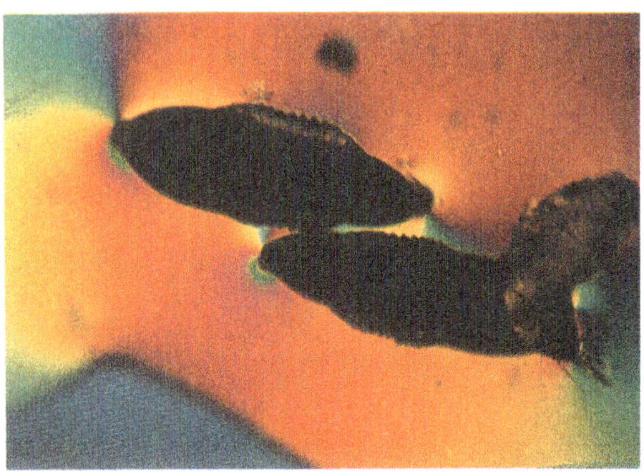

120

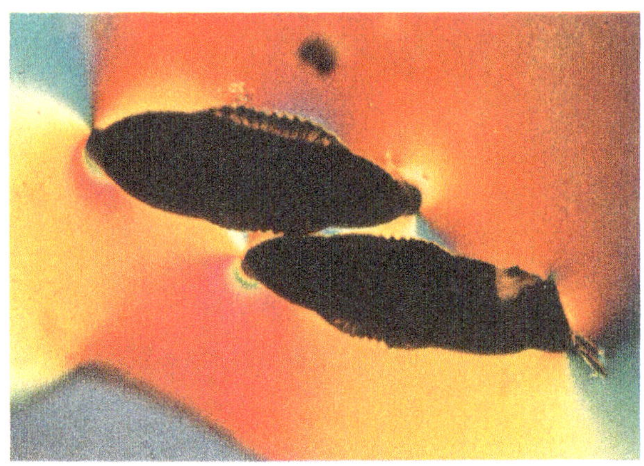

121

122

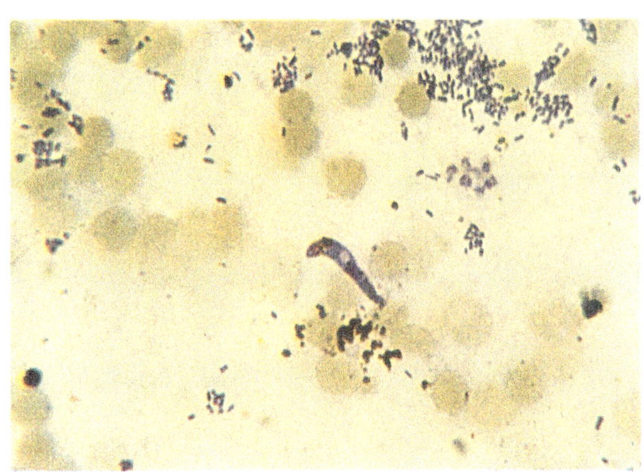

123

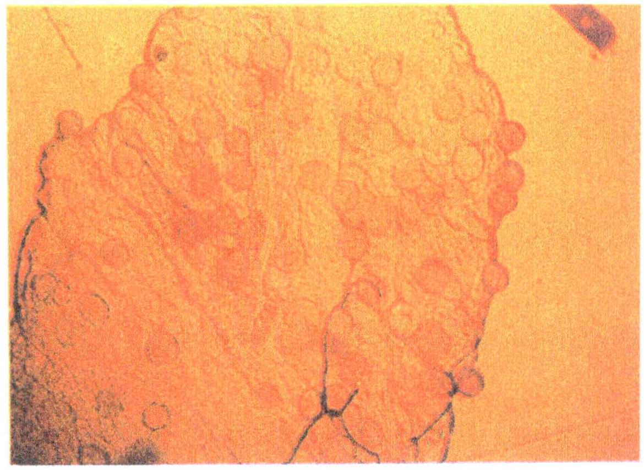

124

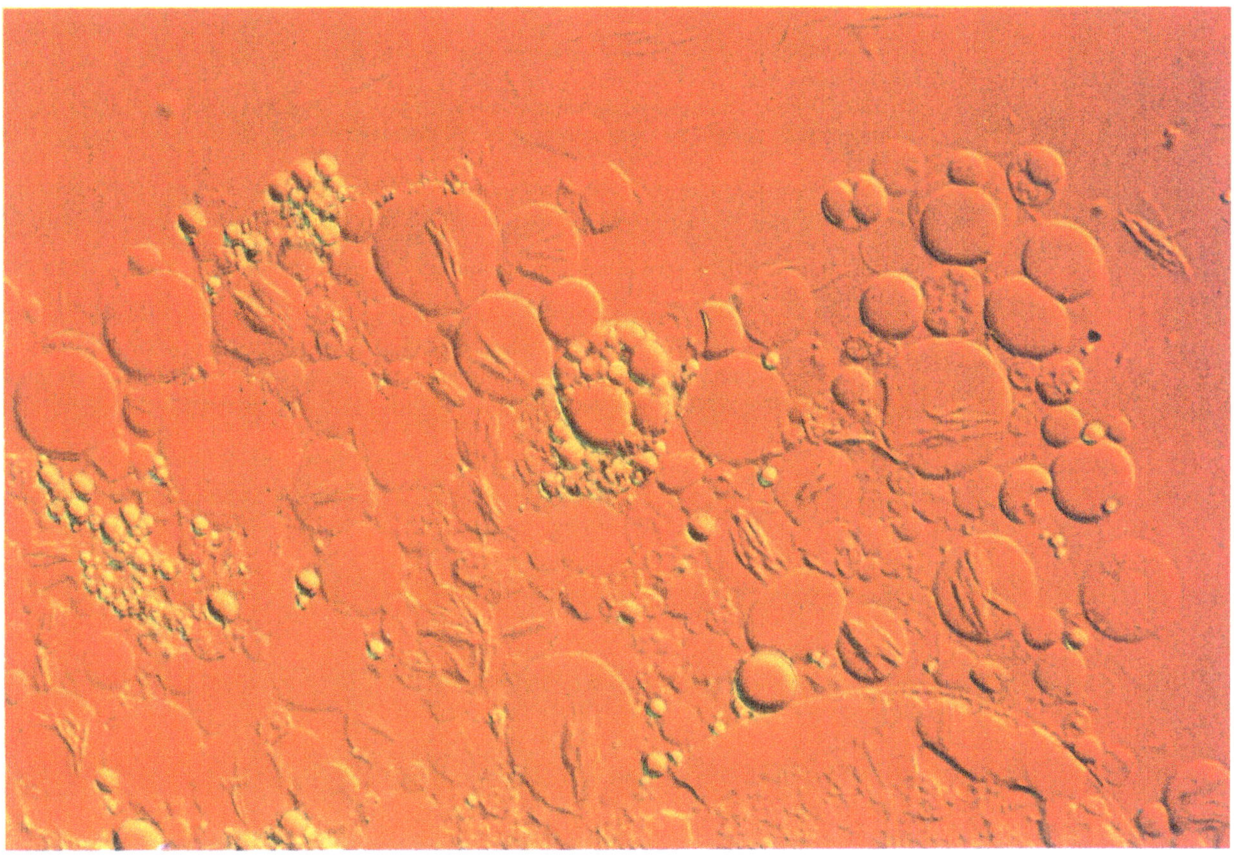

125

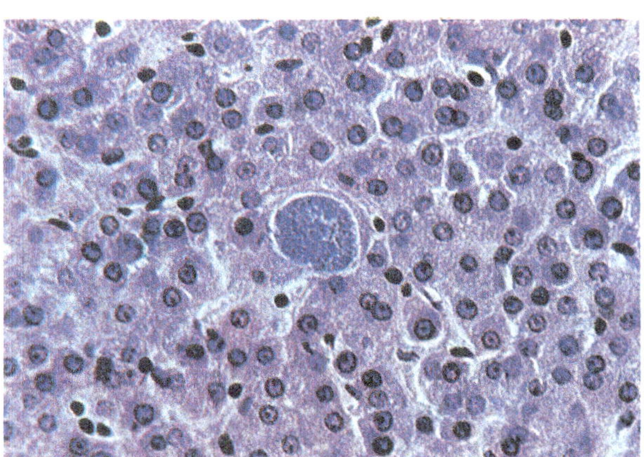

126

**127 P. falciparum thin smear, showing early trophozoites.** In *P. falciparum* schizonts rarely appear in the blood and this is the stage most commonly seen. Giemsa. X1000. Enlarged by 5.4.

**128 P. falciparum thick smear, showing early trophozoites.** The red cells are lysed. Giemsa. X1000. Enlarged by 5.4.

**129 P. falciparum thick smear, showing two gametocytes.** The red cells are lysed. Giemsa. X1000. Enlarged by 5.4.

**130 P. falciparum thin smear.** Showing early trophozoites stained with acridine orange. The nucleus is yellowish (DNA) and the cytoplasm reddish (RNA). At about 6 o'clock a white cell can be seen. X1000. Enlarged by 5.4.

**131 P. vivax thin smear, showing early trophozoites.** The infected red cells are enlarged and show some stippling. Giemsa. X1000. Enlarged by 5.4.

**132 P. vivax thin smear.** Showing late trophozoites (amoeboid stage). The infected red cells are enlarged and show marked stippling. Giemsa. X1000. Enlarged by 5.4.

**133 P. vivax thin smear.** A mature schizont about to rupture. A clump of malarial pigment can be seen in the centre. Giemsa. X1000. Enlarged by 5.4.

**134 P. vivax thin smear.** Merozoites lying free. Malarial pigment is seen as a clump on one side. Giemsa. X1000. Enlarged by 5.4.

**135 P. vivax thick smear, early trophozoites.** The red cells are lysed. Giemsa. X1000. Enlarged by 5.4.

**136 P. malariae thick smear, mature schizont.** This stage is frequently seen in *P. malariae* infection. The number of merozoites formed are less than other species. (In this case 8). Giemsa. X1000. Enlarged by 5.4.

127

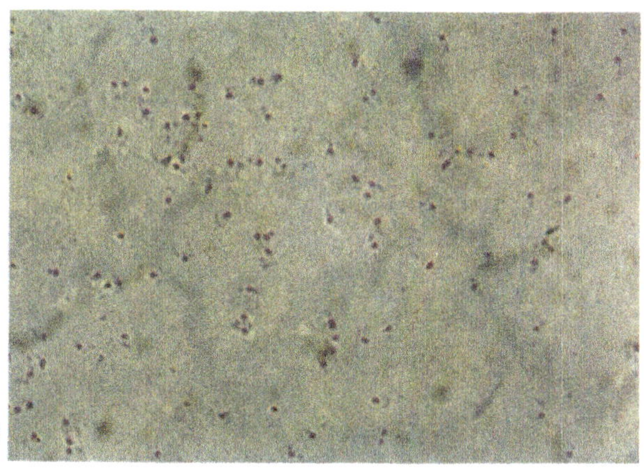
128

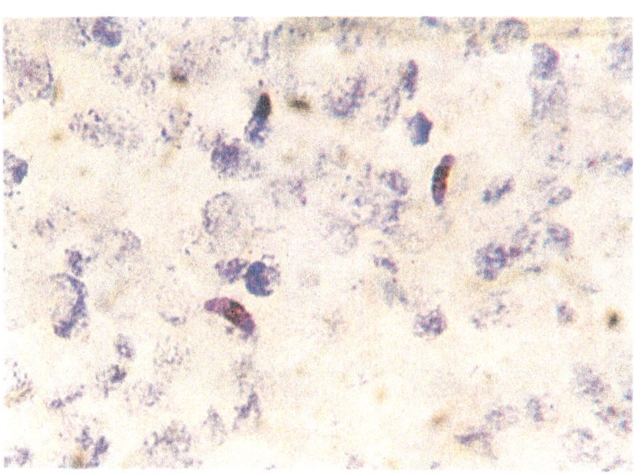
129

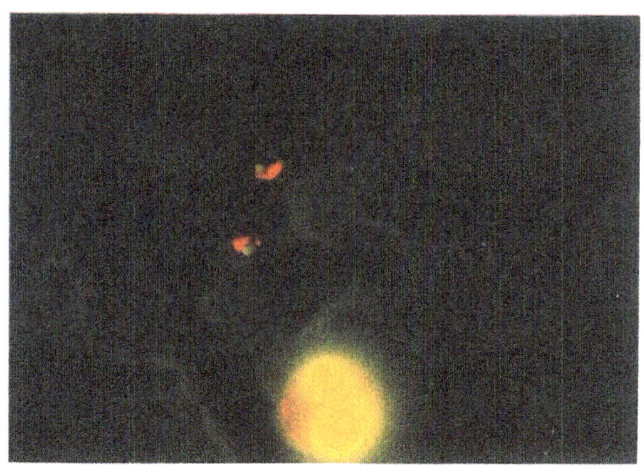
130

131 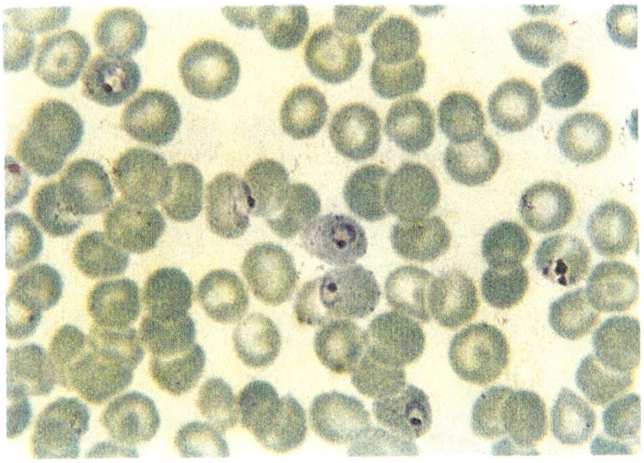

132 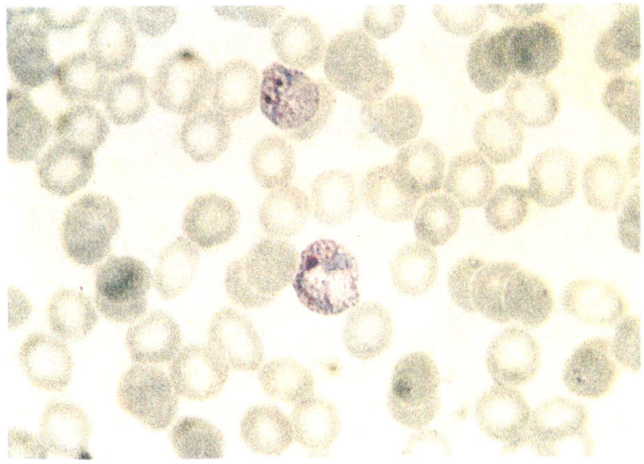

133 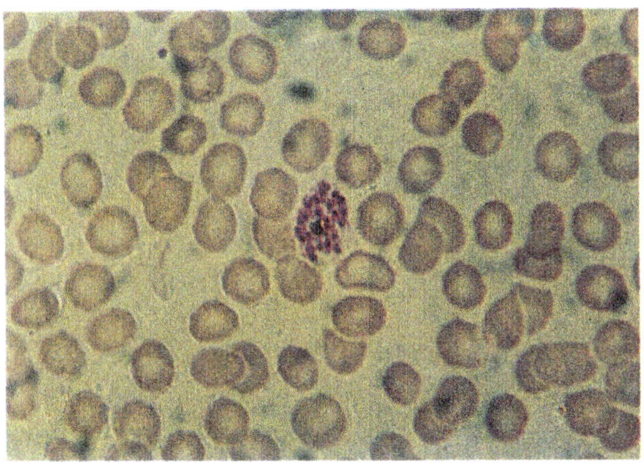

134 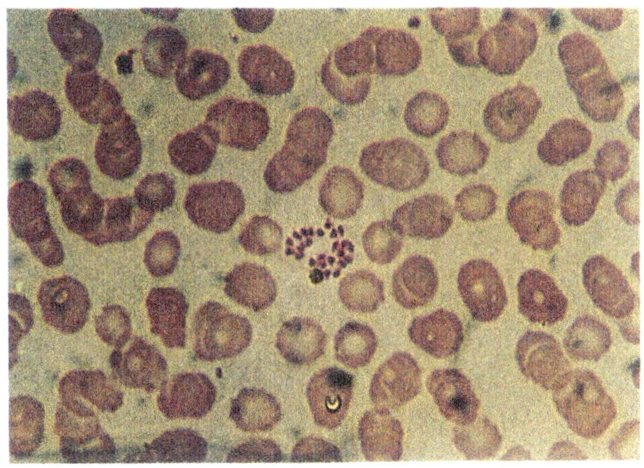

135 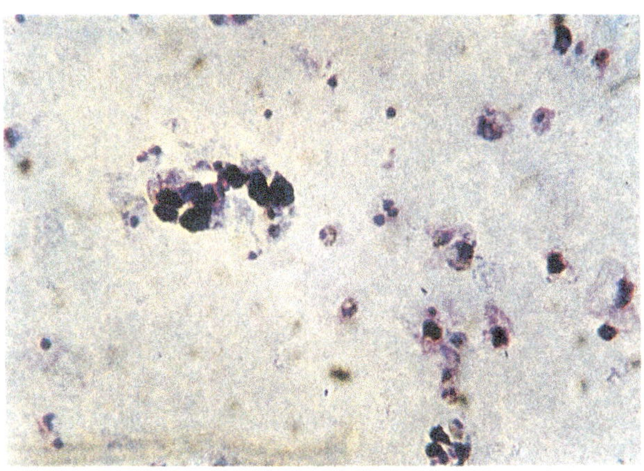

136 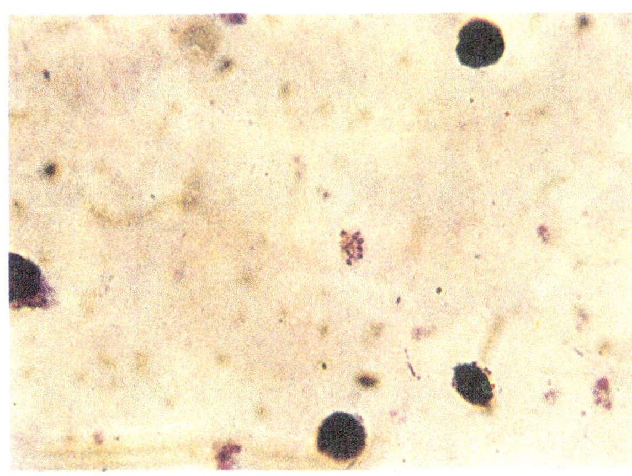

**137 Section of liver from a case of P. falciparum infection.** The Kupffer cells (macrophages) are loaded with black pigment, which is malarial pigment (haemozoin). H and E. X400. Enlarged by 23.4.

**138 Section of brain from a case of P. falciparum infection.** The infected cells adhere to the endothelial lining of the blood vessels. In small blood vessels this leads to occlusion and necrosis. H and E. X400. Enlarged by 23.4.

**139 Same tissues observed in polarized light.** The malarial pigment inside the red cells shines making it easily visible. H and E. X400. Enlarged by 23.4.

137

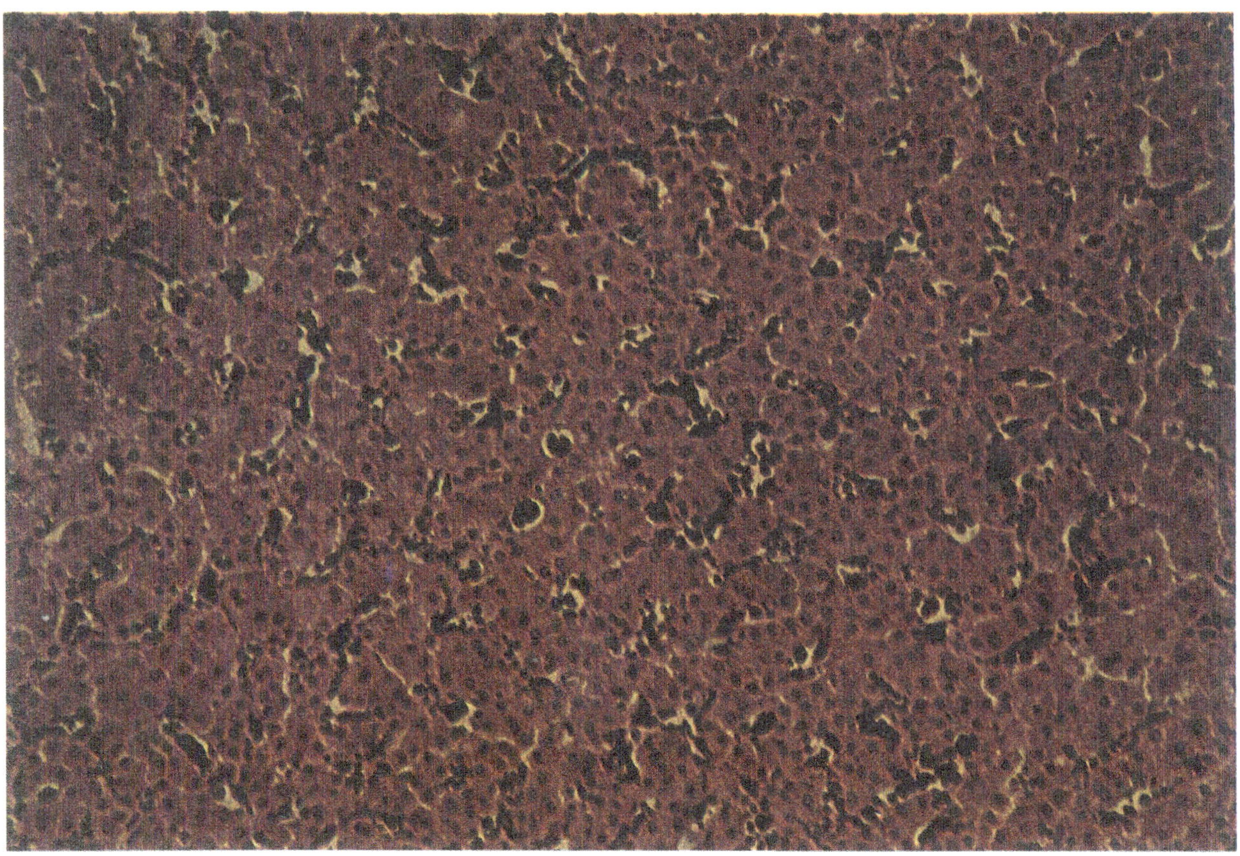

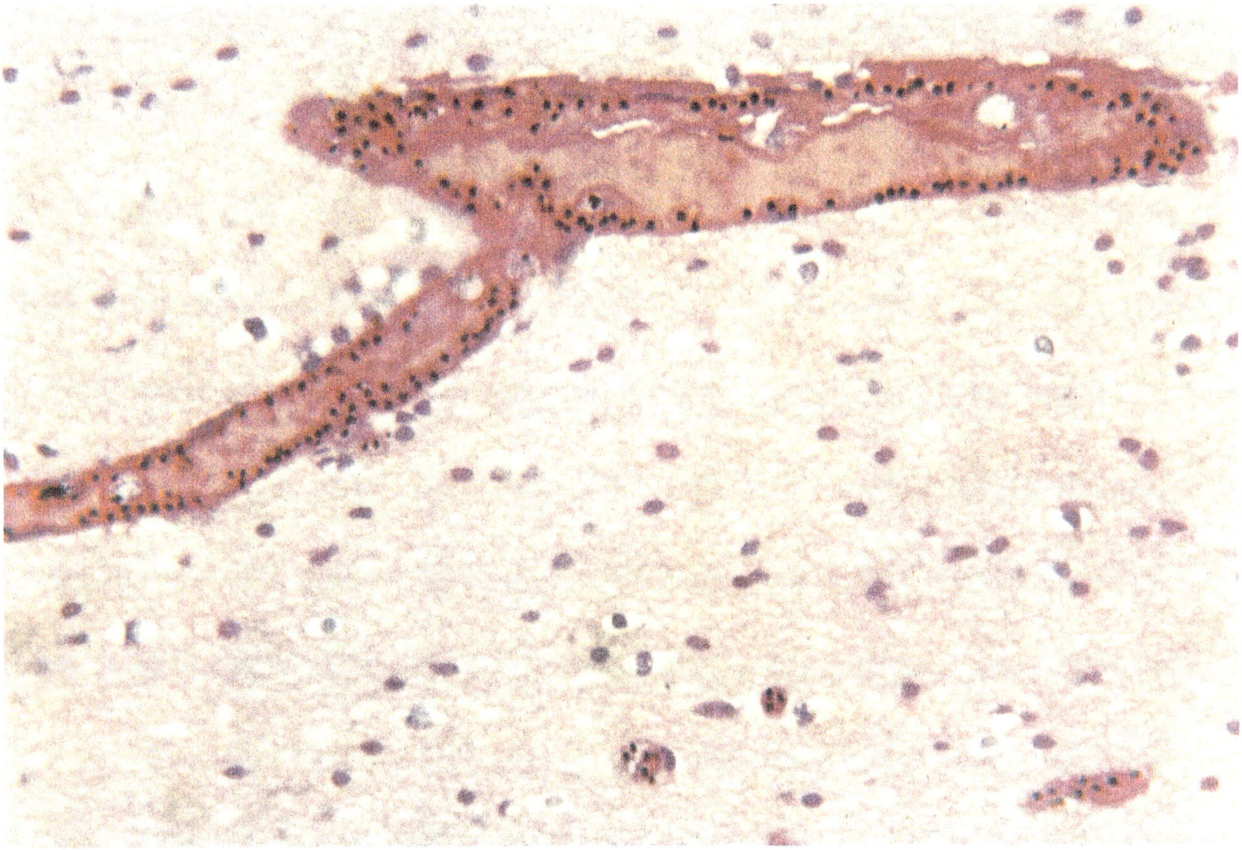

138

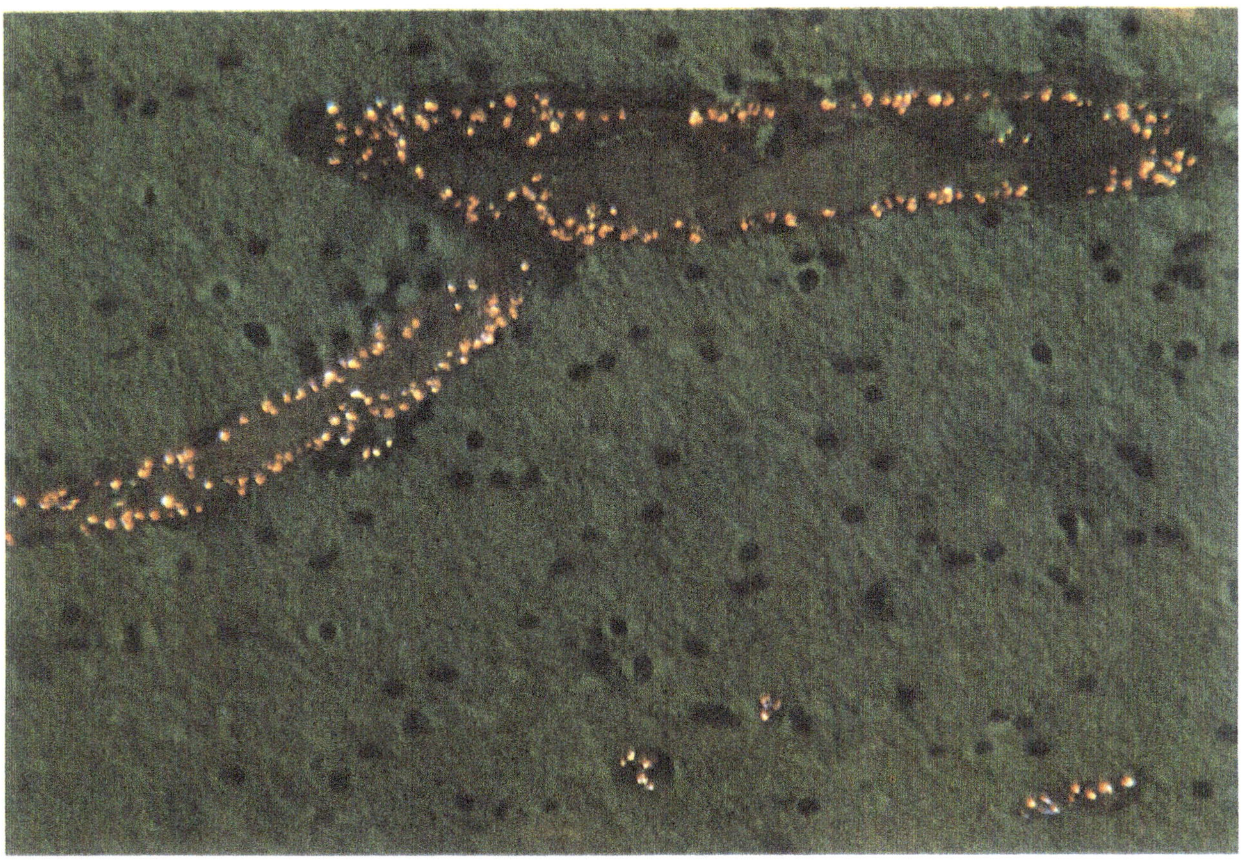

139

**140 Section of placenta from a case of P. falciparum infection.** The maternal blood shows red cells containing parasites and pigment. The foetal blood shows red cells without any parasites. It is generally believed that congenital transmission of malaria does not occur and that malarial parasites are unable to pass the placental barrier H and E. X400. Enlarged by 5.4.

**141 Spleen from a case of P. falciparum infection.** The deposition of malarial pigment makes it almost black in colour.

**142 Brain from a case of P. falciparum infection.** The cortex has a brownish colour and minute haemorrhages can be seen in the region of the basal ganglia. In this case death occurred due to haemorrhages and necrosis in the CNS (cerebral malaria).

**143 Tropical splenomegaly syndrome.** This young Malaysian girl had a protuberant abdomen due to a large spleen extending to the pelvis. She showed high antibody levels to *P. falciparum* and raised IgM. It is believed that the condition results from repeated infection with malarial parasite along with an abnormal immunolgical response.

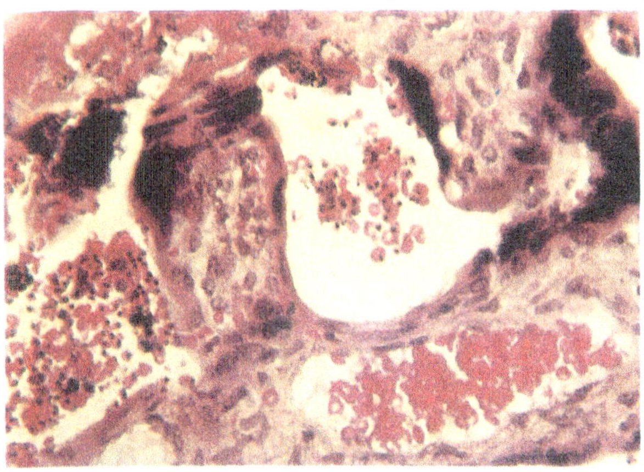

140

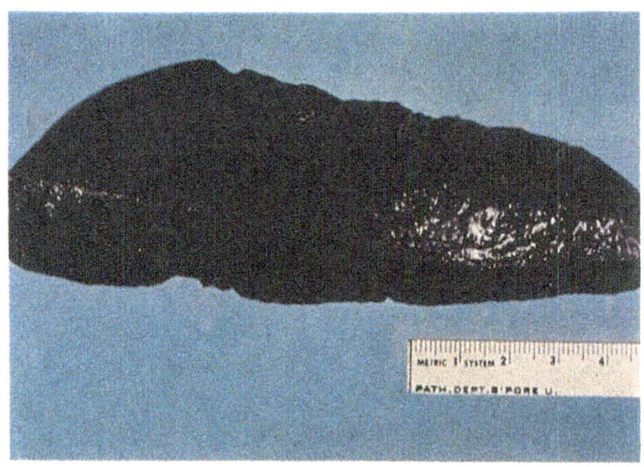

141

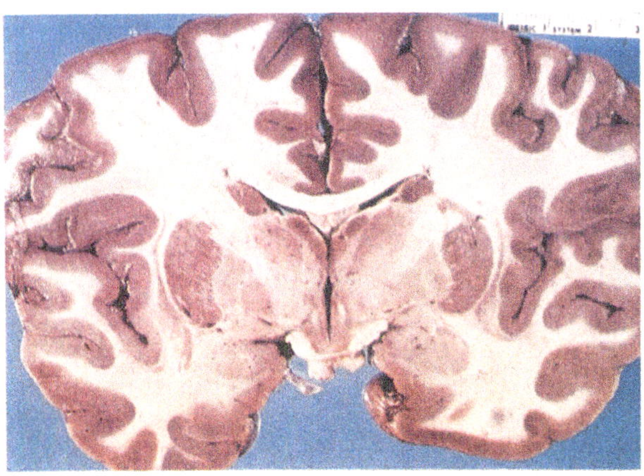

142

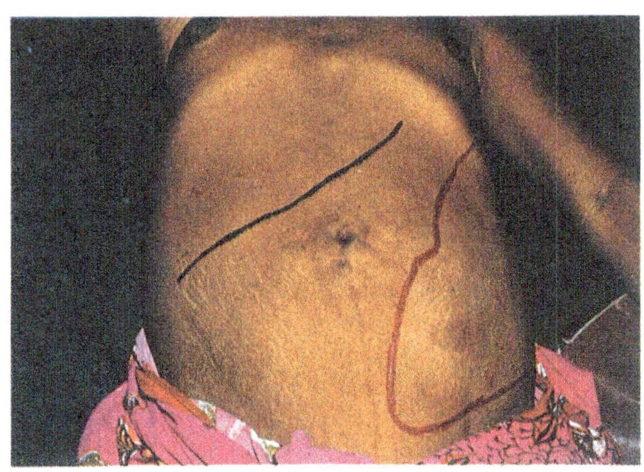

143

# Section 7
# Toxoplasma gondii

## Terms Used in Relation to Toxoplasma and Sarcocystis

**Endozoites** — Stages found in the macrophages and other host cells. As these multiply rapidly they are also known as tachyzoites. Formerly they were known as trophozoites or the proliferative forms. These stages measure between 4 to 7 by 2 to 4μm. The shape is generally crescentic.

**Pseudocysts** — These consist of a large number of endozoites enclosed in a macrophage or some other host cell. The parasites are bound by host cell tissue.

**Cystozoites** — Stages found in the cysts. As these multiply slowly they are also known as bradyzoites. These are morphologically similar to endozoites but contain a large amount of polysaccharide material and, therefore, react strongly to staining by the periodic acid-Schiff (PAS) method.

**Cyst** — Consists of a large number of parasites enclosed within a well defined cyst wall, made up of parasitic tissue. The cyst may be located in any organ of the body and is generally spherical in shape. The cyst wall is argyrophilic.

**Merozoites** — Are the stages formed as a result of asexual development by multiple fission in the epithelial cells of the intestine. Basic structure of this stage resembles endozoites.

**Schizont** — Is the stage which gives rise to merozoites. This process of division is known as schizogony.

**Microgametocyte** — Is the male stage of the parasite found in the epithelial cells of the intestine. The nucleus undergoes many divisions leading to the formation of microgametes.

**Macrogametocyte** — Is the female stage of the parasite found in the epithelial cells of the intestine. The nucleus does not divide and gives rise to only a single macrogamete.

**Gametogony** — Is the sexual form of division which involves the formation of micro and macrogametocytes.

**Oocyst** — Is the resistant stage of the parasite which is formed as a result of gametogony. *Toxoplasma* oocysts measure approximately 9 x 13μm.

**Sporogony** — Is the process of development of sporozoites, which are the infective stages of the parasite. This form of development occurs outside the host.

**Endodyogeny** — Is an asexual form of division in which 2 daughter cells form within a mother cell. Both endozoites and cystozoites multiply by this method.

**Endopolygeny** — Is an asexual form of division seen in the cat intestine. The process resembles endodyogeny but differs in the formation of more than 2 offspring with multiple nuclear divisions occurring simultaneously.

**Conoid** — Is the hollow cone-like structure at the anterior end of the cystozoites, endozoites and merozoites. It is covered by spirally arranged fibrillar structures. It is generally believed that this organ is used by the parasite to penetrate host cells.

**Rhoptries** — Also known as the paired organelles. These are electron dense club-like structures located at the anterior end of endozoites, cystozoites and merozoites. It is postulated that these structures produce enzymes required for the penetration of the parasite into the host.

**Micronemes** — Also known as taxonemes. These are electron dense, small, cord-like structures located near the conoid. In cross-section they appear as oval or spherical bodies. Their function is not known.

## Toxoplasmosis

*Toxoplasma gondii* is a very common parasite of man and animals. The infection is known as Toxoplasmosis. It is estimated that at least one-third of the world's adult population contracts toxoplasmosis.

### *Life cycle of Toxoplasma gondii* [144]

**Development in the cat**

The parasites undergo gametogony only in the family Felidae. The developmental stages in the domestic cat have been studied in detail and seem to follow the same basic pattern of development as other coccidia.

The infective stages are the sporozoites, cystozoites and endozoites which on ingestion, penetrate the epithelial cells of the intestines. Within the host cells they round up and grow. Asexual form of division occurs first, leading to the formation of merozoites. The merozoites enter fresh host cells and initiate up to 5 different cycles (Frenkel, 1973). Some of the merozoites are transferred into the sexual stages, initiating gametogony. A macrogamete is fertilised by a motile microgamete resulting in the formation of a zygote. The zygote secretes a protective coat and is transformed into an oocyst. The oocysts are passed in the faeces after the disintegration of the host cell epithelium. During sporogony 2 sporoblasts form from a single cell. These then become sporocysts by acquiring a cyst wall. Each sporocyst gives rise to 4 sporozoites.

*Toxoplasma* differ from other coccidia in that extra-intestinal development can also occur simultaneously with the intestinal phase. This takes place by the passage of merozoites from the intestines into the lymphatics and the blood stream. The extra-intestinal stages consist of cysts and pseudocysts.

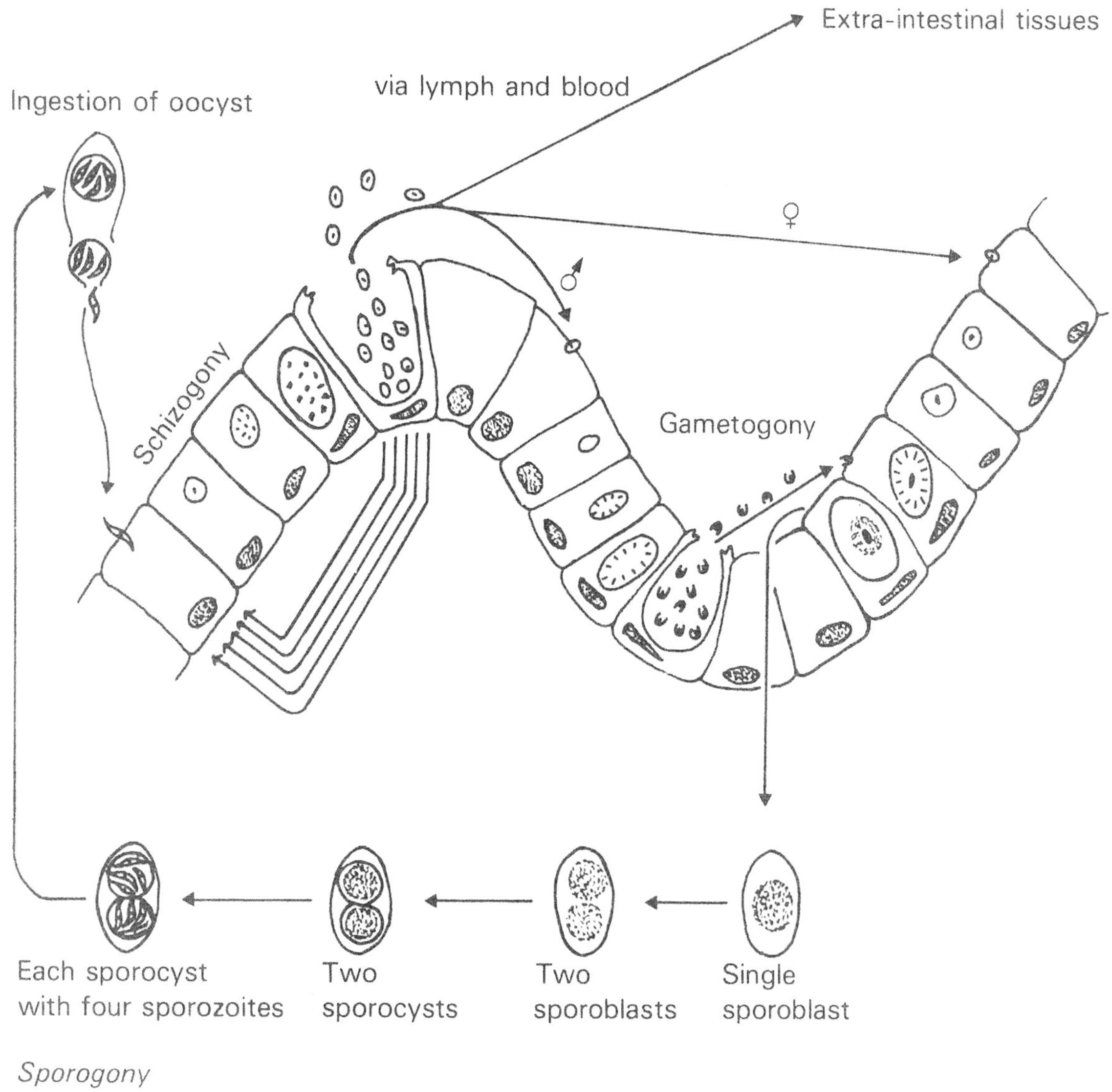
Extra-intestinal tissues
via lymph and blood
Ingestion of oocyst
♀
♂
Schizogony
Gametogony
Each sporocyst
with four sporozoites
Two
sporocysts
Two
sporoblasts
Single
sporoblast
Sporogony

144

### Development in man

Infection occurs following ingestion of oocysts from the cat or by eating improperly cooked meat containing cysts or pseudocysts. Meat of various domestic animals may be infected and these include pork, mutton, beef and poultry. Only asexual development occurs in man and oocysts are not formed in the intestine. Merozoites arising from the asexual development enter into the lymphatics and the blood and give rise to the formation of pseudocysts and cysts in various organs of the body. Human to human transmission can occur through the placenta giving rise to congenital toxoplasmosis.

## *Invasion of tissues by Toxoplasma*

*Toxoplasma* are essentially intracellular parasites. They enter host cells either by rupturing the membrane or by invaginating it. The entry into macrophages occurs mainly by phagocytosis. Within the macrophage the parasite multiplies by repeated endodyogeny giving rise to large colonies. The macrophage finally ruptures liberating the endozoites.

The cyst formation occurs mostly in the central nervous system, eyes, cardiac and skeletal muscle. The cyst is surrounded by a distinct wall which is argyrophilic. The cyst does not produce an inflammatory reaction when it is intact.

## *Clinical aspects*

### Acquired Toxoplasmosis

Toxoplasmosis is less severe in its acquired form, although the infection itself is very common. Individuals with clinical toxoplasmosis may show symptoms involving the eye or the lymphatic system. The eye lesion may present as uveitis, choroiditis and choroidoretinitis. The involvement of the lymphatic system may produce lymphadenopathy with or without fever. Rarely, myocarditis or myositis may also be seen.

### Congenital Toxoplasmosis

Toxoplasmosis is more severe in its congenital form. Congenital infection may lead to abortion or still birth and give rise to abnormalities of the central nervous system such as hydrocephalus or mental retardation. Less serious consequences of congenital infection such as eye lesions and hepatosplenomegaly are generally missed at birth and may be detected much later in life.

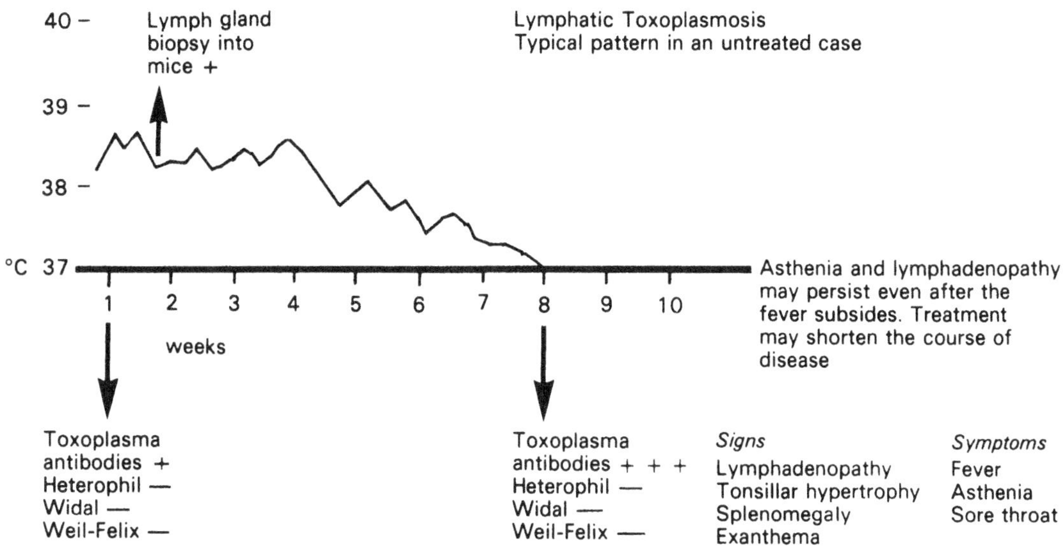
Lymphatic Toxoplasmosis
Typical pattern in an untreated case
40
39
38
°C 37
Lymph gland biopsy into mice +
1 2 3 4 5 6 7 8 9 10
weeks
Asthenia and lymphadenopathy may persist even after the fever subsides. Treatment may shorten the course of disease
Toxoplasma antibodies +
Heterophil —
Widal —
Weil-Felix —
Toxoplasma antibodies + + +
Heterophil —
Widal —
Weil-Felix —
*Signs*
Lymphadenopathy
Tonsillar hypertrophy
Splenomegaly
Exanthema
*Symptoms*
Fever
Asthenia
Sore throat

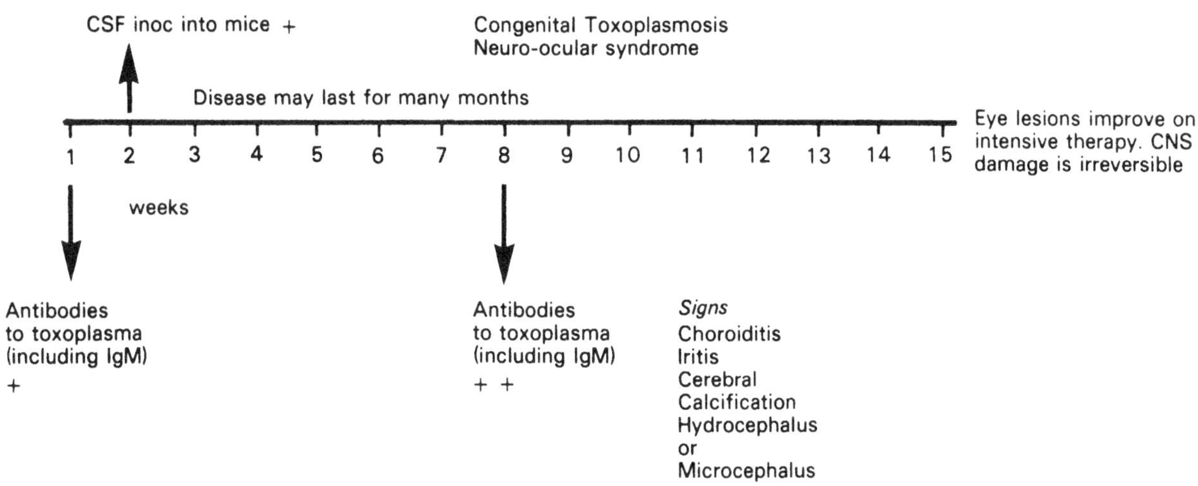
Congenital Toxoplasmosis
Neuro-ocular syndrome
CSF inoc into mice +
Disease may last for many months
1 2 3 4 5 6 7 8 9 10 11 12 13 14 15
weeks
Eye lesions improve on intensive therapy. CNS damage is irreversible
Antibodies to toxoplasma (including IgM) +
Antibodies to toxoplasma (including IgM) + +
*Signs*
Choroiditis
Iritis
Cerebral
Calcification
Hydrocephalus
or
Microcephalus

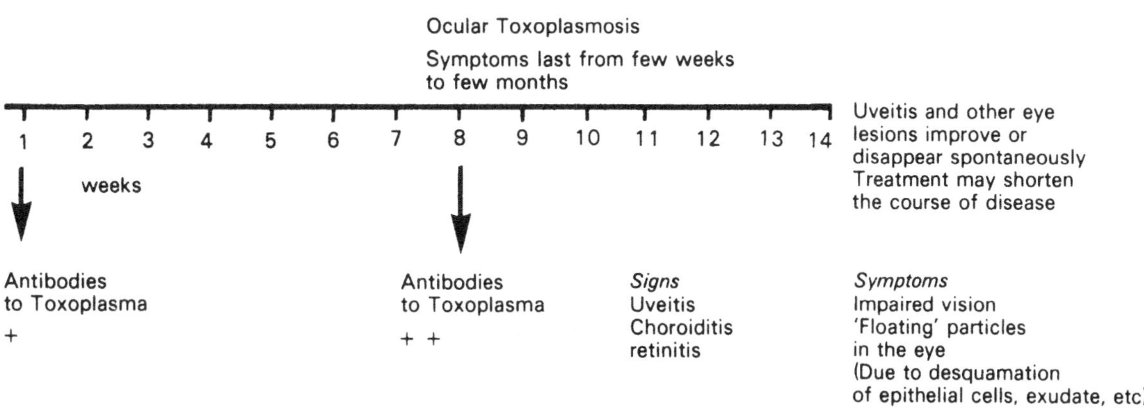
Ocular Toxoplasmosis
Symptoms last from few weeks to few months
1 2 3 4 5 6 7 8 9 10 11 12 13 14
weeks
Uveitis and other eye lesions improve or disappear spontaneously Treatment may shorten the course of disease
Antibodies to Toxoplasma +
Antibodies to Toxoplasma + +
*Signs*
Uveitis
Choroiditis
retinitis
*Symptoms*
Impaired vision
'Floating' particles in the eye
(Due to desquamation of epithelial cells, exudate, etc)

*Laboratory Diagnosis*

**Isolation of parasite**

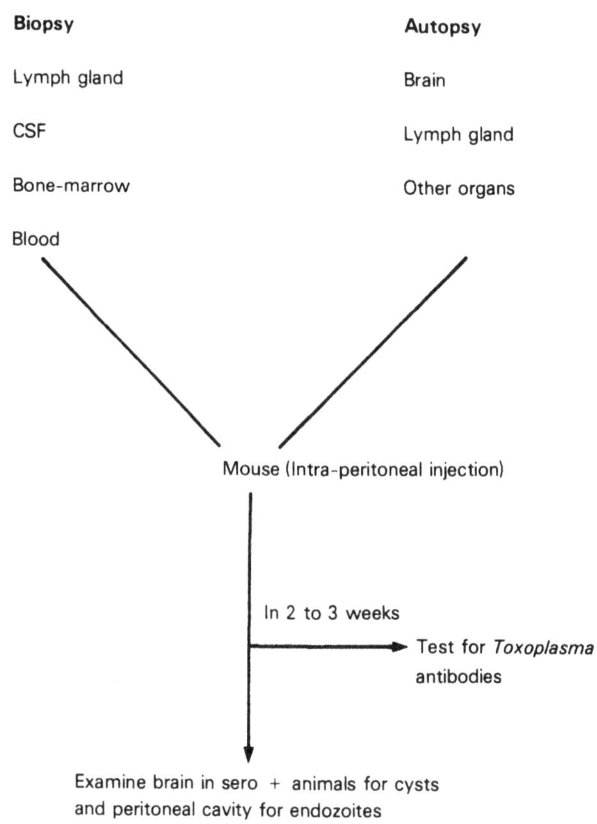


*Serological tests*

Three most commonly used tests are:
**Dye test of Sabin and Feldman** (Beverley and Beattie, 1952) — Depends on cytoplasmic lysis of the endozoites when exposed to antibody, in the presence of a heat sensitive non-specific substance found in the serum of certain individuals known as the 'accessory' factor. Modified parasites appear unstained or clear when treated with Methylene blue.

**Fluorescent antibody test** (Fletcher, 1965) — Is a sensitive test and has the advantage over the dye test in not requiring live parasites or the 'accessory' factor. Fluorescein labelled anti-IgM can be used to diagnose infection in the new born as passive transfer of IgM from mother to foetus does not occur. Detection of high titre of IgM antibodies in acquired toxoplasmosis, in the presence of low titres of indirect haemagglutination reaction, is highly suggestive of acute infection as IgM antibodies appear earlier than IgG (Karim and Ludlam, 1975).

**Indirect haemagglutination test** (Jacobs and Lunde, 1957) — Is a very sensitive test but has a disadvantage in that it takes longer to become positive as compared to the dye test and the fluorescent antibody test. Once it becomes positive it remains positive for many years. It is used widely for conducting antibody surveys.

## References

Beverley, J.K.A. and Beattie, C.P.: Standardization of the dye test for toxoplasmosis. Journal of Clinical Pathology 5: 350 (1952).

Fletcher, S.: Indirect fluorescent antibody technique in the serology of Toxoplasma gondii. Journal of Clinical Pathology 18: 193 (1965).

Frenkel, J.K.: Toxoplasmosis: Parasite life cycle, pathology, and immunology; in The Coccidia, Hammond and Long (Eds), University Park Press, Baltimore (1973).

Jacobs, L. and Lunde, M.N.: A haemagglutination test for toxoplasmosis. Journal of Parasitology 43: 308 (1957).

Karim, K.A. and Ludlam, G.B.: The relationship and significance of antibody titres as determined by various serological methods in glandular and ocular toxoplasmosis. Journal of Clinical Pathology 28: 42 (1975).

**145 Endodyogeny as seen in the electron microscope.**
1) Structure of *Toxoplasma* prior to division.
2) The nucleus changes its shape and becomes crescentic in appearance.
3) Two conoids develop anterior to the two poles of the crescentic nucleus.
4) The nucleus begins to divide and the conoids enlarge.
5) Two fully formed daughter parasites appear inside the mother parasite.
6) The mother parasite disintegrates liberating the daughter parasites. R = rophtries. G = golgi body. N = nucleus. M = mitochondria. C = conoid. V = vacuole.

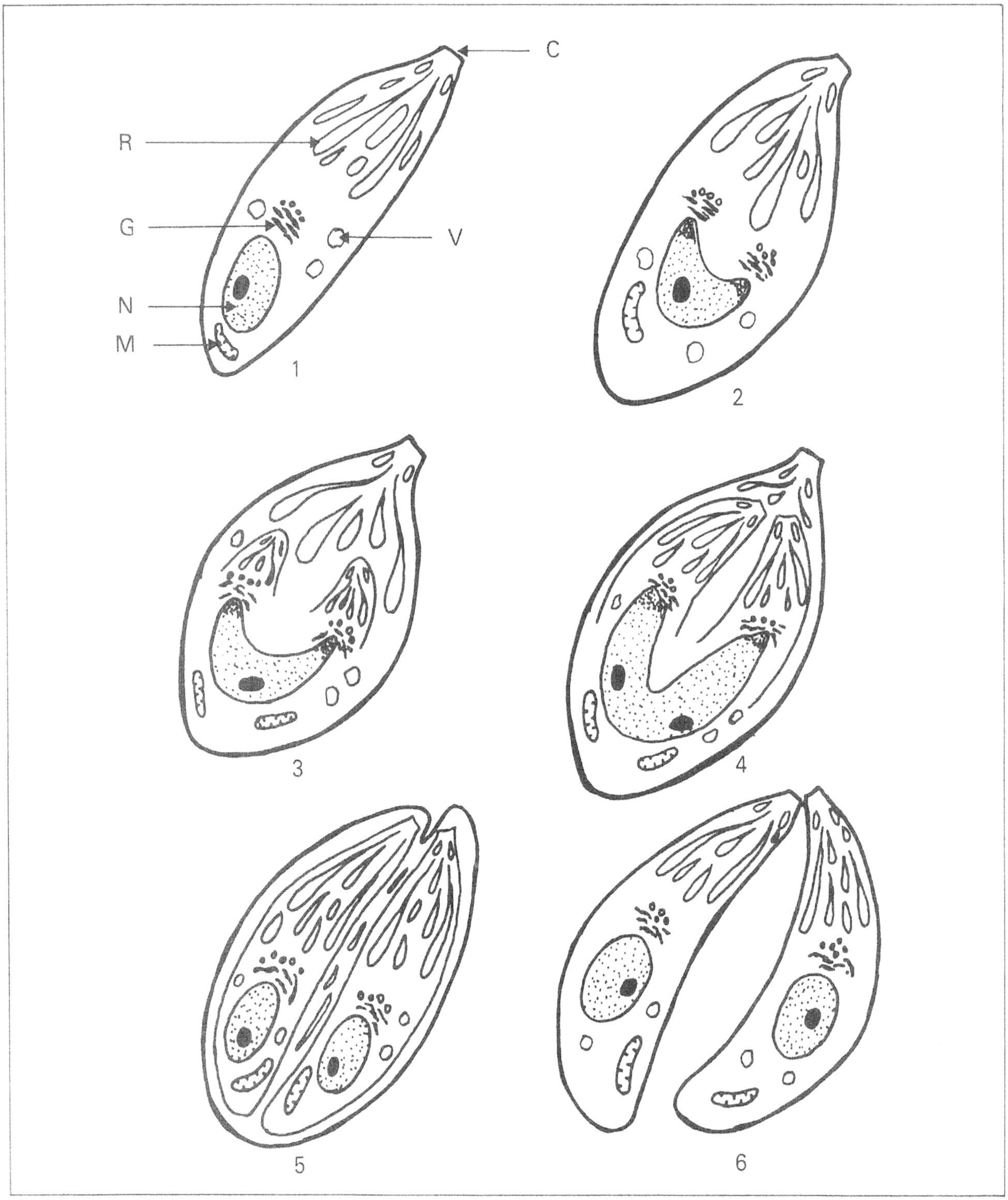


145

**146 Endozoite entering a macrophage.** The parasite enters with the conoid directed towards the host cell. During this process pseudopods are thrown around the parasite by the macrophage and it is phagocytosed. M = macrophage. C = conoid. P = pseudopods. X36000. Electron micrograph.

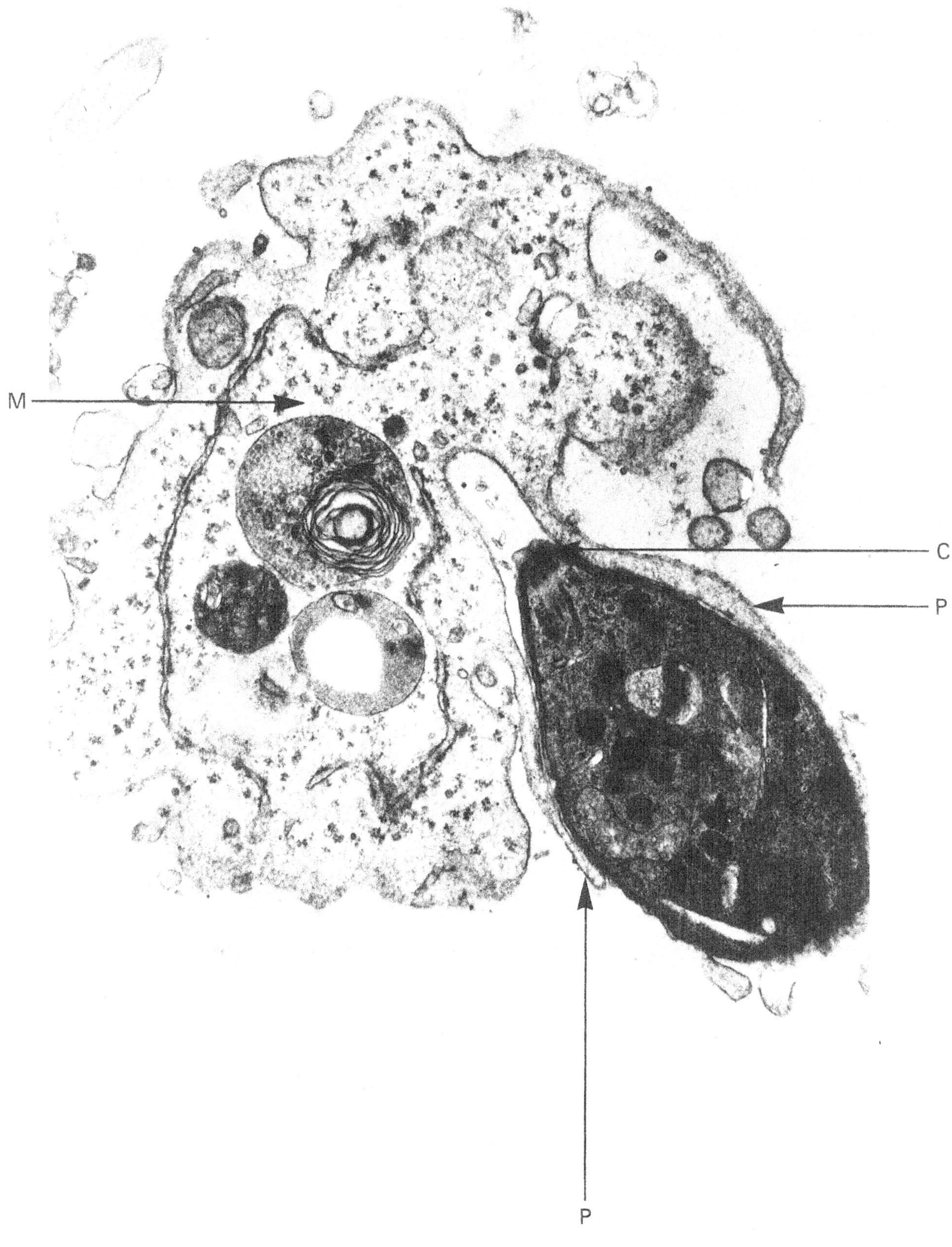


146

**147 Cyst in the brain.** Shows a large number of cystozoites enclosed by a distinct cyst wall. CZ = cystozoites. CW = cyst wall. B = brain. R = rhoptries. PG = polysaccharide granules. N = nucleus. X28,000. Electron micrograph.

**148 Cystozoites high magnification.** Rhoptries, micronemes and polysaccharide granules can be clearly seen. It is the presence of these polysaccharides that gives a strong PAS reaction. B = brain. CW = cyst wall. M = micronemes. R = rhoptries. PG = polysaccharide granules. X56,000. Electron micrograph.

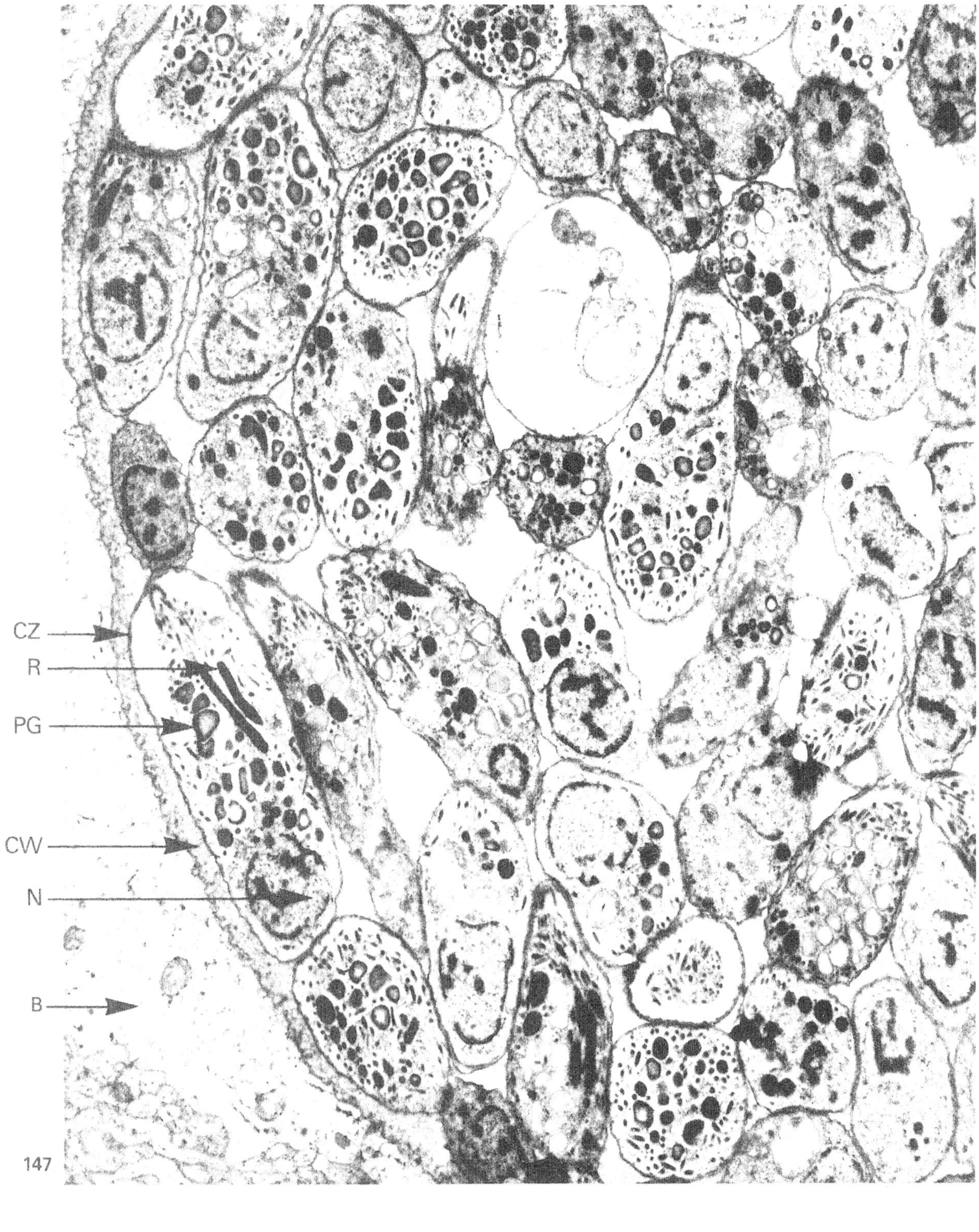


147

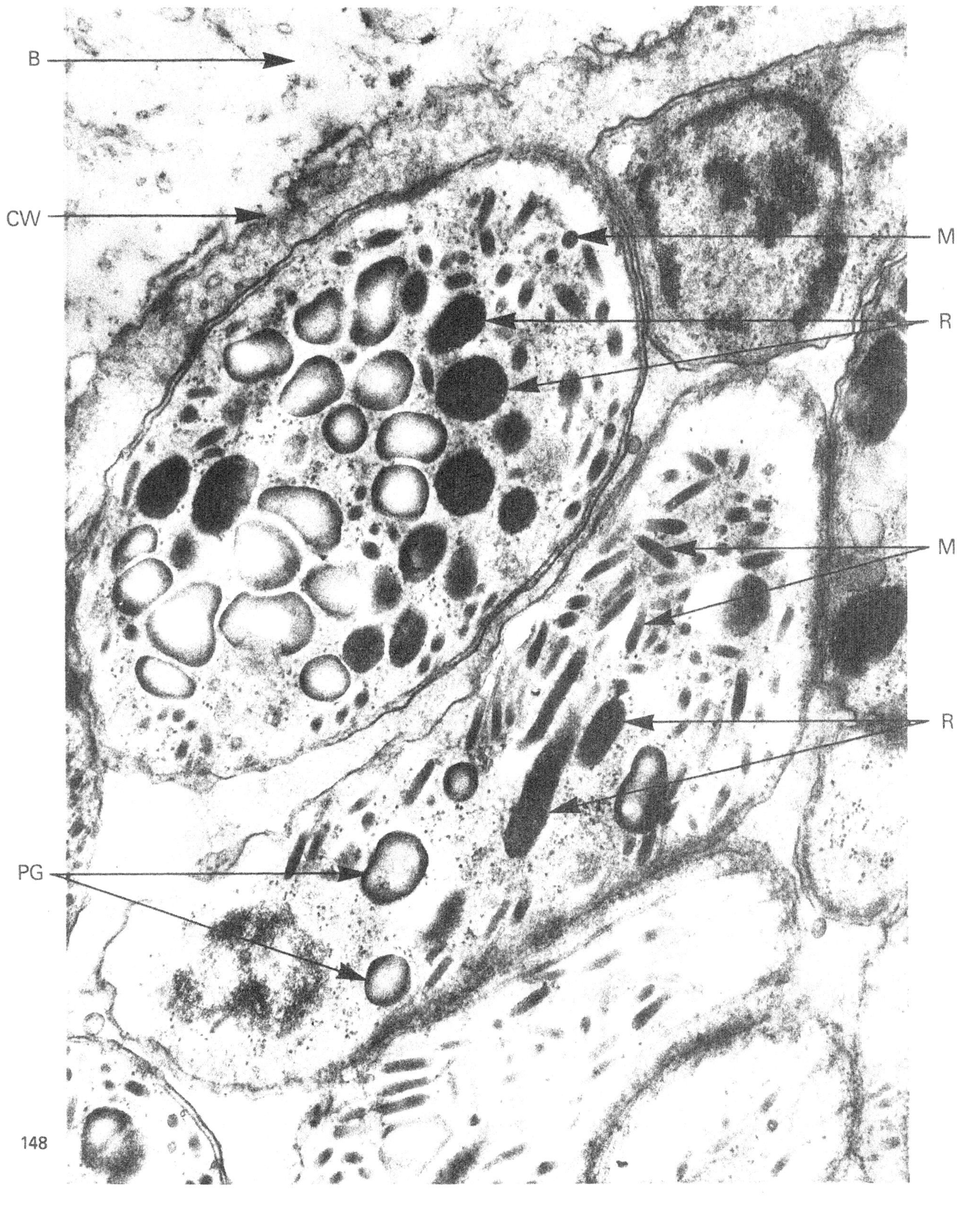


148

**149 Merozoite high magnification.** A distinct conoid with spiral markings at the anterior end. The parasite has elongated rhoptries. C = conoid. R = rhoptries. V = vacuoles. X56,000. Electron micrograph.

**150 Endozoite in a macrophage.** A macrophage with a clump of parasites in its cytoplasm can be seen. The macrophage membrane is stretched to accommodate the parasites. Two free endozoites are lying nearby. X6000. Scanning electron micrograph.

**151 Endozoite in terminal stage of division.** The parasites are attached at the posterior end, which indicates the terminal stage of endodyogeny. X22,500. Scanning electron micrograph.

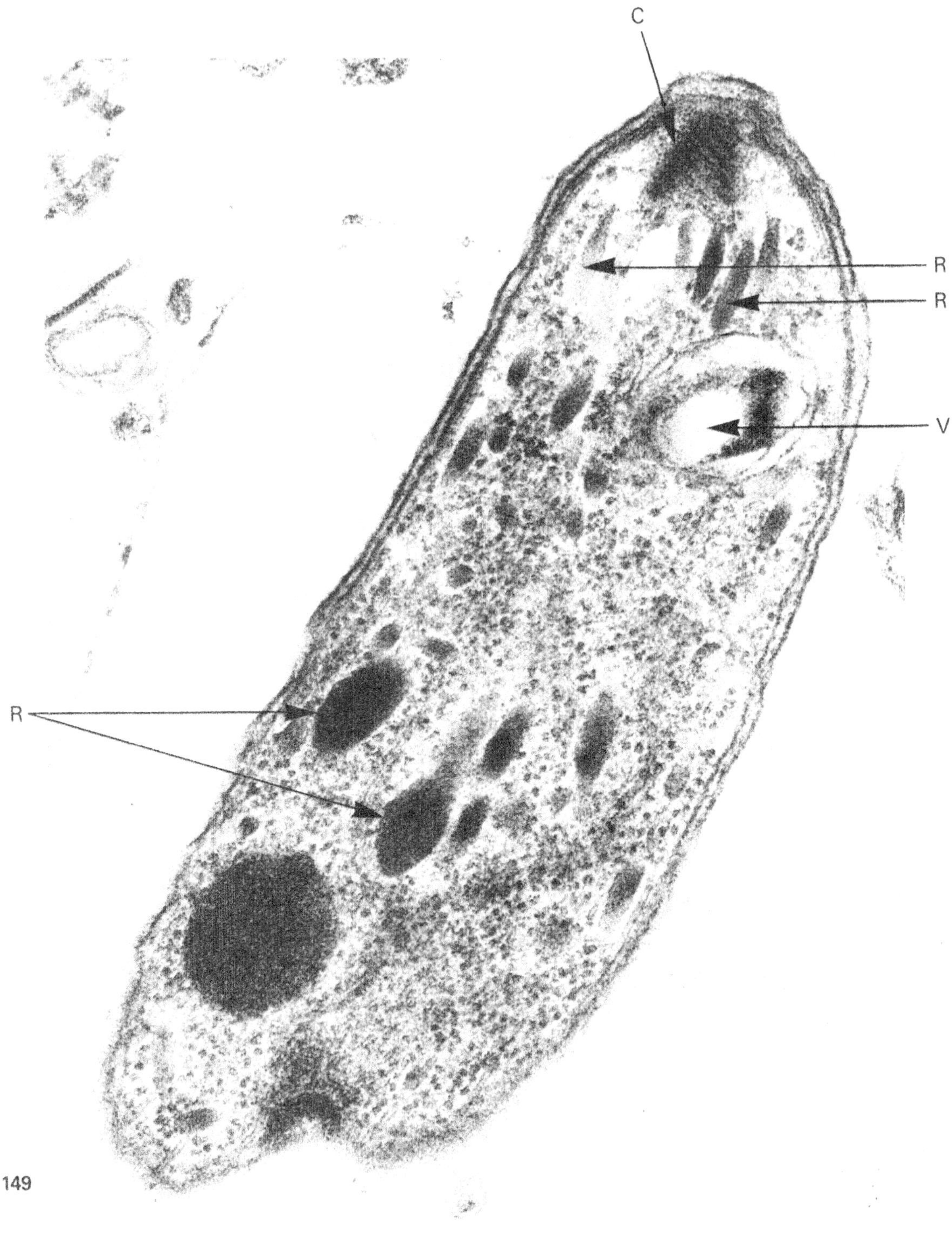


149

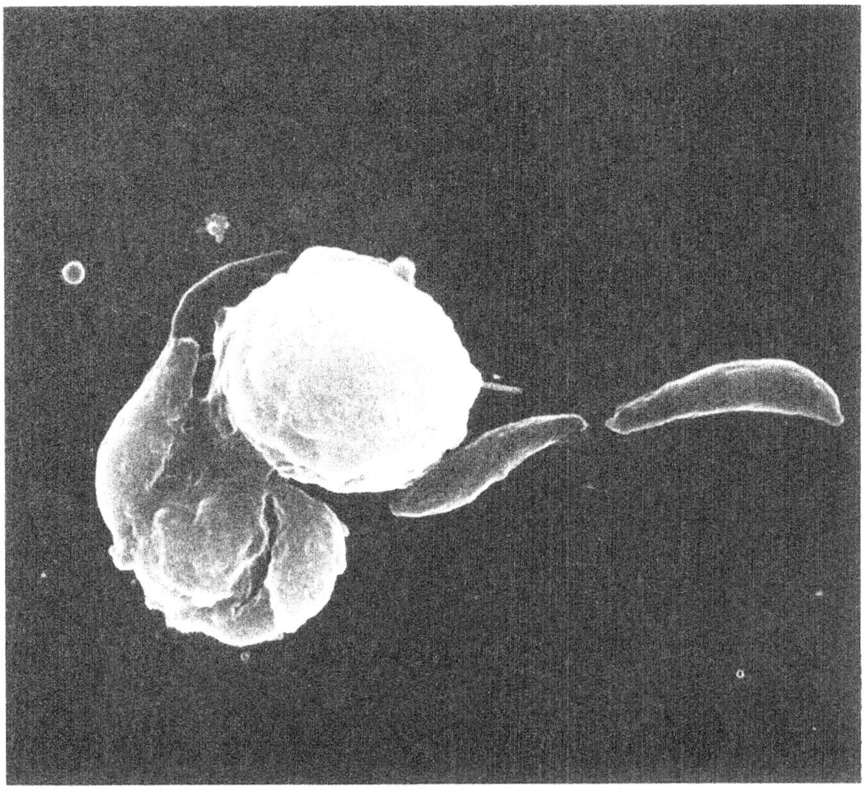

150

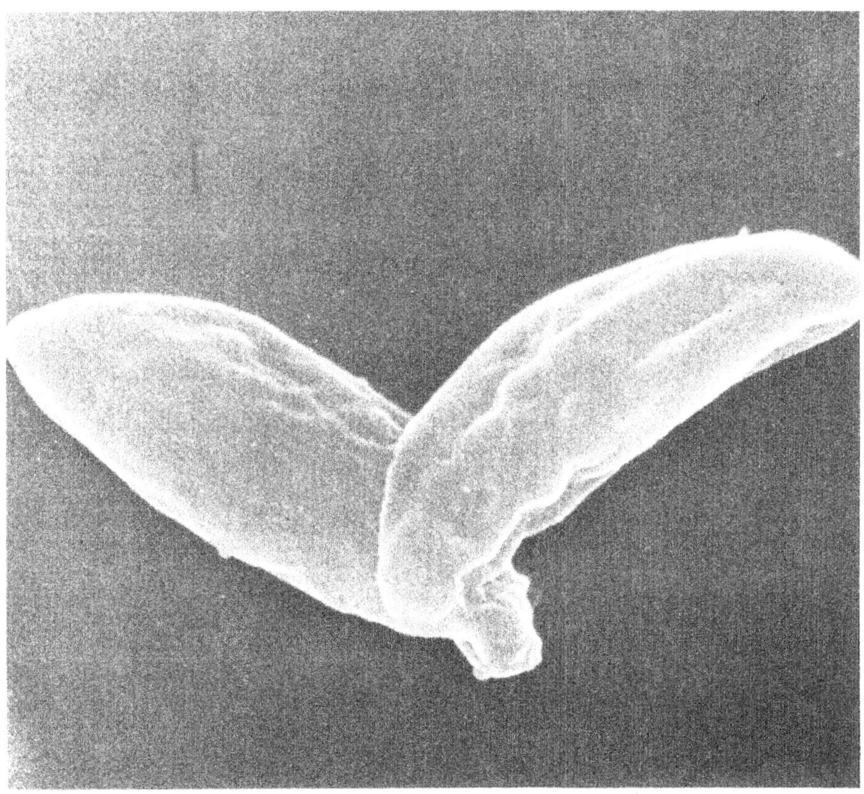

151

**152 Indirect fluorescent antibody test.** Slide shows endozoites exposed to antibody containing serum and then treated with fluorescein conjugated anti-human globulin. There is a zone of light green fluorescence around each organism. The cytoplasm appears reddish as it was counterstained with Evan's blue. X1000. Enlarged by 5.4.

**153 Freshly passed Toxoplasma oocysts.** Most of the oocysts have a single sporoblast. Interference contrast. X400. Enlarged by 5.4.

**154 Toxoplasma oocysts after incubation in vitro for 12 hours.** Most of the oocysts have now sporulated. Interference contrast. X400. Enlarged by 23.4.

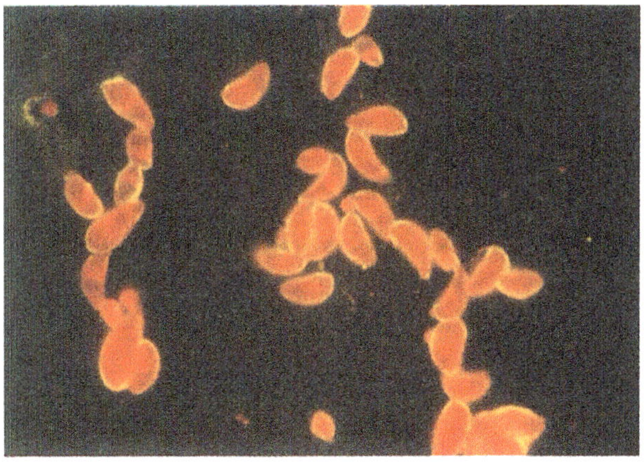
152

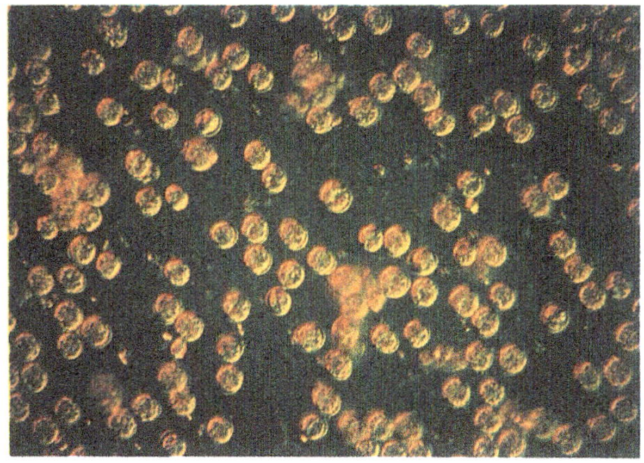
153

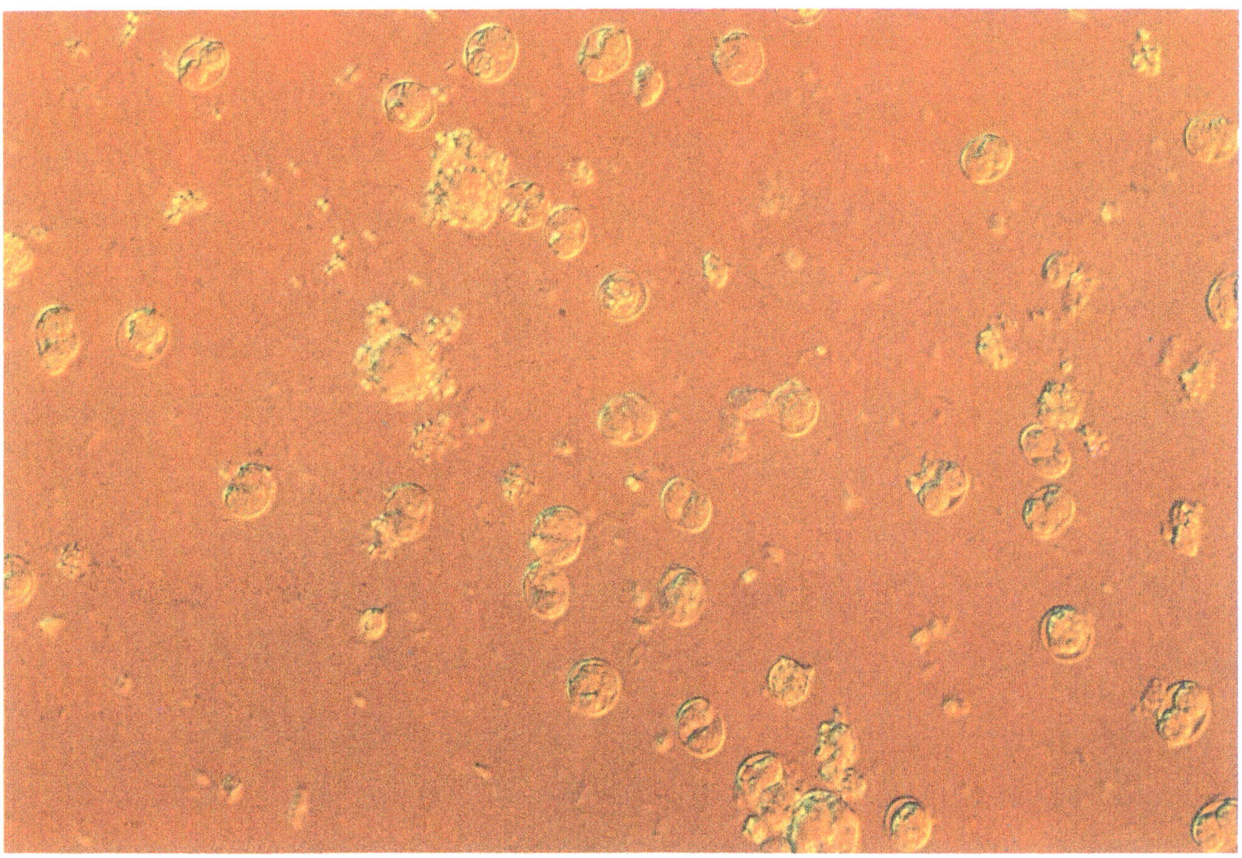
154

**155 Section of cat intestine infected with Toxoplasma.** Macrogametocytes appear as round bodies with granular cytoplasm lining the anterior part of the epithelial cells. Iron-haematoxylin satin. X400. Enlarged by 9.6.

**156 Section of cat intestine infected with Toxoplasma.** A microgametocyte is lying in the centre. It is irregular in shape because of the clumping of the flagellae of the microgametes. X400. Enlarged by 9.6.

**157 An impression smear from the Toxoplasma infected cat intestine.** Showing a large number of merozoites. The merozoites are morphologically similar to endozoites. RNA of the cytoplasm takes reddish colour and the DNA of the nucleus yellowish-green. Acridine orange stain. X1000. Enlarged by 9.6.

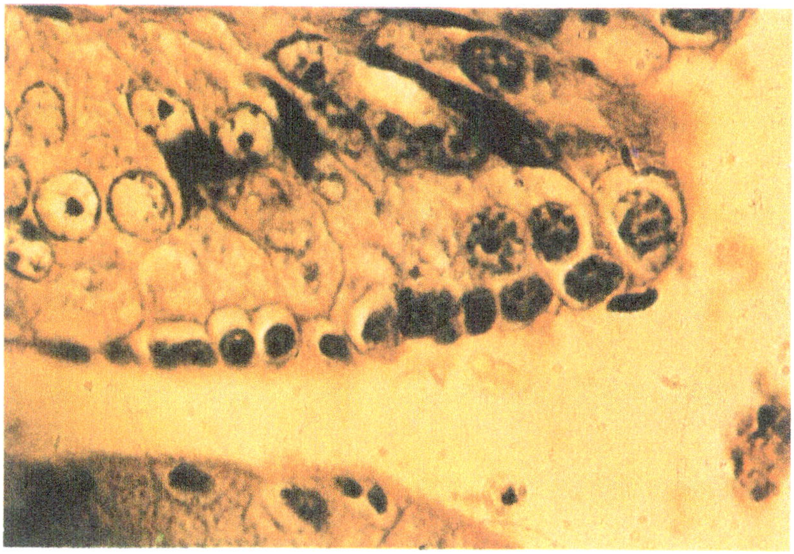
155

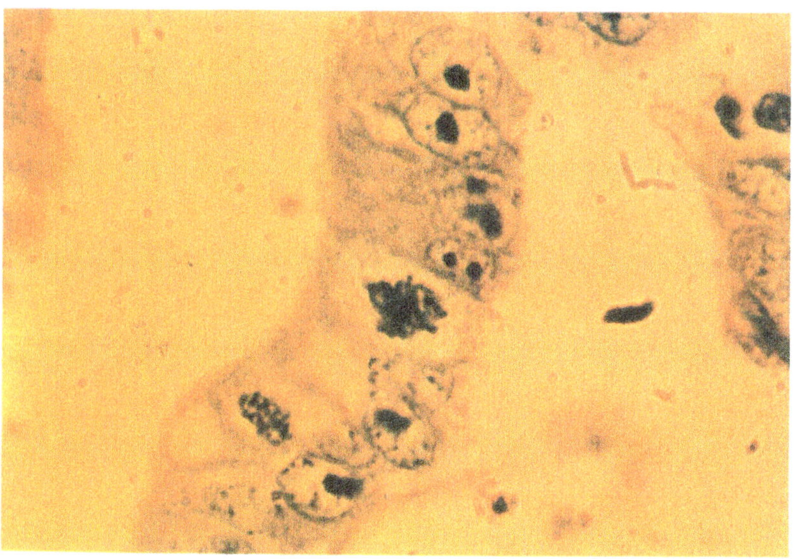
156

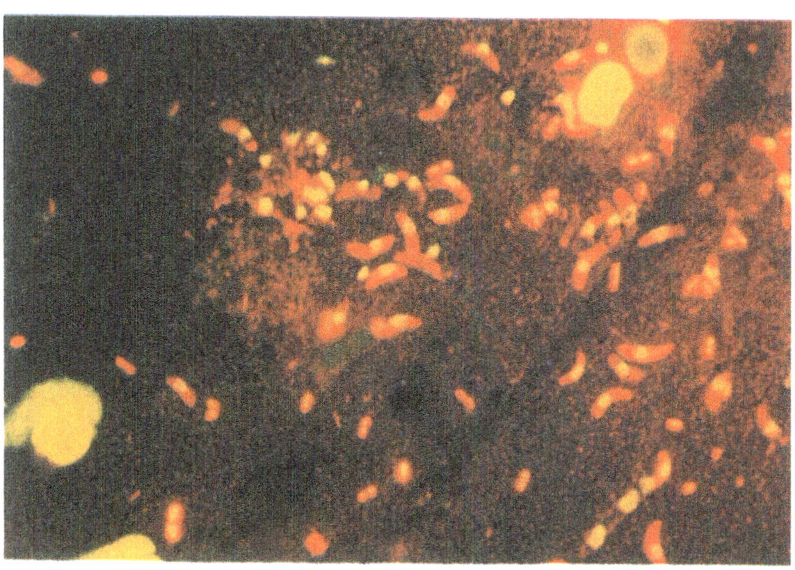
157

**158 Toxoplasma cyst from mouse brain.** Cystozoites appear as crescentic bodies inside a well defined cyst wall. Interference contrast. X1000. Enlarged by 23.4.

158

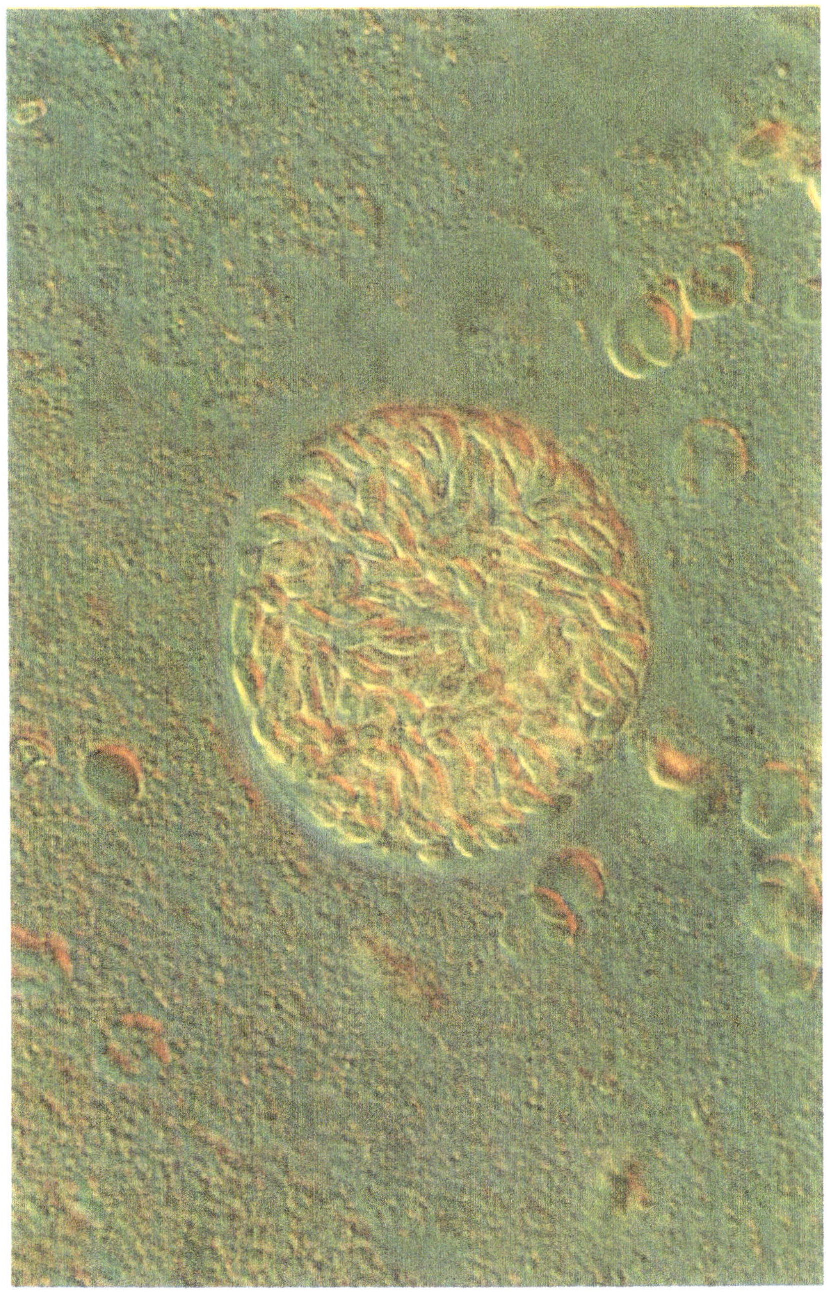

**159 Toxoplasma cyst from mouse brain.** The cyst has partly detached from the brain tissue. Phase contrast. X400. Enlarged by 5.4.

**160 Toxoplasma cyst from mouse brain.** The cyst has completely detached from the brain tissue. Phase contrast. X400. Enlarged 5.4.

**161 Toxoplasma cyst from mouse brain.** Showing the argyllophilic nature of the cyst wall after silver staining. This is the best method for demonstrating the cyst wall. X1000. Enlarged 5.4.

**162 Toxoplasma cyst from mouse brain.** Showing the PAS positive cystozoites. The cyst wall is faintly visible. This is the best method for demonstrating the cystozoites. X1000. Enlarged by 5.4.

**163 Toxoplasma cyst from mouse brain after H and E stain.** The cyst wall is faintly visible and the nuclei appear as dark bodies inside the cyst. X1000. Enlarged by 5.4.

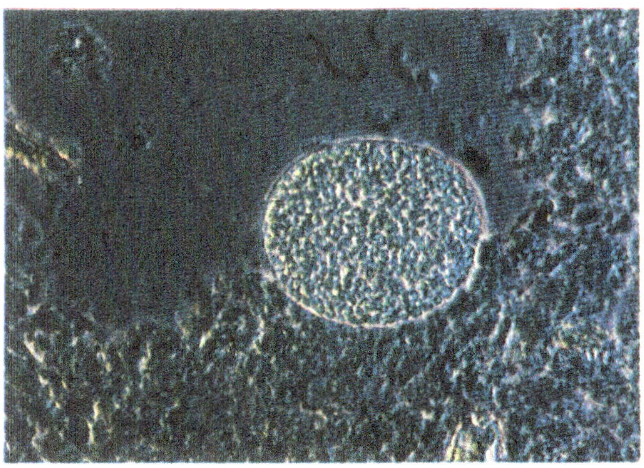

159

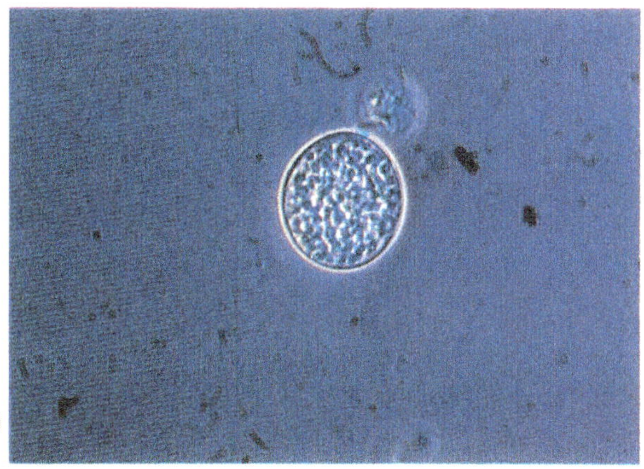

160

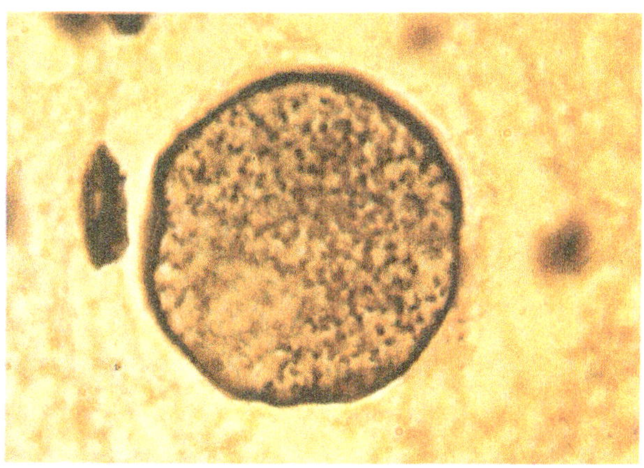

161

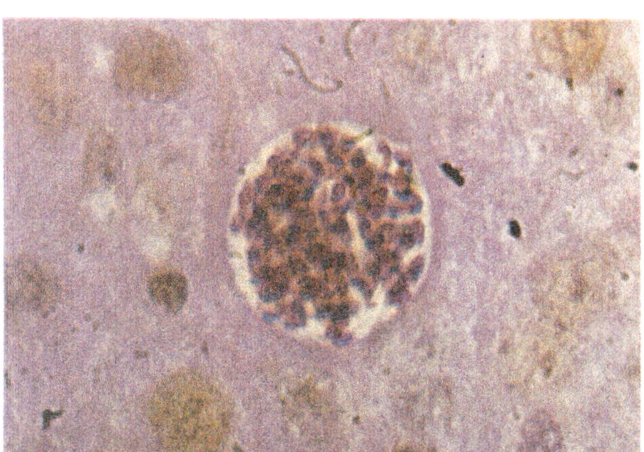

162

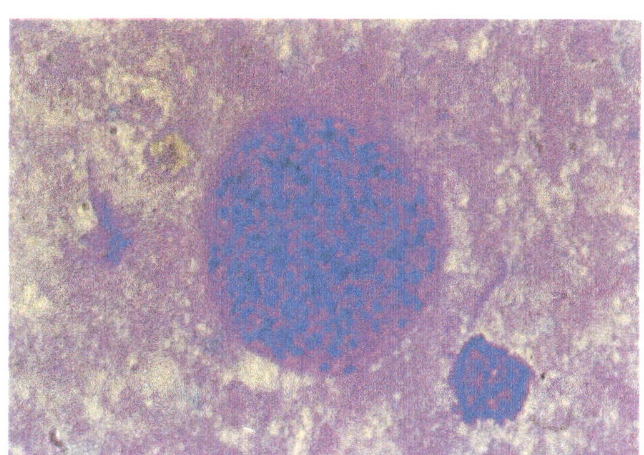

163

**164 In the centre is a macrophage filled with endozoites.** This stage is also known as the pseudocyst. Giemsa stain. X1000. Enlarged by 9.6.

**165 A ruptured macrophage which has liberated endozoites.** Many are still lying close to the macrophage nucleus. Giemsa stain. X1000. Enlarged by 9.6.

**166 A ruptured macrophage with the endozoites dispersed in the background.** Some parasites are undergoing division. The large dense body is the macrophage nucleus. Giemsa stain. X1000. Enlarged by 5.4.

**167 A macrophage containing endozoites and some free endozoites, stained with acridine orange.** The cytoplasm of the endozoites consisting mostly of RNA takes up a red colour and the nucleus consisting mainly of DNA takes up a yellowish green colour. X1000. Enlarged by 5.4.

**168 Photograph of fundus.** Showing a heavily pigmented central choroidal lesion in a young person due to acquired toxoplasmosis. The whitish patches inside the lesion are areas where the sclera is exposed due to the necrosis of the retina. A sharply demarcated lesion, as seen in this photograph, is typical of both acquired and congenital toxoplasmosis. However, in some cases a more diffused lesion may be observed. (Courtesy Dr A.S.M. Lim).

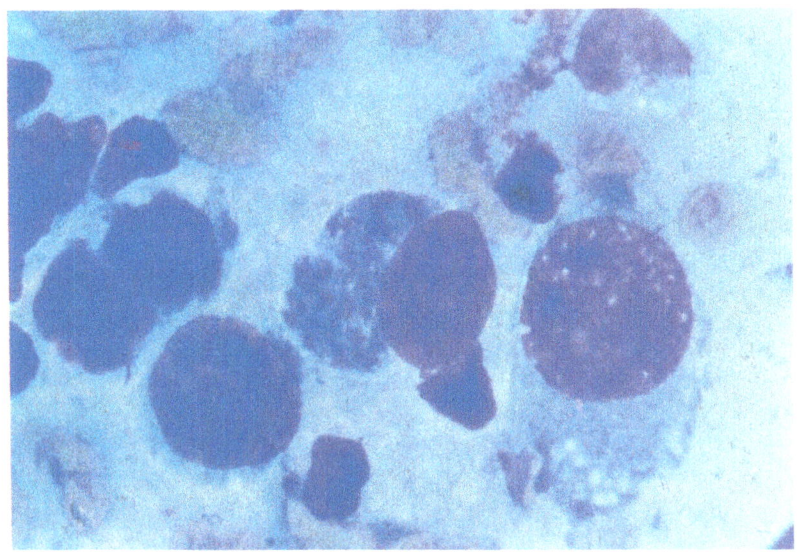

164

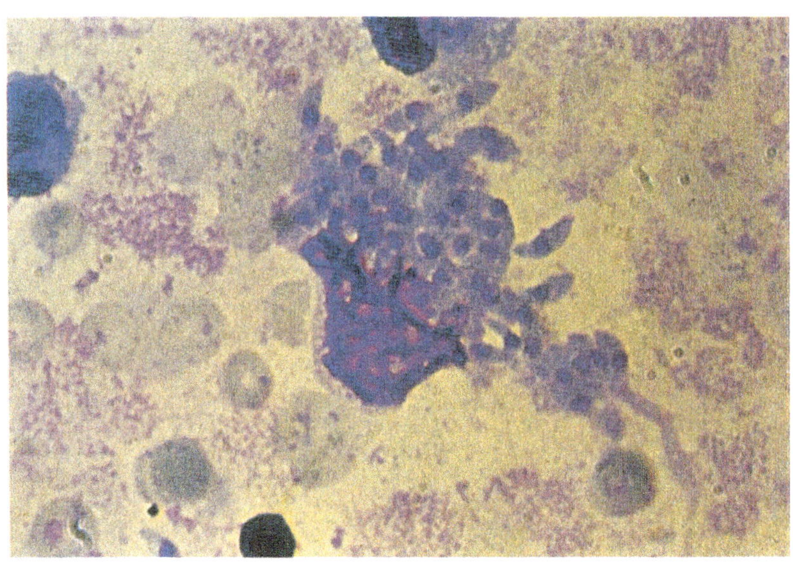

165

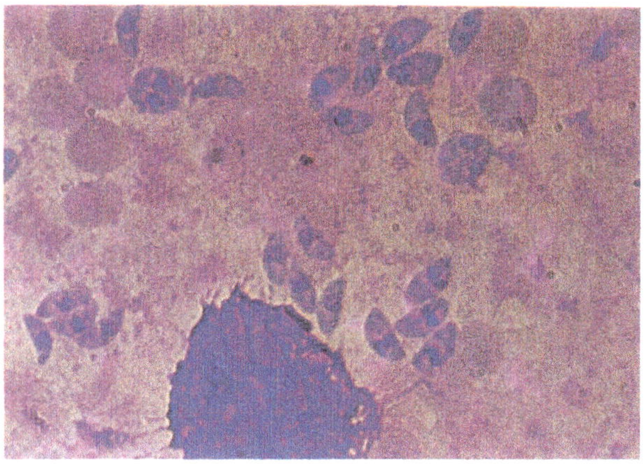
166

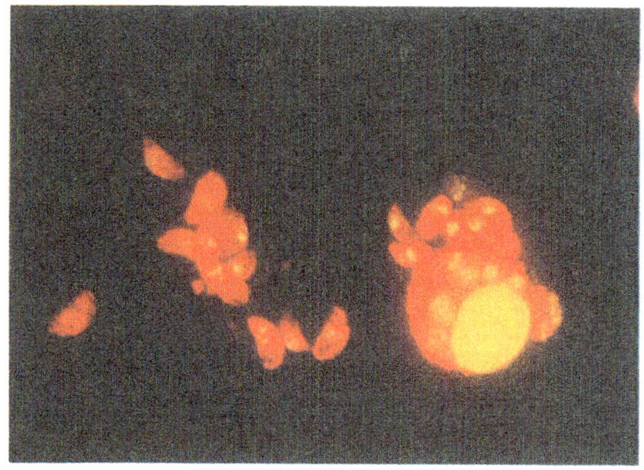
167

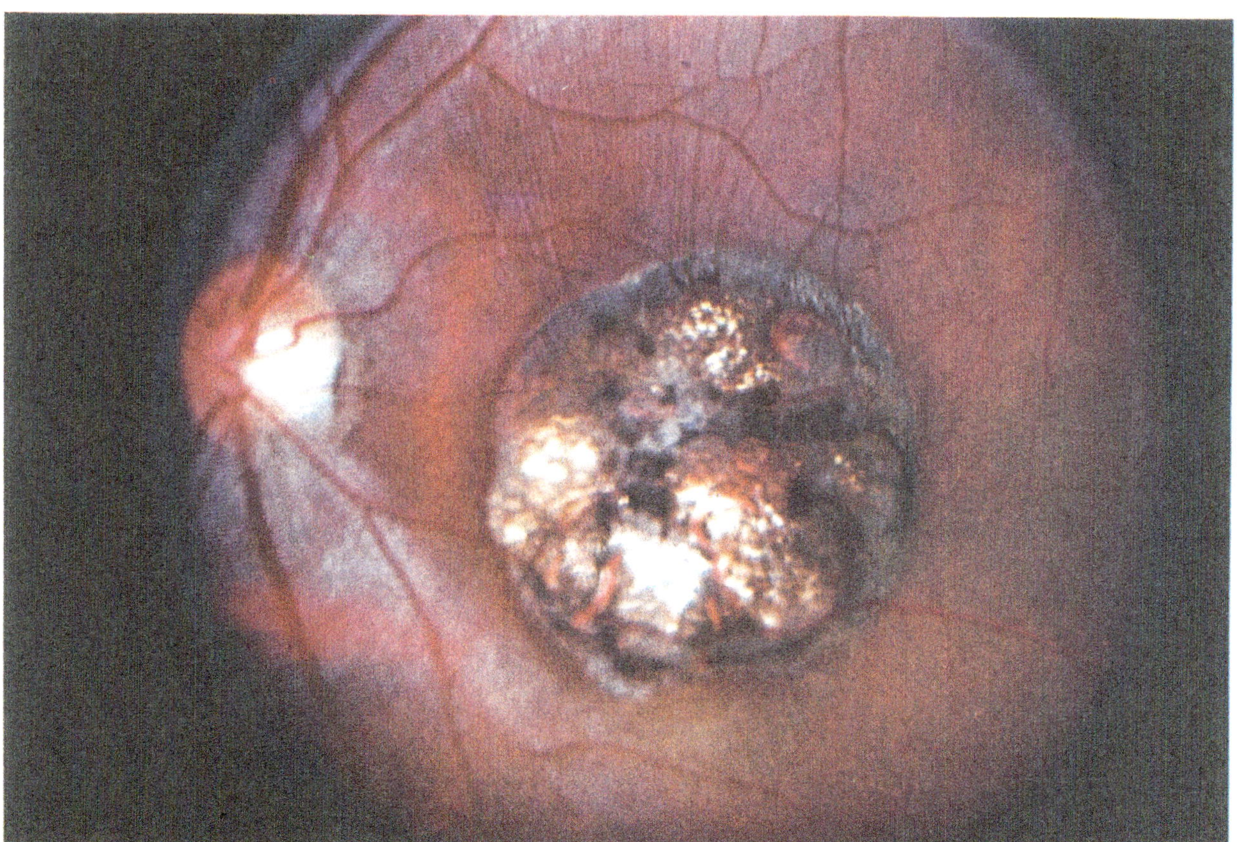
168

# Section 8
# Sarcocystis and Isospora

*Sarcoystis* is the name given to a coccidian parasite belonging to the family Sarcocystidae, which forms characteristic cysts in the muscle of infected animals and humans. Infection with this parasite is common in animals but rare in humans. The infection in humans is generally diagnosed in biopsy or autopsy specimens, as in the majority of cases it is asymptomatic. Occasionally, the parasite may give rise to pain and swelling in the muscles.

*Table I.* Differential diagnosis of important *Isospora* type oocysts

| | *Toxoplasma gondii* | *Sarcocystis spp* | *Isospora belli* | *Isospora hominis* |
|---|---|---|---|---|
| *Size* | | | | |
| Oocyst | 13 x 9μm | Oocyst wall usually absent | 30 x 12μm | Oocyst wall usually absent |
| Sporocyst | 6 x 5μm | Varies in different spp 9 to 13μ x 7 to 9μm | 11 x 9μm | 15 x 10μm |
| *Shape* | Ovoid oocysts | Ovoid sporocyst generally free | Ovoid or slightly elongated oocysts | Ovoid sporocysts generally coupled |
| *Maturity at the time of passage* | Immature generally single sporoblast stage | Mature generally with sporozoites (sporulated) | Immature generally single sporoblast stage | Mature generally with sporozoites (sporulated) |
| *Host* (shedding oocysts in faeces) | Felidae only | Many predators | Man<br>Other animals (?) | Man<br>Other animals (?) |

## *Life cycle of Sarcocystis* [169]

*Sarcocystis* — Life cycle involves 2 hosts — a predator and a prey. The sexual stages develop in the intestinal tract of a predator and asexual stages in various tissues of a prey. It appears that the sporozoites (9) liberated from the sporocysts in the intestinal tract of the prey develop in the intestinal cells (1) and the blood vessels (2), before entering the muscle cells. In the muscle cells they form cysts (3). These cysts rupture in the intestinal tract of the predator (4), liberating cystozoites (5), which enter the intestinal cells, producing male (6) and female gametocytes (7). Fertilisation of the female gametocyte results in the formation of oocysts and sporocysts (8). The sporozoites are, therefore, infective to prey and cystozoites (bradyzoites) to predator.

The ultrastructure of the cystozoites of *Sarcocystis* resemble the cystozoites of *Toxoplasma.* However, the number of micronemes (taxonemes) in the cystozoites of *Sarcocystis* are greater than *Toxoplasma.* The cyst wall of *Sarcocystis* is generally thicker and structurally more complex than that of *Toxoplasma.* Adjacent to the *Sarcocystis* cyst wall are irregular cells known as metrocytes, which are absent in the cysts of *Toxoplasma.*

On ingestion of *Sarcocystis* infected meat humans pass *Isospora* oocysts. Two human species of *Isospora* are generally recognised. These are *I. belli* and *I. homonis.* However, mature *I. belli* resemble *I. hominis.* Human to human transmission also occurs by the ingestion of oocysts. The infection is, therefore, seen more often in hostels, mental institutions and other places where people live in close proximity. The parasite during its development in the intestine may give rise to mild gastrointestinal symptoms. Clinically the disease is not important and no treatment is indicated.

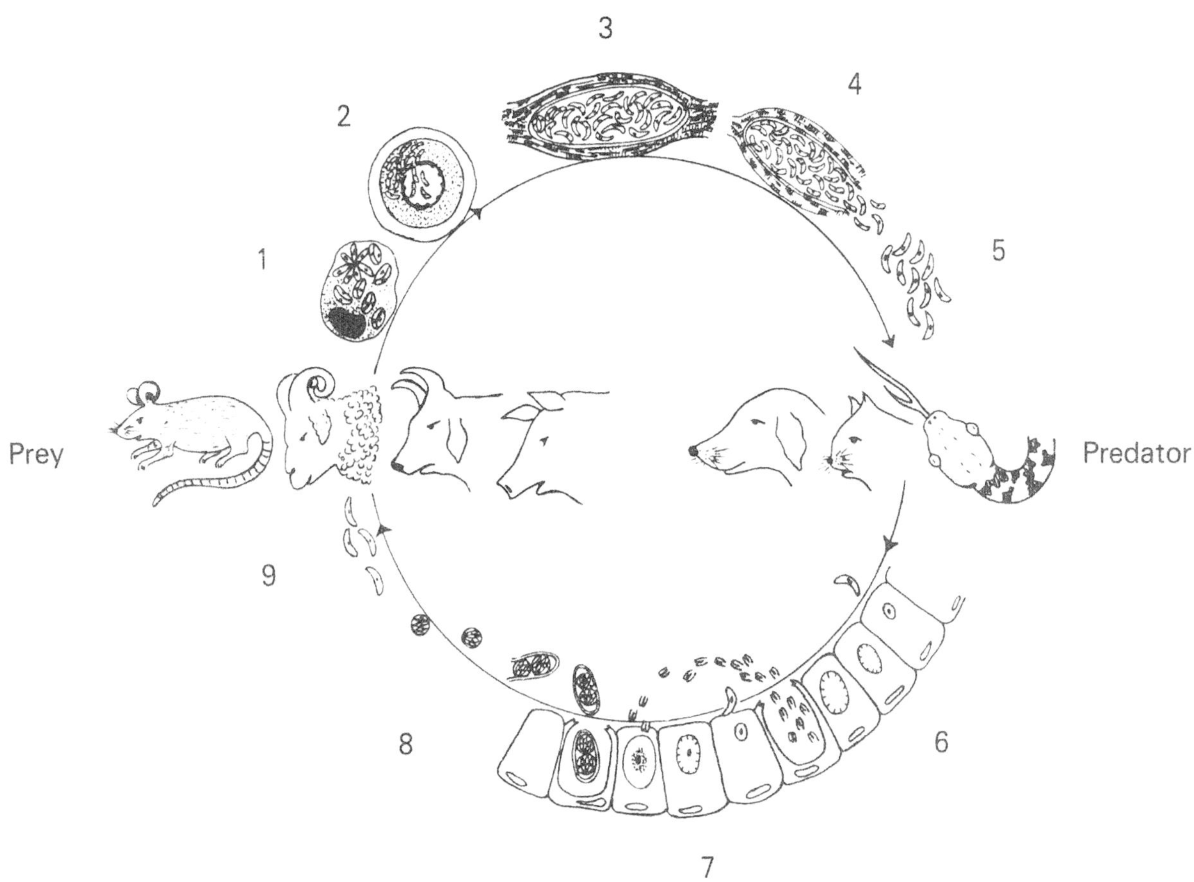


169

**170 Section of a rat Sarcocystis.** The cyst wall is thick and made up of finger like protrusions, which appear 4 sided in cross sections. Immediately adjacent to the cyst wall are metrocytes. The cyst is full of cystozoites cut at different levels. M = metrocytes. CZ = cystozoites. CW = cyst wall. HM = host muscle. X4500. Electron micrograph.

**171 Section of a cystozoite.** Showing a large number of micronemes. M = micronemes. R = rhoptries. PG = polysaccharide granules. N = nucleus. X32,000. Electron micrograph.

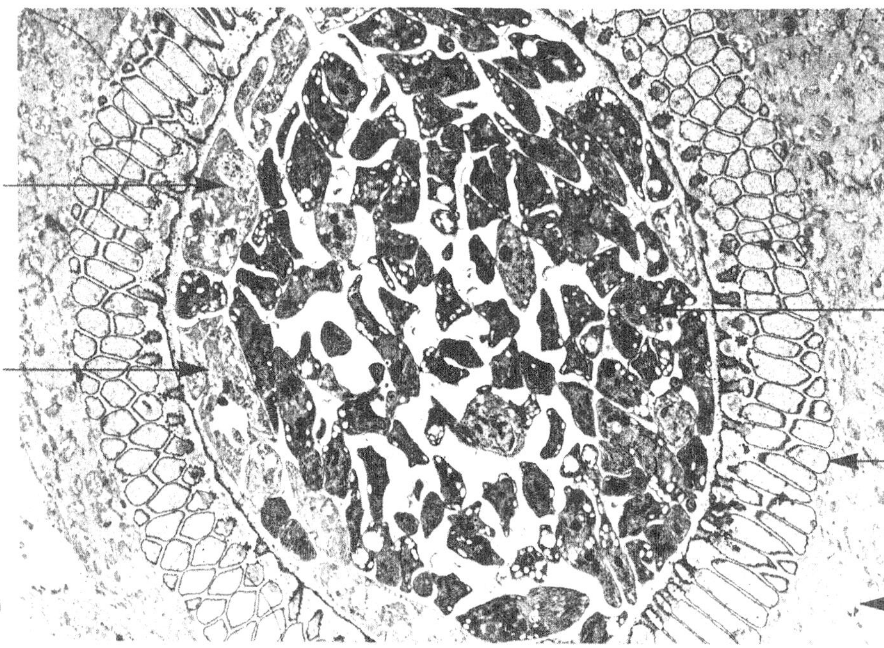


170

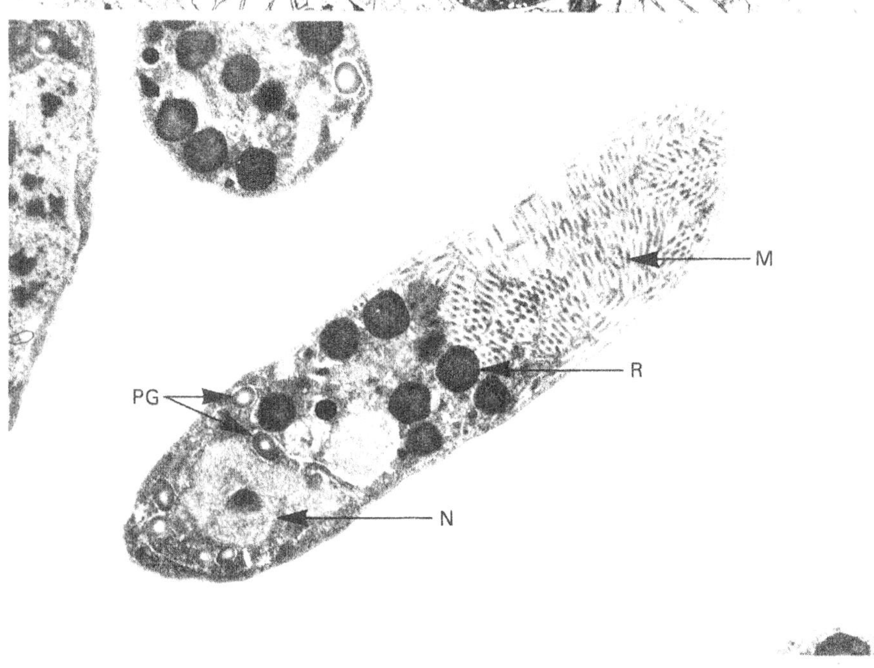


171

**172 Stages in the development of 'Isospora' type oocysts (T. gondii, Sarcocystis spp, I. belli and I. hominis).** These oocysts pass through 4 phases of development.
1) The single cell or the zygote stage. The contents appear granular and almost fill the oocyst.
2) The single sporoblast stage in which the zygote shrinks to form a circular body.
3) The 2 sporoblast stage resulting from the division of the single sporoblast.
4) The sporocyst stage in which a cyst wall forms around each sporoblast, and at the same time 4 sporozoites develop in each sporocyst. A granular mass appears in each sporocyst known as the sporocyst residium. The residium acts as a food store.

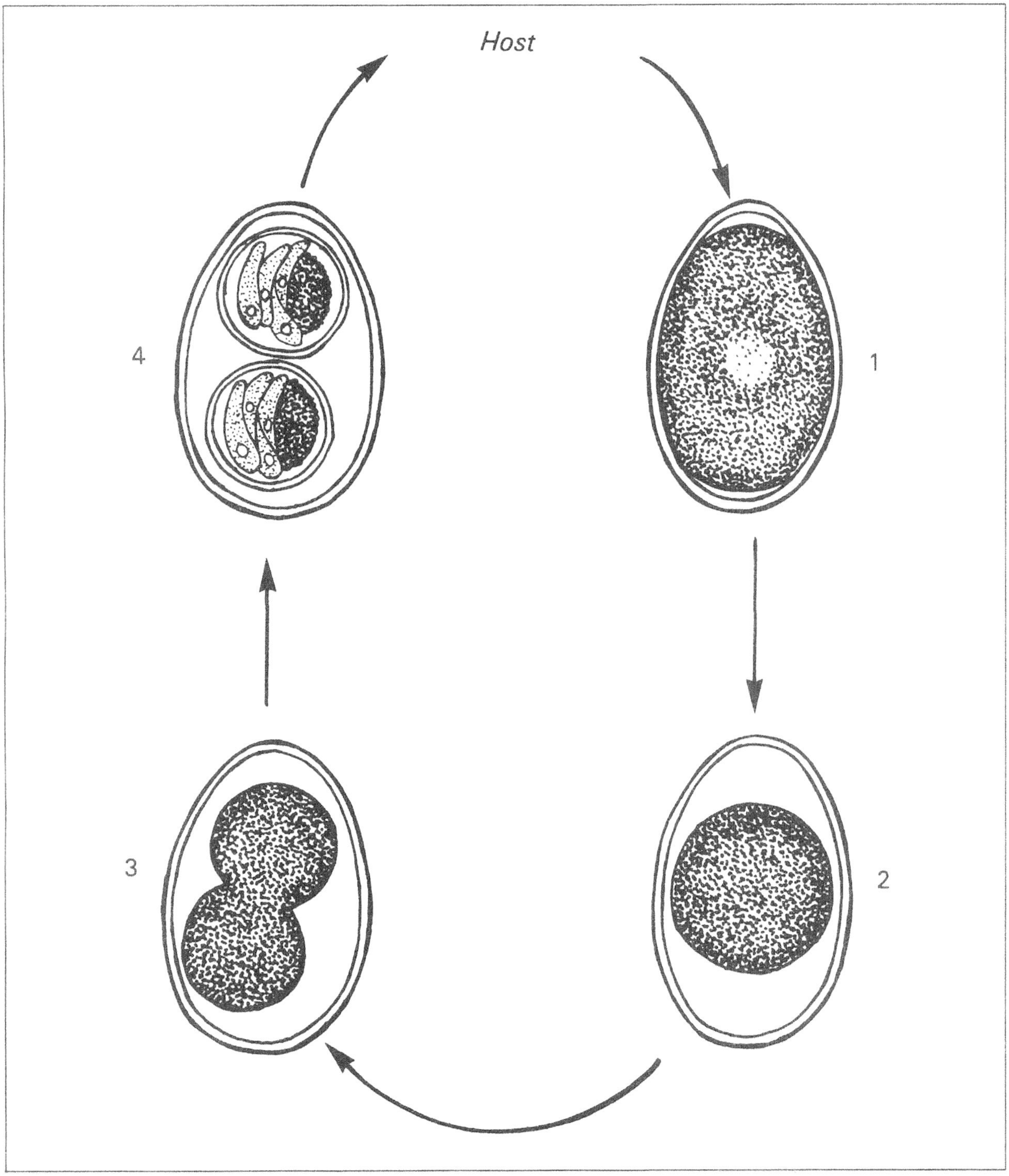


172

**173 A piece of rat muscle with a cyst.** Unfixed. Interference contrast. X1000. Enlarged by 5.4.

**174 Cyst after rupture.** Showing a large number of free cystozoites. Unfixed. Interference contrast X100. Enlarged by 5.4.

**175 Free cystozoites.** Unfixed. Interference contrast. X400. Enlarged by 5.4.

**176 Giemsa stained smear of cystozoites.** The posterior end is broader than the anterior end and approximately 3 zones can be seen. The anterior end represents the area of micronemes and the posterior end that of the nucleus. X800. Enlarged by 5.4.

**177 Oocysts of a Sarcocystis species.** Squash preparation from the infected gut of a python. The sporocyst formation has already occurred. In this respect they differ from the *Toxoplasma* oocysts. Interference contrast X400. Enlarged by 5.4.

**178 Biopsy of a muscle.** From a human case of *Sarcocystis*. The cyst wall is thin. H and E. X400. Enlarged by 9.6.

**179 Section of a Sarcocystis infected muscle from rat.** The cyst wall is thick. PAS. X400. The structure of the cyst wall is useful in identifying the species. Enlarged by 5.4.

**180 Freshly passed oocyst of Isospora belli.** At this stage it has a single cell (sporoblast). X1000. Enlarged by 5.4.

173

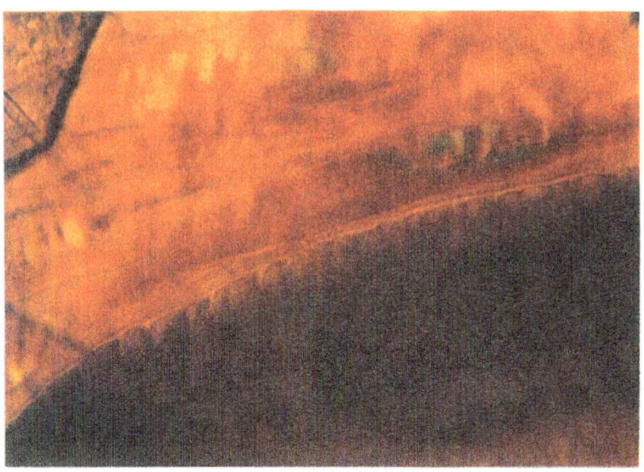

174

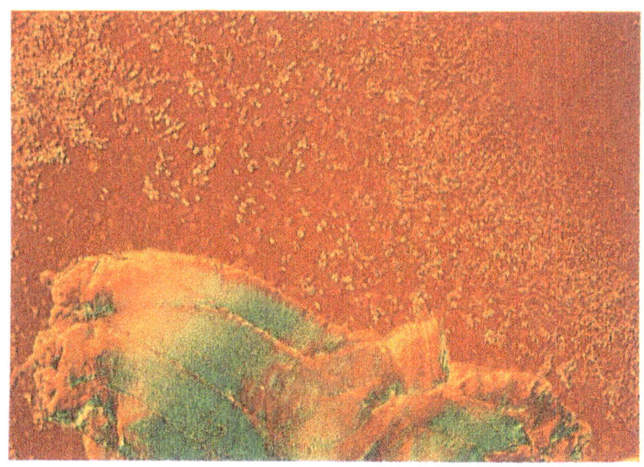

175

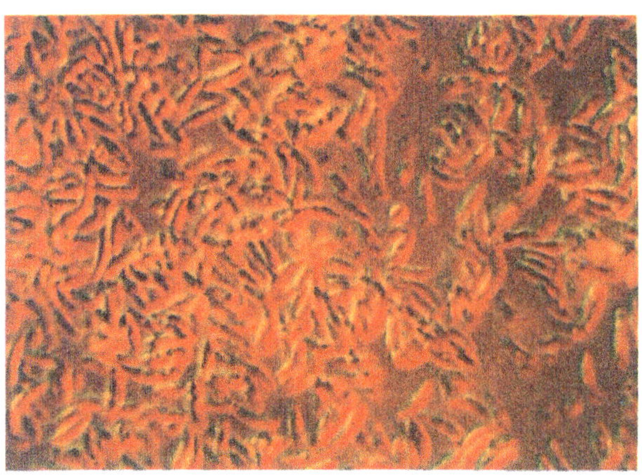

176

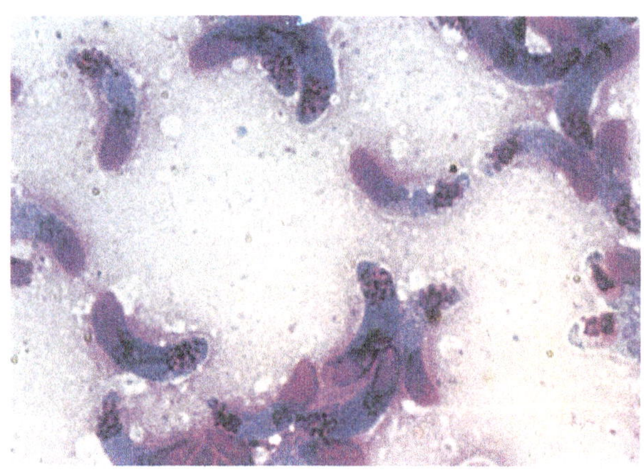

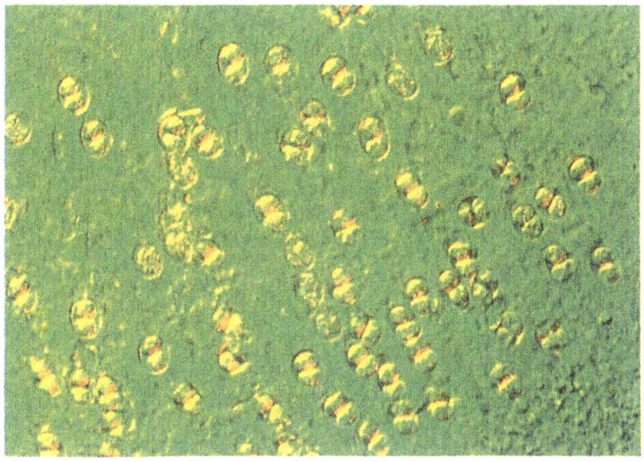

177

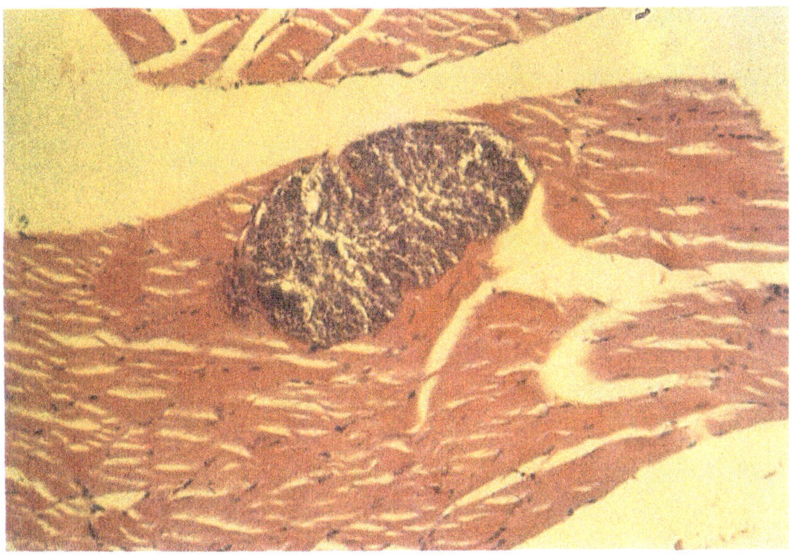

178

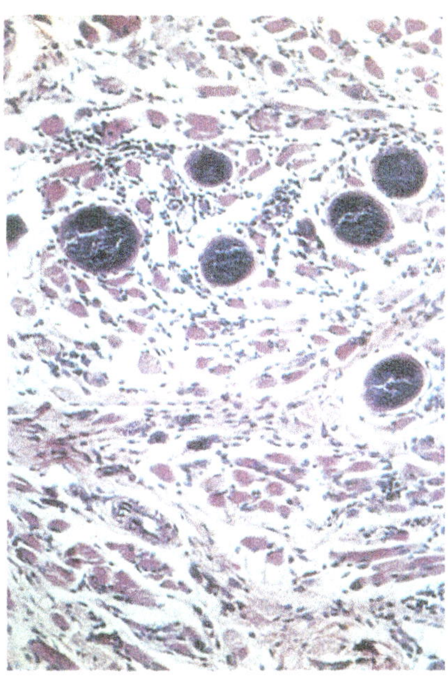

179

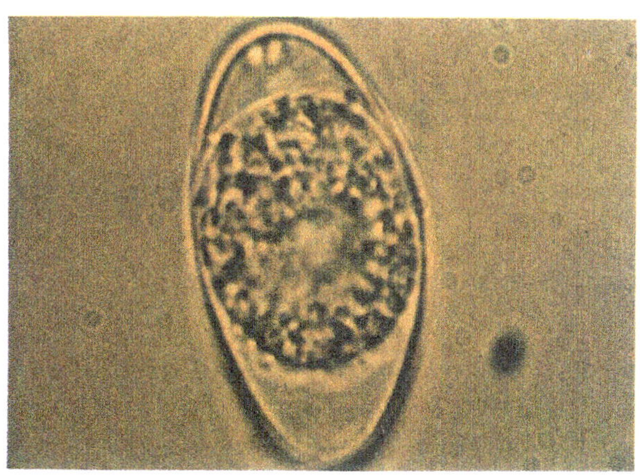

180

# Section 9
# Pneumocystis carinii

This is a common parasite of animals. Human infection occurs in individuals who are on immunosuppressive therapy or suffer from congenital hypogammaglobulinaemia. The parasite produces severe pneumonitis and the course of the disease is usually fatal unless treated. The histological examination of lung biopsy is the only definite method of confirming the diagnosis. This can be done by a transbronchoscopic approach (Hodgkin et al., 1973).

## Reference

Hodgkin, J.E.; Andersen, H.A. and Rosenow, E.C.: Diagnosis of *Pneumocystis carinii* pneumonia by Transbronchoscopic lung biopsy. Chest 64: 551 (1973).

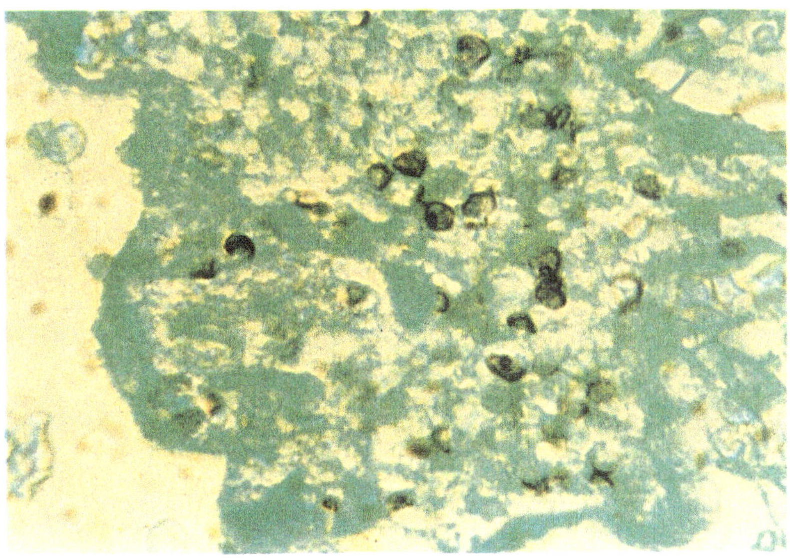

181

**181 Pneumocystis carinii infection of human lung.** The cyst wall is strongly argyrophilic and, therefore, appears dark in colour. The outline of the individual cysts is often irregular. Methenamine silver. X1000. Enlarged by 9.6.

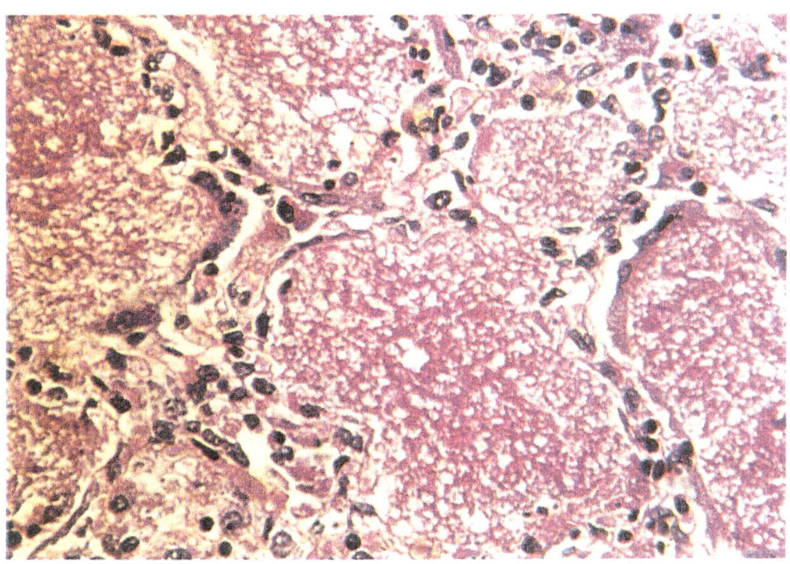

182

**182 Pneumocystis carinii infection of human lung.** The alveoli are packed with 'foamy' material. This appearance is characteristic of *Pneumocystis* pneumonia. There is, in addition, thickening of the alveolar septa with infiltration by inflammatory cells, particularly plasma cells. PAS. X1000. Enlarged by 9.6.

# Part II: Helminths

# Section 10 Cestodes

## Classification of Medically Important Cestodes

| | |
|---|---|
| Phylum | Platyhelminthes |
| Class | Cestoda |
| Order | Pseudophyllidea |
| | Cyclophyllidea |

Pseudophyllidea

| Family | Genus | Species |
|---|---|---|
| Diphyllobothriidae | *Diphyllobothrium* | *D. latum* |

Cyclophyllidea

| Family | Genus | Species |
|---|---|---|
| Taeniidae | *Taenia* | *T. saginata*<br>*T. solium* |
| | *Echinococcus* | *E. granulosus*<br>*E. multilocularis* |
| | *Multiceps* | *M. multiceps* |
| Hymenolepididae | *Hymenolepis* | *H. nana*<br>*H. diminuta* |
| Dilepididae | *Dipylidium* | *D. caninum* |

## Terms Used in Relation to Cestodes

**Rostellum** — the protuberant anterior part of the scolex of certain tapeworms.

**Scolex** — the anterior organ of a tapeworm used for attachment to host tissues. Also known as the holdfast.

**Bothrium** — a longitudinal groove in the scolex of pseudophyllideans.

**Neck** — the connecting tissues between the scolex and strobila of a tapeworm. This part is unsegmented.

**Proglottid** — single segment of a tapeworm.

**Strobila** — the body of a tapeworm.

**Strobilisation** — the process of producing or growing new segments (proglottids). This happens near the neck region.

**Protoscolex** — the scolex of a larval stage. Morphologically it resembles the adult scolex.

**Hexacanth** — a larval stage of the tapeworm having 6 hooks.

**Hydatid** — larval stage of *Echinococcus,* generally containing a large number of protoscoleces.

**Hydatid sand** — free protoscoleces lying inside a hydatid.

**Daughter cyst** — a cyst formed by endogenous or exogenous budding from the germinal layer of a hydatid.

**Endogenous budding** — inward development from the germinal layer of a hydatid resulting in the formation of a daughter cyst or a brood capsule.

**Exogenous budding** — outward or external development from the germinal layer of a larval cestode.

**Brood capsule** — a small cyst attached to the germinal layer of the hydatid, containing many protoscoleces.

**Multilocular hydatid** — larval stage of *Echinococcus multilocularis* in which exogenous development occurs resulting in infiltration of tissues.

**Cysticercus** — a larval form of a tapeworm having a fluid filled bladder.

**Cysticercoid** — a larval form of a tapeworm which has a solid body and no bladder.

**Oncosphere** — a hexacanth embryo.

**Coracidium** — a ciliated oncosphere which develops in the ova of pseudophyllidea.

**Procercoid** — the second stage larva of pseudophyllidea which bears 6 hooks near the posterior end.

**Plerocercoid** — the third stage larva of pseudophyllidea which has a solid body.

**Egg capsule** — a membranous structure containing eggs of a tapeworm, in the absence of a uterus.

**Other terms** — see page 164 for terms relating to reproductive system.

## Introduction

This group of parasites are also known as tapeworms. Their body consists of a row of segments or proglottids. In between the scolex and the first segment is a narrow zone known as the neck. This part of the parasite contains the germinal cells responsible for the process of strobilization. The scolex is the principal means of locomotion and attachment to the host tissues. It is provided with grooves, suckers or hooks.

The nervous system consists of longitudinal cords running laterally in each segment. The principal organs for osmoregulation are the flame cells which are connected to a series of collecting tubules.

All medically important tapeworms are hermaphroditic and each segment contains a complete set of male and female reproductive organs. The male reproductive system consists of a number of testes scattered in the proglottid. Each testis is connected with a vas efferens which unite to form a vas deferens which leads into the cirrus pouch, containing a cirrus. The cirrus opens into the common genital atrium. The female reproductive system consists of a single lobed or unlobed ovary which is located at the posterior part of the segment. An oviduct arises from the ovary which leads to the ootype. The vitelline duct and the duct from the seminal receptacle also open at the ootype. The ootype leads into the uterus. In Cyclophyllidea the uterus ends blindly, so the eggs can only be dispersed by the disintegration of the segment. In Pseudophyllidea the uterus opens to the outside of the body by a uterine pore. In gravid proglottids of Cyclophyllidea the uterus is distended with eggs which throws out lateral branches to accommodate them. In some cases such as *Dipylidium caninum,* the uterus breaks into small sacs known as egg capsules.

The nutrition of cestodes is obtained through the integument as they lack a digestive system. The calcareous corpuscles are characteristic inclusions in the parenchyma of all cestodes. Their ultrastructure has been studied in *Taenia taeniaeformis* (Nieland and von Brand, 1969).

## Taenia saginata [183] and Taenia solium [184]

Man is the only definitive host of these animals although gibbons can be experimentally infected. The adults are located in the small intestine and the size of *T. saginata* is approximately 5 metres with 1,000 to 2,000 proglottids. Immediately behind the scolex is the unsegmented neck which is followed by a row of immature proglottids, mature proglottids and the gravid proglottids. The gravid proglottids become separated from the strobila and migrate out of the anus or are discharged in the faeces. The evacuated proglottids disintegrate in the soil freeing the eggs which find their way into the intermediate host. In the intermediate host, the hexacanth embryos enter various organs of the body via the circulation where they transform into the larva (cysticercus) stages. The larval stage in the case of *T. saginata* is known as *Cysticercus bovis* and in the case of *T. solium* as *Cysticercus cellulosae.* Humans are infected by *C. cellulosae.*

### *Clinical aspects*

The majority of cases are symptomless. In some patients there may be vague intestinal discomfort, vomiting or diarrhoea. Appendicitis and intestinal obstruction has also been reported.

### *Diagnosis*

This is based on detection of eggs or segments in the faeces. The eggs of *T. saginata* and *T. solium* are indistinguishable. They consist of an outer vitelline layer, which is a membranous covering. The inner layer is a thick brown embryophore which is made up of longitudinal blocks giving the eggs a radially striated appearance. Inside the egg is the hexacanth embryo or oncosphere which carries 6 hooks.

## Cysticercus cellulosae

The infection is called cysticercosis. Humans acquire the parasite by autoinfection or by accidental swallowing of cysticerci with food or water. The hexacanth embryos usually settle down in the striated muscles, the central nervous system and the eye.

### *Clinical aspects*

The most common symptoms are palpable subcutaneous nodules and epileptiform fits. In 3 to 6 years' time cysts begin to die and are calcified. The calcified cysts are visible on X-ray examination.

### *Diagnosis*

This is based on biopsy of nodules and X-ray examination. Various serological tests also become positive (Proctor et al., 1966).

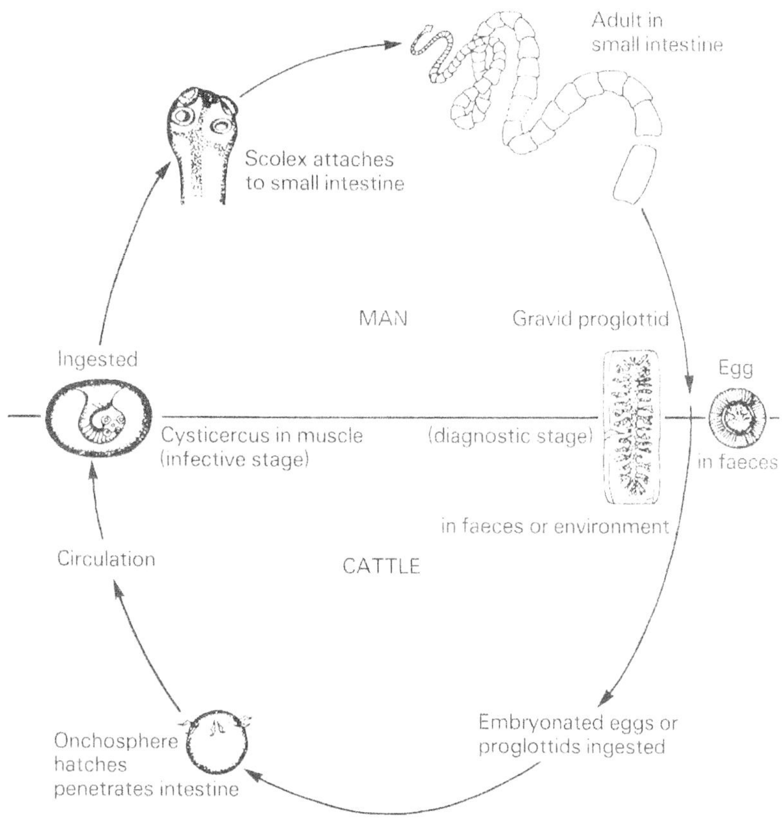
Adult in small intestine
Scolex attaches to small intestine
MAN
Gravid proglottid
Ingested
Egg
Cysticercus in muscle (infective stage)
(diagnostic stage)
in faeces
in faeces or environment
Circulation
CATTLE
Embryonated eggs or proglottids ingested
Onchosphere hatches penetrates intestine

183

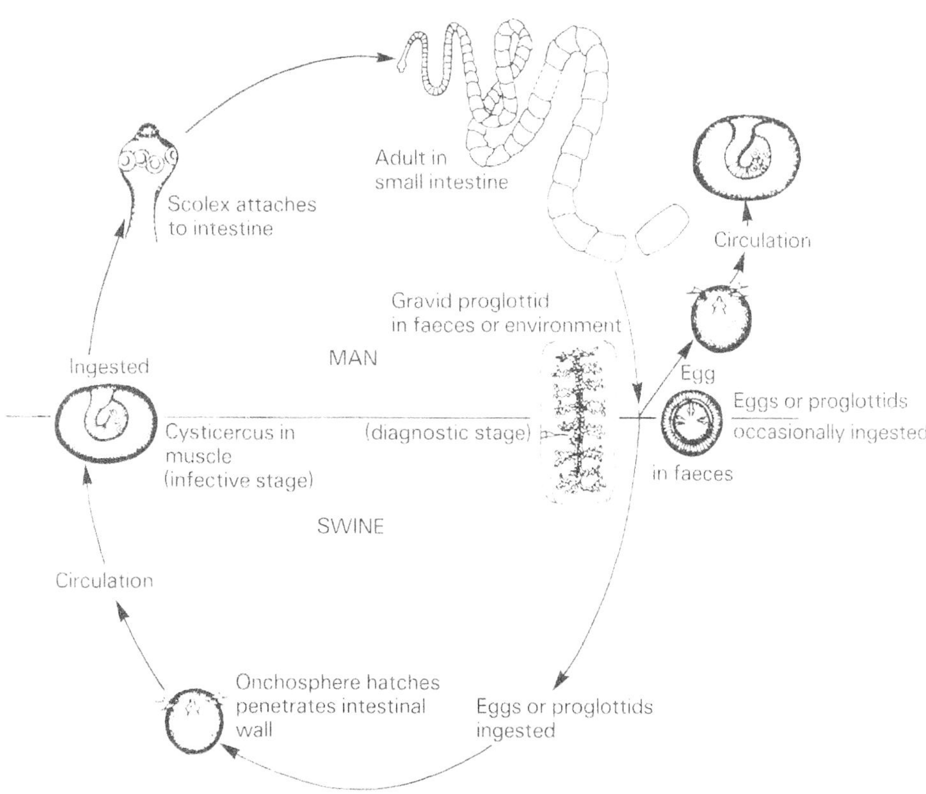
Adult in small intestine
Scolex attaches to intestine
Circulation
Gravid proglottid in faeces or environment
MAN
Ingested
Egg
Eggs or proglottids occasionally ingested
Cysticercus in muscle (infective stage)
(diagnostic stage)
in faeces
SWINE
Circulation
Onchosphere hatches penetrates intestinal wall
Eggs or proglottids ingested

184

**185 Comparative morphology of T. saginata and T. solium**

| | *T. saginata* | *T. solium* |
|---|---|---|
| *Scolex* | 4 suckers<br>no hooks | 4 suckers<br>rostellum<br>with hooks |
| *Egg* | Radially striated embryophore | Radially striated embryophore |
| *Mature Segment* | | |
| Ovary | 2 large lobes | 1 small and 2 large lobes |
| Testes | Small follicular testes 300 to 400 | Small follicular testes 150 to 200 |
| *Gravid Segment* | | |
| Uterine branches | 15 to 30 | 7 to 12 |

h = hooks
r = rostellum
s = suckers
n = neck
e = embryophore
he = hexacanth embryo
t = testes
u = uterus
vd = vas deferens
go = genital opening
v = vagina
ot = ootype
vi = vitelline gland
osl = small lobe of ovary
o = ovary

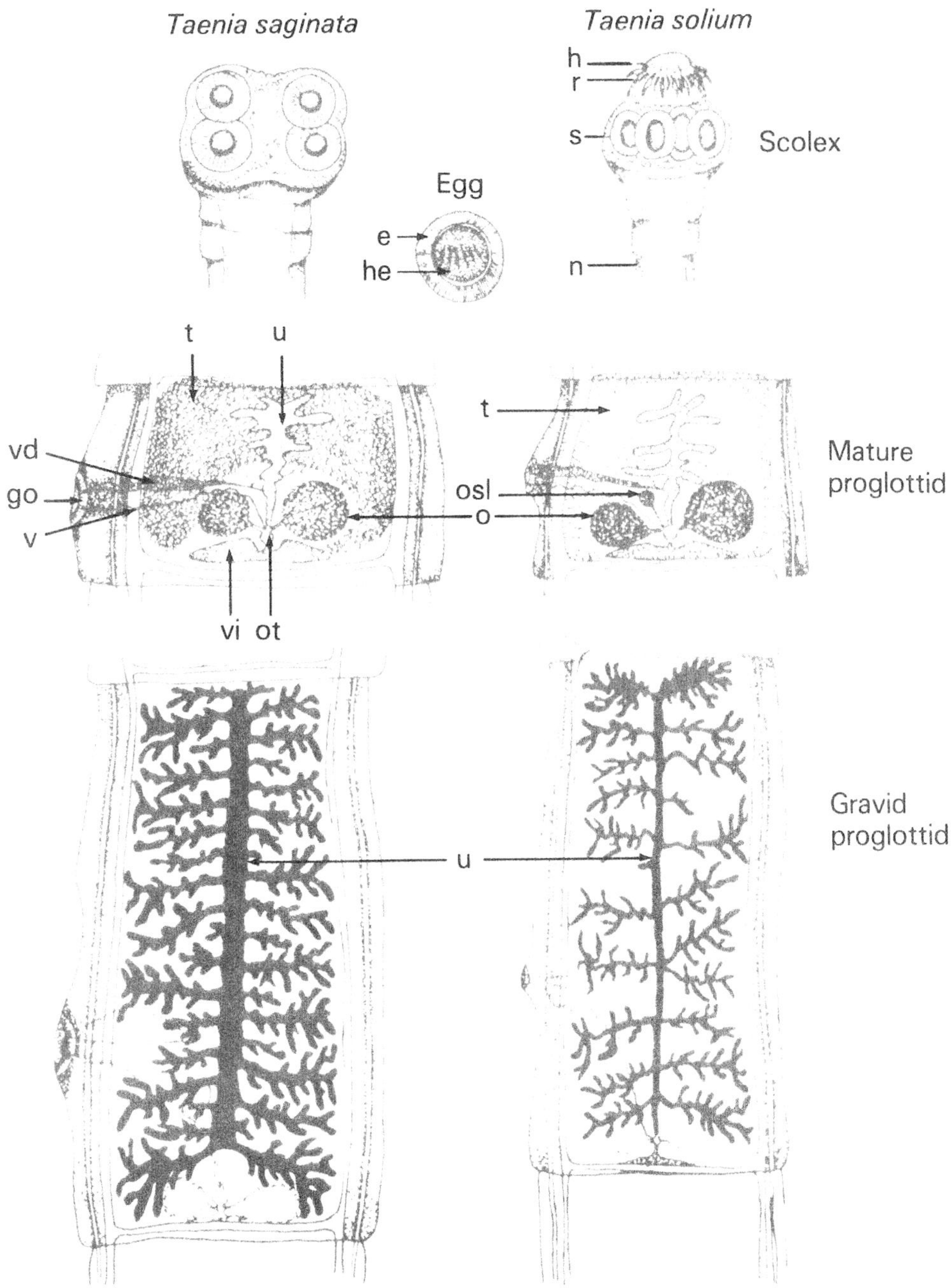
Taenia saginata
Taenia solium
h
r
s
Scolex
Egg
e
he
n
t
u
t
vd
go
v
osl
o
Mature proglottid
vi ot
Gravid proglottid
u

185

### Echinococcus granulosus [186]

The adults are located in the small intestine of the dog and other carnivores and are 3 to 6mm long with 3 to 4 proglottids. When eggs are swallowed by herbivores or humans, the hexacanth embryos enter various organs of the body where they develop into hydatid cysts.

#### *Clinical aspects*

This depends on the site and size of the hydatid cyst. The early stage of infection is generally asymptomatic. As the cyst grows in size, symptoms of a space occupying lesion develops. Allergic reactions and sometimes anaphylactic shock may occur if the cyst ruptures.

#### *Diagnosis*

Radiological examination will show the cyst, if it is calcified. Immunological tests are useful in confirming the diagnosis and as indicators of post operative persistence or of recurrence of infection. The tests include an intradermal reaction (Casoni's test), haemagglutination, complement fixation and fluorescent antibody test (Matossian and Araj, 1975).

### Echinococcus multilocularis

Is usually found in wild carnivores and the adult is approximately half the size of *E. granulosus.* Man is occasionally infected and the cyst forms a proliferating growth with exogenous budding.

### Multiceps multiceps

This is a parasite of dogs and wild carnidae and the normal intermediate host is the sheep. The larval stage normally develops only in the central nervous system and is known as a coenurus. In humans the cysts are most often found in the subarachnoid space giving rise to meningitis. The symptoms are mainly caused by meningeal irritation and include headache, vomiting and neck rigidity.

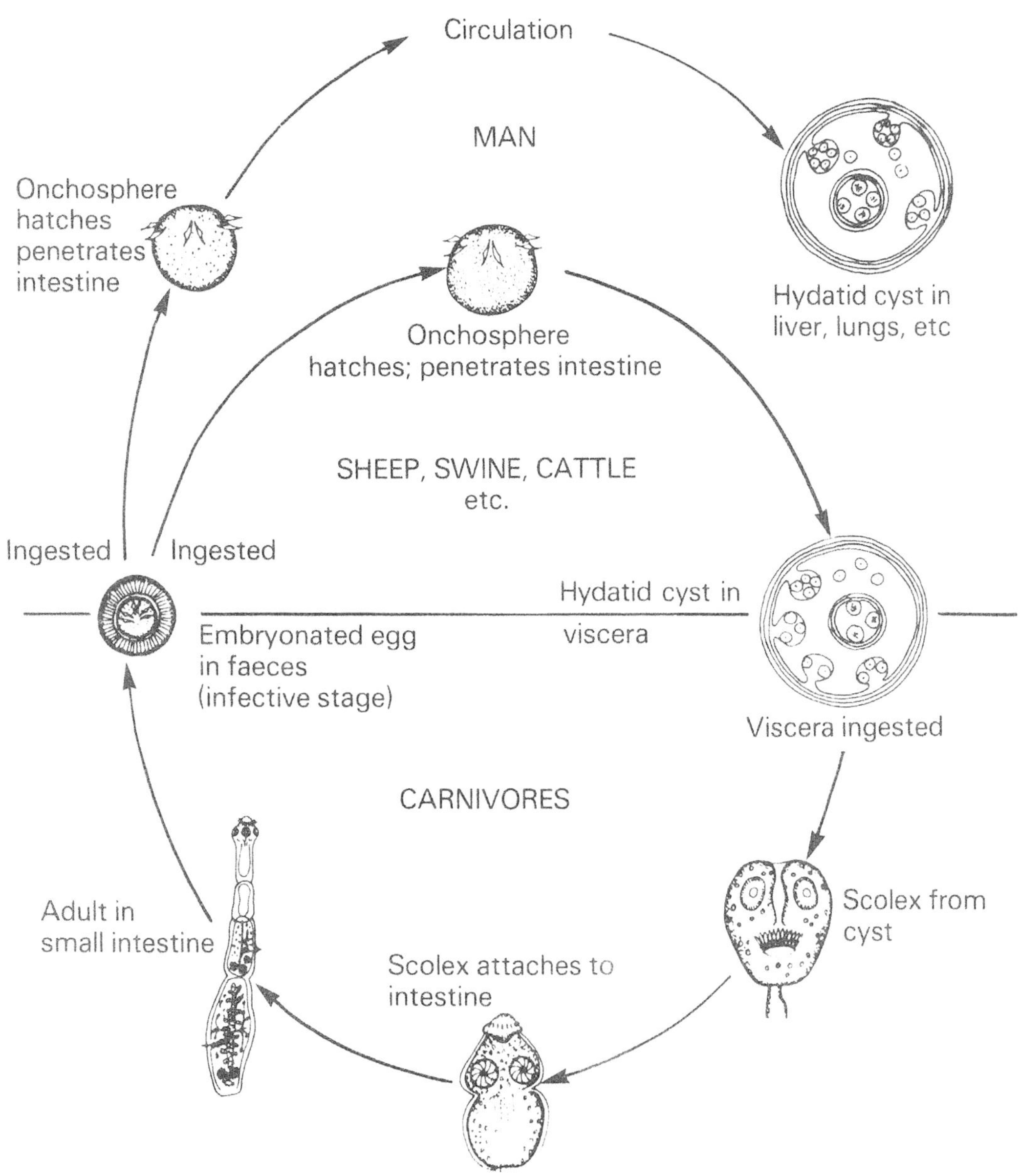
Circulation
MAN
Onchosphere
hatches
penetrates
intestine
Hydatid cyst in
liver, lungs, etc
Onchosphere
hatches; penetrates intestine
SHEEP, SWINE, CATTLE
etc.
Ingested
Ingested
Hydatid cyst in
viscera
Embryonated egg
in faeces
(infective stage)
Viscera ingested
CARNIVORES
Scolex from
cyst
Adult in
small intestine
Scolex attaches to
intestine

186

**187 Internal structure of Hydatid Cyst.** The diagram shows host connective tissue (hct), on the outside of the cyst followed by the laminated layer (ll), formed by the parasite. Inside the ll is the basal layer (bl) on which rests the germinal layer (gl) or the proliferative cell layer (pcl). Brood capsules (bc) are formed by the thickening and protrusion of the pcl towards the cystic cavity to form 'clusters' or 'buds'. These 'buds' grow, become stalked and are vacuolated (1,2,3). A similar budding process leads to the formation of protoscolices inside each brood capsule (Thompson, 1976). The brood capsule may separate from the main cyst if the stalk (st) breaks. In which case, it may swim freely in the cavity or rupture to liberate the protoscolices. The protoscolices generally settle down at the bottom of the cyst and are known as hydatid sand (hs). The brood capsule may also develop a laminated layer when it becomes a daughter cyst (dc).

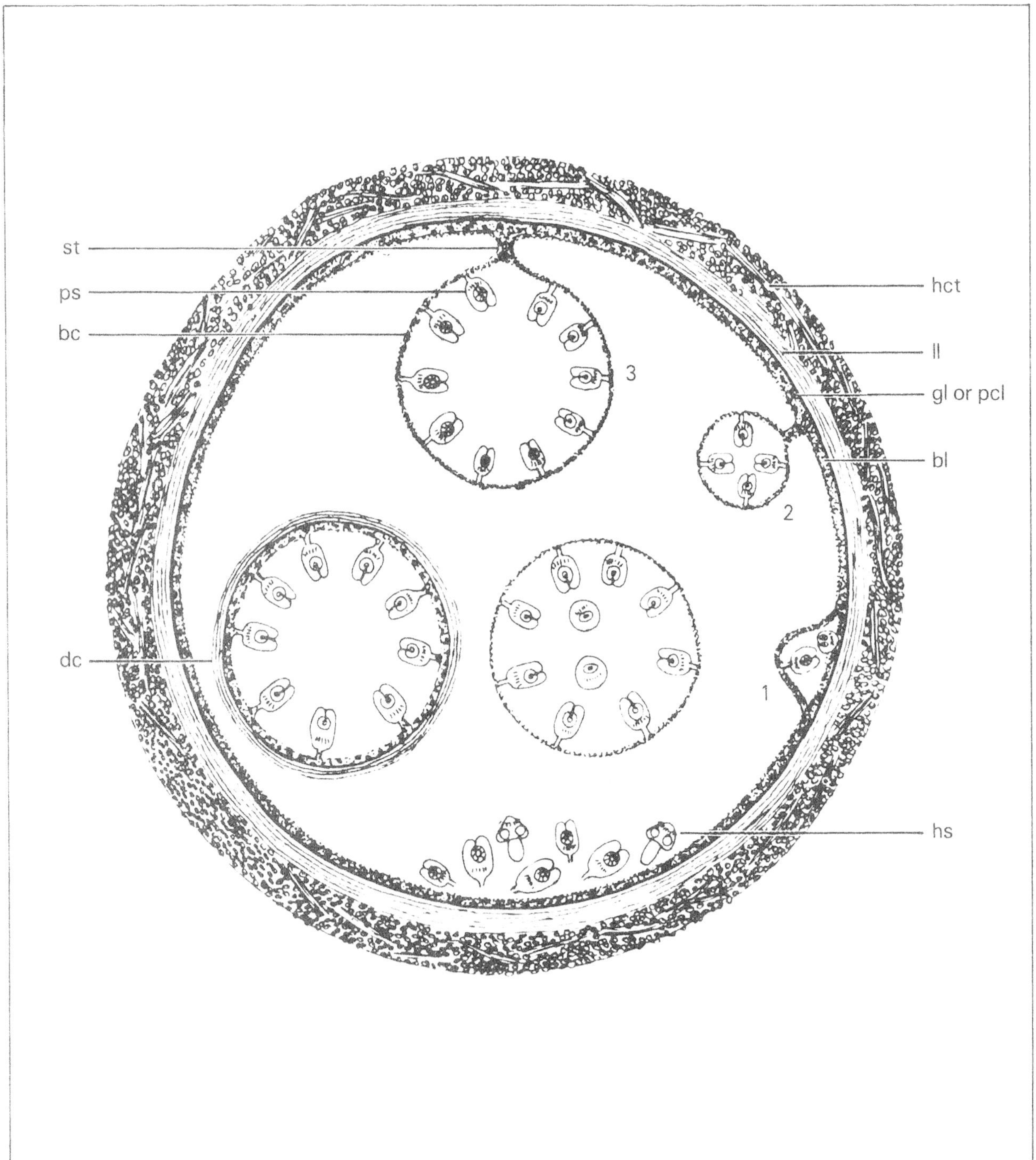


187

**188 Larval Tapeworms**

*Coenurus* — Protoscolices develop from the proliferative cell layer, each on a separate stalk. Brood capsules are not formed and the laminated layer is not present.

*Multilocular hydatid* — There is extensive exogenous budding giving the cyst an irregular appearance. It infiltrates the tissues unlike the hydatid cyst which is localised.

*C. cellulosae* — The cyst is generally ovoid with the scolex invaginated from the broad side. The bladder is distended with fluid. The scolex has a rostellum with hooks.

*C. bovis* — Is similar in structure to *C. cellulosae* excepting the scolex which is generally broader and does not possess hooks.

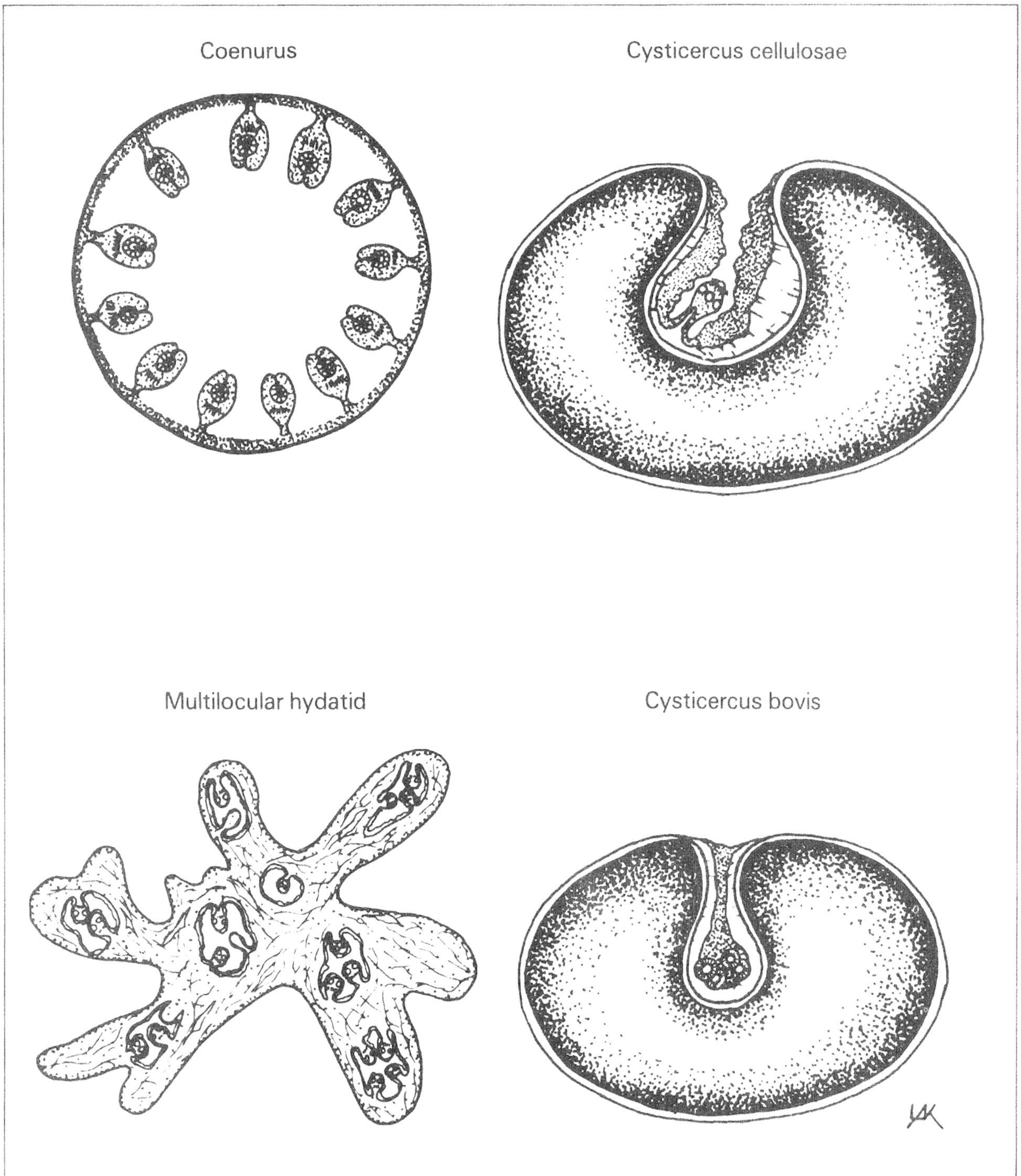


188

## Hymenolepis nana [189]

This is the smallest tapeworm found in man. It varies in length from 25 to 40mm, with approximately 200 proglottids. The eggs are hyaline and ovoidal, 30 to 47μm in diameter. They have 2 membranous shells. The inner shell has 2 poles. From each pole 4 to 8 filaments arise and lie in the space between the 2 shells. Inside the egg is the oncosphere which carries 6 hooks.

Infection is established by the swallowing of the egg. The hexacanth embryo penetrates the villi of the small intestine where it transforms into a cysticercoid larva. The larva re-enters the intestinal lumen and develops into a complete tapeworm.

### *Clinical aspects*

In light infections the infection is asymptomatic. In heavy infections the parasite may give rise to intestinal disturbances such as diarrhoea, vomiting and anorexia.

### *Diagnosis*

This is based on finding the eggs in the faeces.

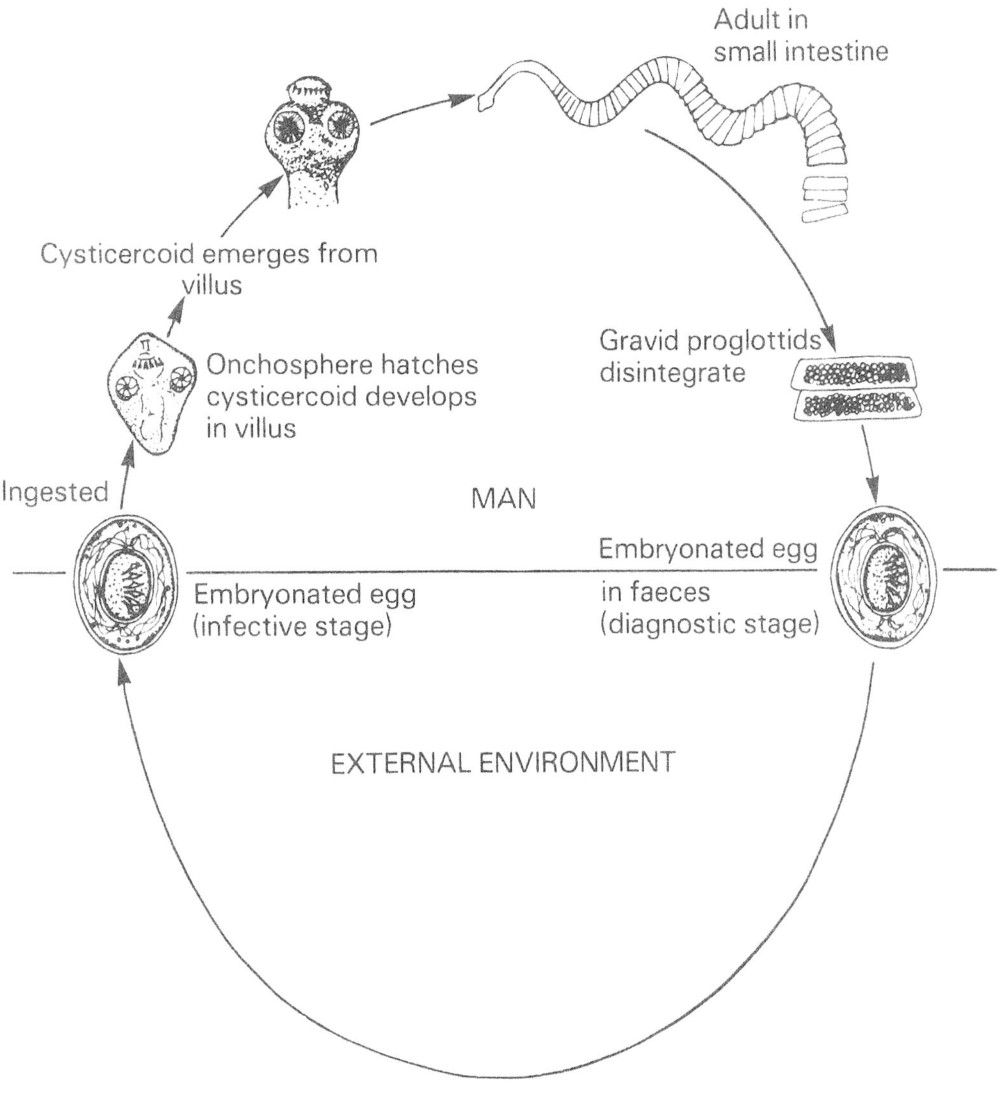
Adult in
small intestine
Cysticercoid emerges from
villus
Onchosphere hatches
cysticercoid develops
in villus
Ingested
Gravid proglottids
disintegrate
MAN
Embryonated egg
in faeces
(diagnostic stage)
Embryonated egg
(infective stage)
EXTERNAL ENVIRONMENT

189

**Hymenolepis diminuta** [190]

Measures 20 to 60cm in length and may have up to 1,000 proglottids. The eggs are spherical and hyaline. They measure 72 to 86μm and like *H. nana* have 2 membranous shells. The inner shell has polar thickening but no filaments. The eggs which are passed in the faeces are ingested by arthropods, usually a rat flea or a flour beetle. In the arthropod the egg develops into a cysticercoid. If the infected arthropod is ingested by a rodent or a human the worm develops into maturity in the intestine. The infection is often asymptomatic and diagnosis is made by finding the egg in the faeces.

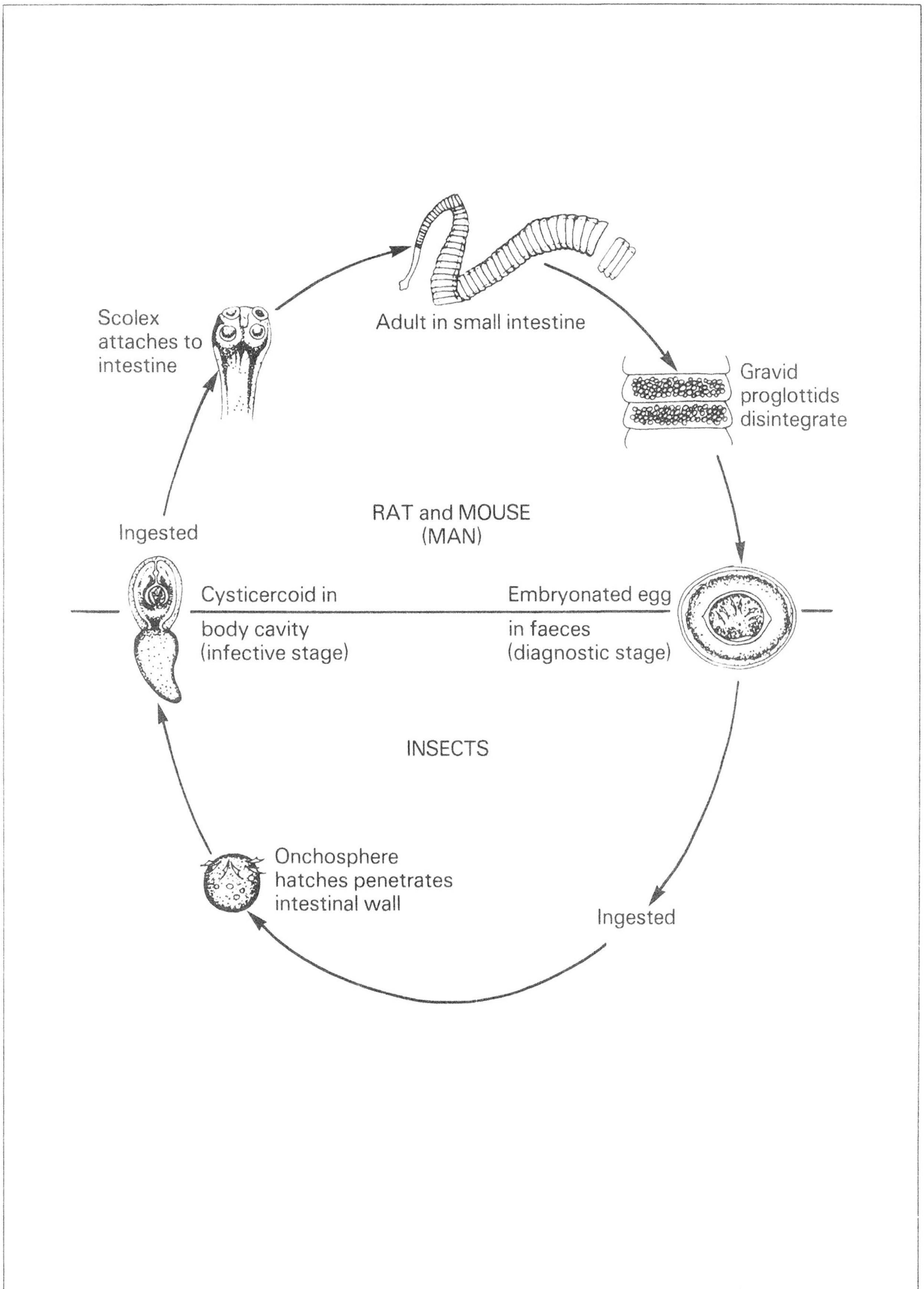
Scolex
attaches to
intestine
Adult in small intestine
Gravid
proglottids
disintegrate
RAT and MOUSE
(MAN)
Ingested
Cysticercoid in
body cavity
(infective stage)
Embryonated egg
in faeces
(diagnostic stage)
INSECTS
Onchosphere
hatches penetrates
intestinal wall
Ingested

190

**191 Comparative morphology of H. nana and H. diminuta**

| | *H. nana* | *H. diminuta* |
|---|---|---|
| *Scolex* | 4 suckers<br>rostellum<br>with hooks | 4 suckers<br>no hooks |
| Egg | Polar thickening<br>with filaments | Polar thickening<br>without filaments |
| *Mature Segment* | | |
| Ovary | Bilobed | Bilobed |
| Testes | 3 globular<br>close | 3 globular<br>widely separated |
| *Gravid Segment* | | |
| Uterus | Sac like<br>irregular | Sac like<br>irregular |

h = hooks
r = rostellum
s = suckers
t = testes
o = ovary
vi = vitelline gland
u = uterus
pf = polar filaments
he = hexacanth embryo

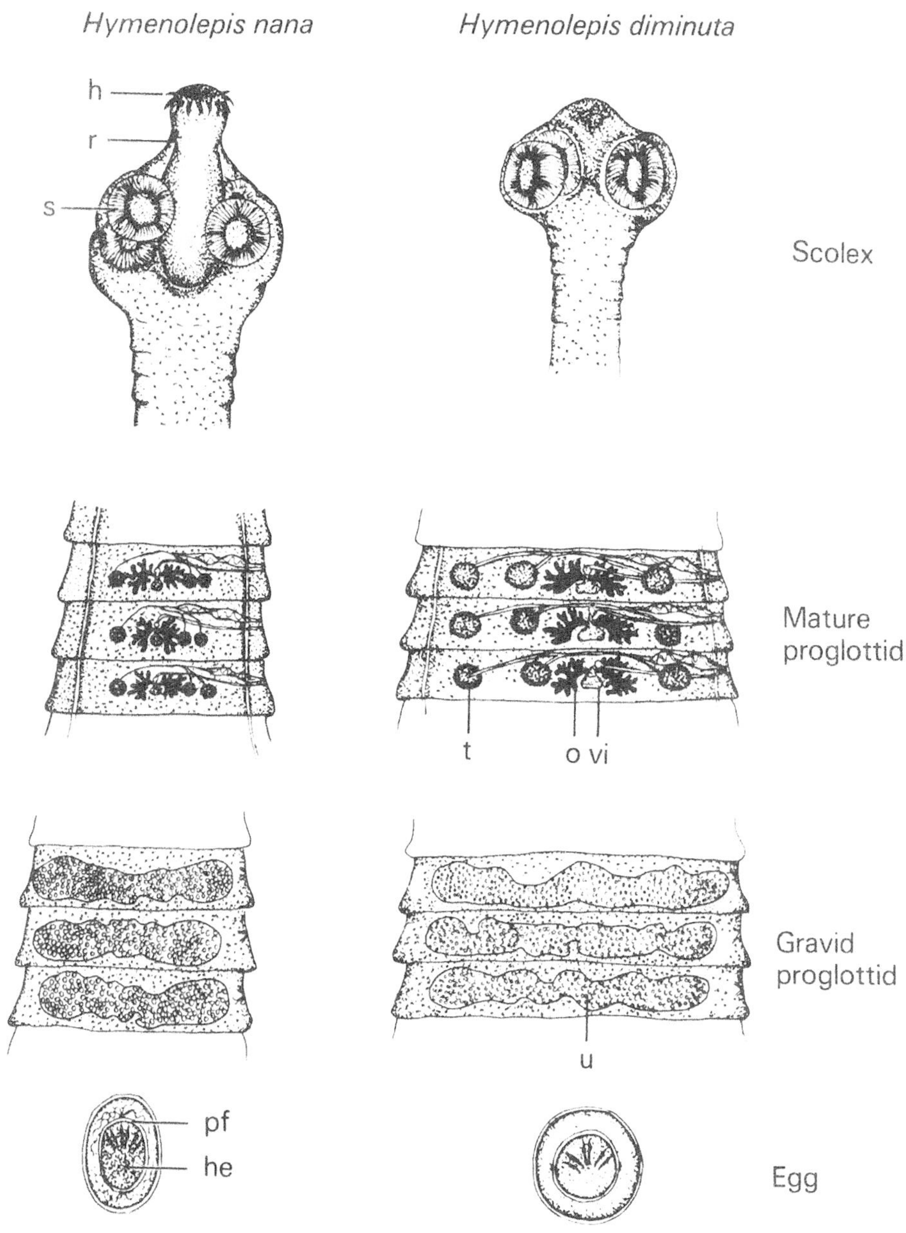
Hymenolepis nana
Hymenolepis diminuta
h
r
s
Scolex
Mature proglottid
t
o vi
Gravid proglottid
u
pf
he
Egg

191

*Dipylidium caninum* [192]

Measures 200 to 400mm and is a common parasite of dogs and cats. It is occasionally seen in man. The gravid proglottids contain egg capsules. These are ingested by the larvae of dog and cat fleas and develop in their body into cysticercoids. Humans are infected by ingesting fleas.

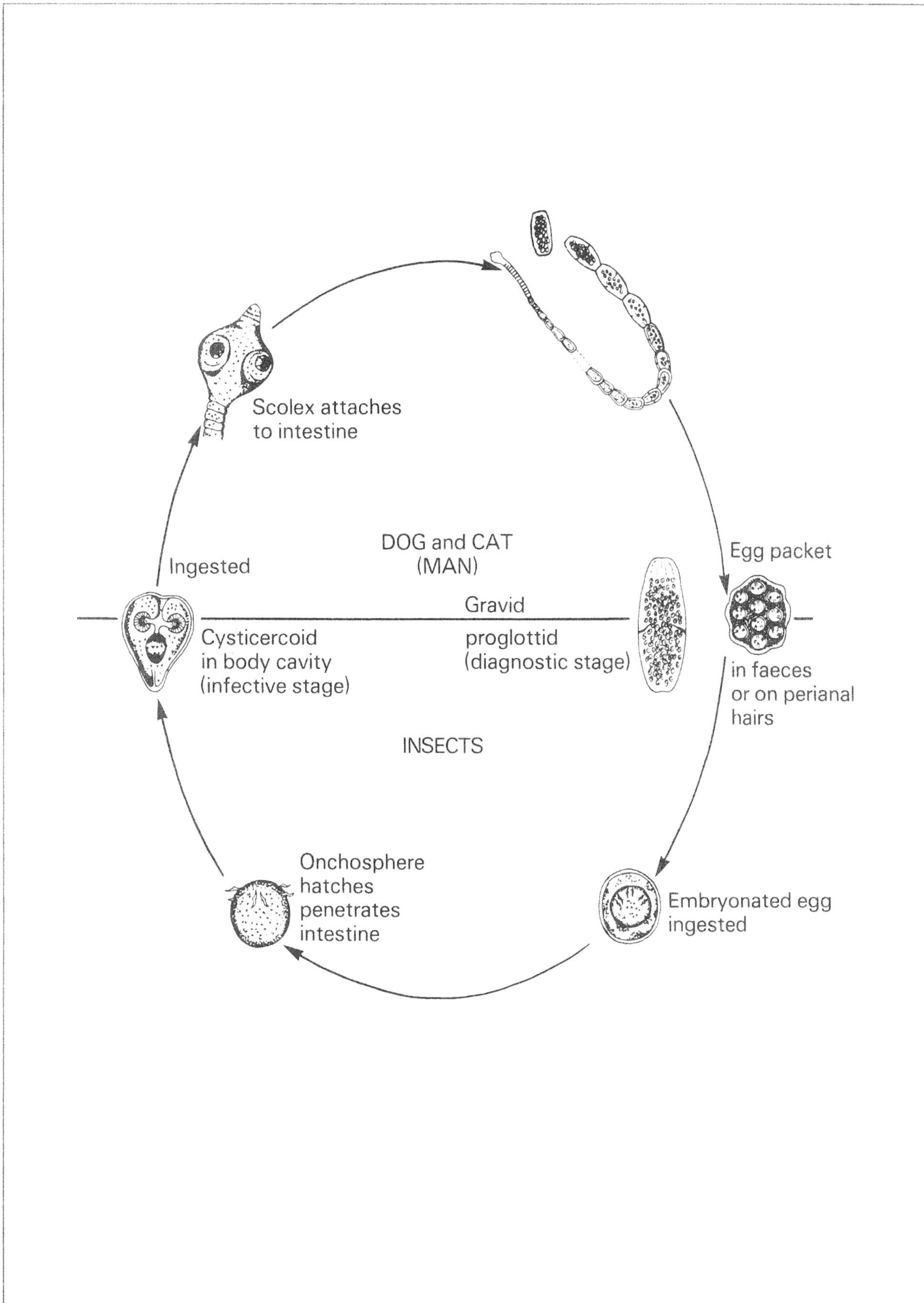
Scolex attaches
to intestine
Egg packet
DOG and CAT
(MAN)
Ingested
Gravid
proglottid
(diagnostic stage)
Cysticercoid
in body cavity
(infective stage)
in faeces
or on perianal
hairs
INSECTS
Onchosphere
hatches
penetrates
intestine
Embryonated egg
ingested

192

## Diphyllobothrium latum [193]

This is a very large tapeworm and may measure up to 10 metres in length. The scolex differs from cyclophyllidean tapeworms in being elongated without any suckers but containing a ventral and a dorsal groove. The eggs are discharged through the uterine pore and are unembryonated at the time of evacuation. A ciliated embryo (coracidium) develops in 11 to 15 days and emerges from the operculated egg. If ingested by a suitable copepod it transforms into a procercoid larva. If the infected copepod is eaten by a fish, the procercoid changes into a plerocercoid larva. Humans may become infected by eating infected fish.

### *Clinical aspects*

The infection may remain asymptomatic. In some cases there are vague digestive disturbances such as diarrhoea, anorexia, nausea and vomiting. A macrocytic hyperchromic anaemia may be produced, apparently due to competitive uptake of vitamin $B_{12}$ by this parasite.

### *Diagnosis*

Is made by finding operculated eggs in the faeces. Proglottids are rarely passed in the faeces unlike other tapeworms.

### References

Matossian, R.M. and Araj, G.F.: Serologic evidence of the postoperative persistence of hydatid cysts in man. Journal of Hygiene 75: 333 (1975).

Nieland, M.L. and von Brand, T.: Electron microscopy of cestode calcareous corpuscle formation. Experimental Parasitology 24: 279 (1969).

Proctor, E.M., Powell, S.J. and Elsdon-Dew, R.: The serological diagnosis of cysticercosis. Annals of Tropical Medicine and Parasitology 60: 146 (1966).

Thompson, R.C.A.: The development of brood capsules and protoscolices in secondary hydatid cysts of *Echinococcus granulosus*. Zeitschrift fur Parasitenkunde 51: 31 (1976).

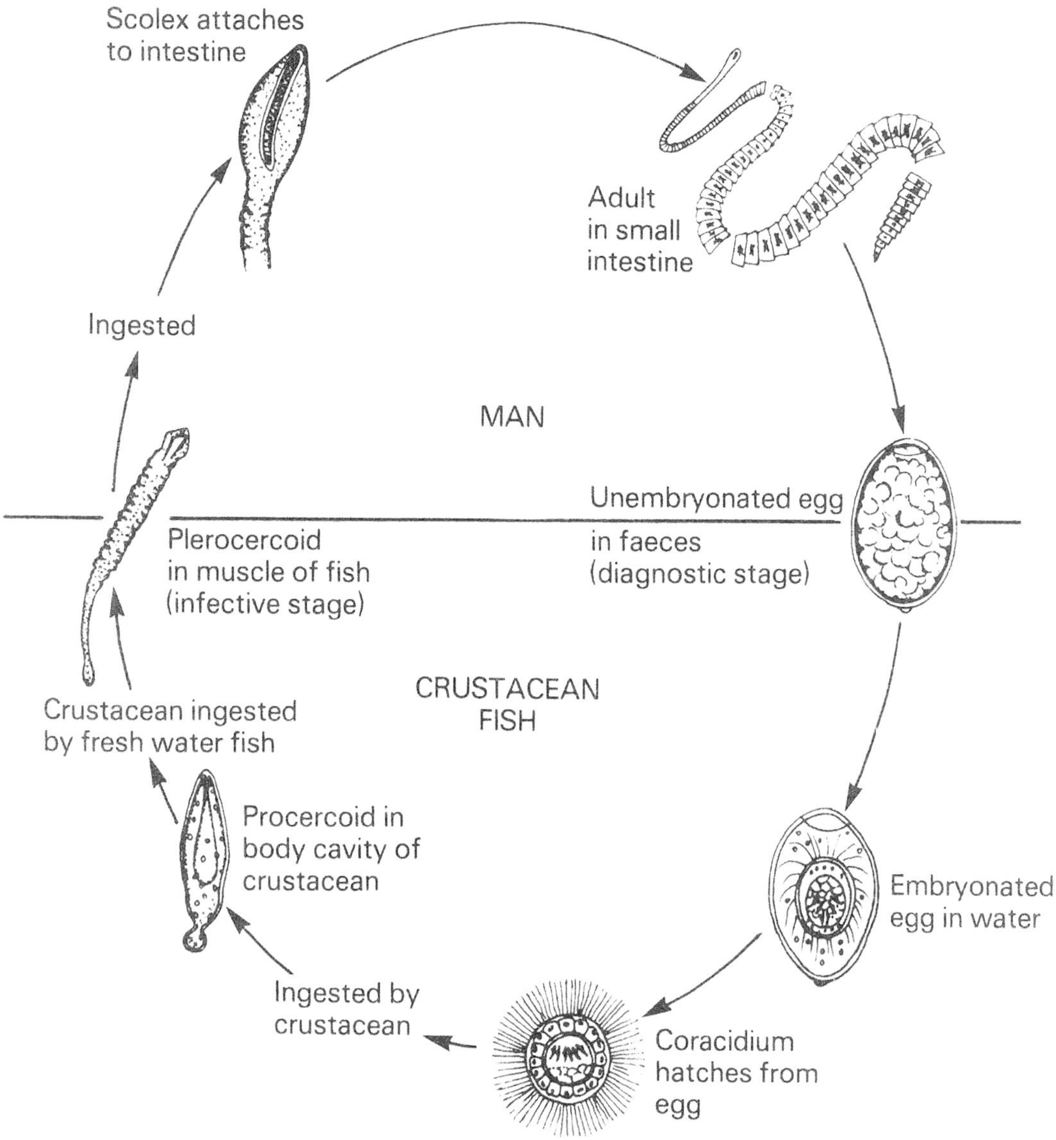
Scolex attaches
to intestine
Adult
in small
intestine
Ingested
MAN
Unembryonated egg
in faeces
(diagnostic stage)
Plerocercoid
in muscle of fish
(infective stage)
CRUSTACEAN
FISH
Crustacean ingested
by fresh water fish
Procercoid in
body cavity of
crustacean
Embryonated
egg in water
Ingested by
crustacean
Coracidium
hatches from
egg

193

**194 Reproductive organs of Diphyllobothrium latum**

| | | | | | |
|---|---|---|---|---|---|
| t | = | testes; | gp | = | genital pore |
| vd | = | vas deferens | u | = | uterus |
| up | = | uterine pore | v | = | vagina |
| vi | = | vitelline gland | ot | = | ootype |
| o | = | ovary | mg | = | Mehlis' gland |

The gravid uterus is easily seen in unstained preparation as a pigmented and convoluted tube. However, it does not spread to cover the whole segment as in *Taenia*.

194

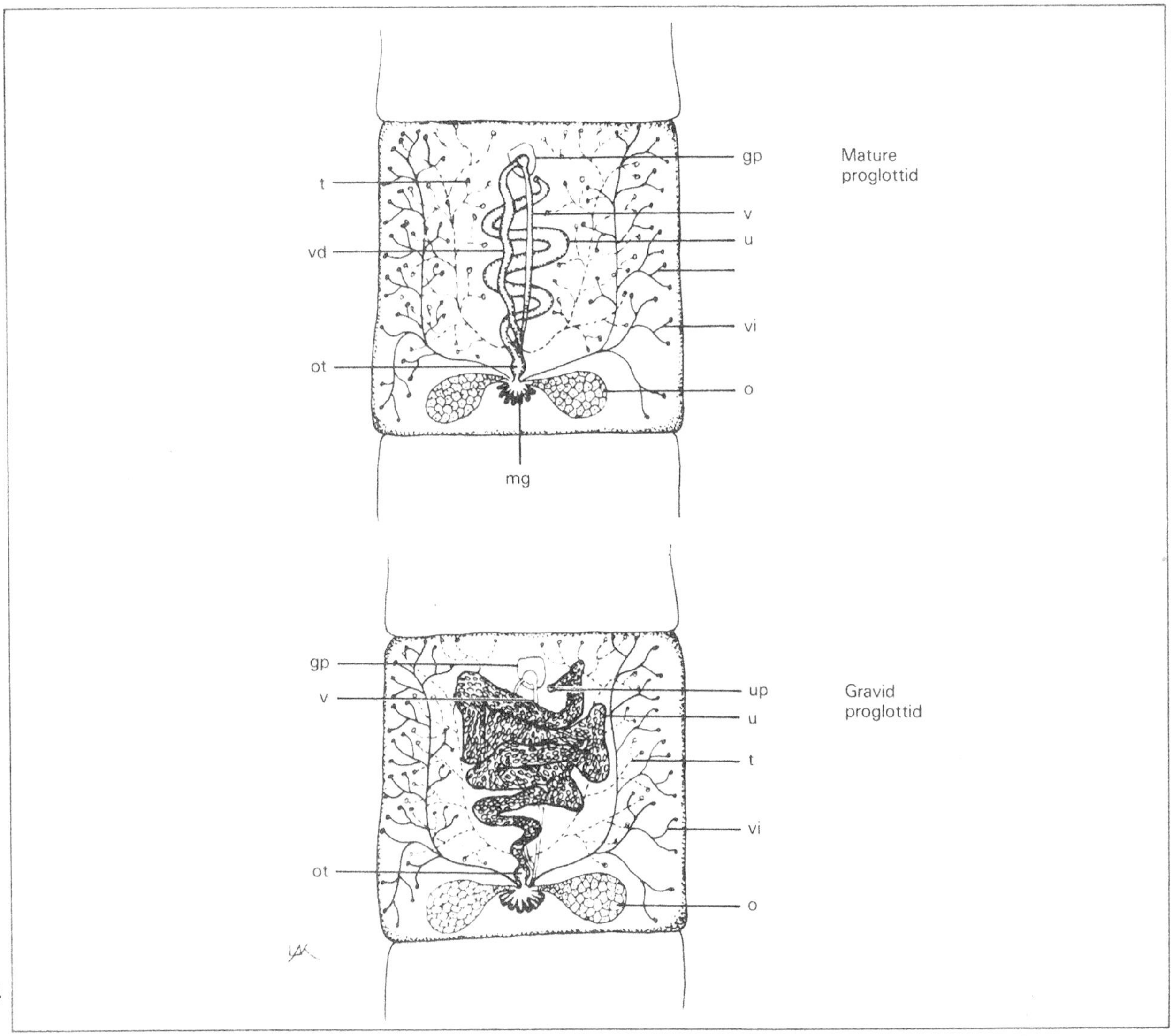

**195 Calcareous Corpuscles of Taenia saginata.** Each corpuscle has a distinct outer wall with a homogenous centre. The homogenous centre is surrounded by a multilayered structure. The multilayered structure may extend to the centre of the corpuscle (b), possibly indicating the age difference between the corpuscles. The background is made up of cells containing a large amount of glycogen. Electron micrograph. X4,500.

a

b

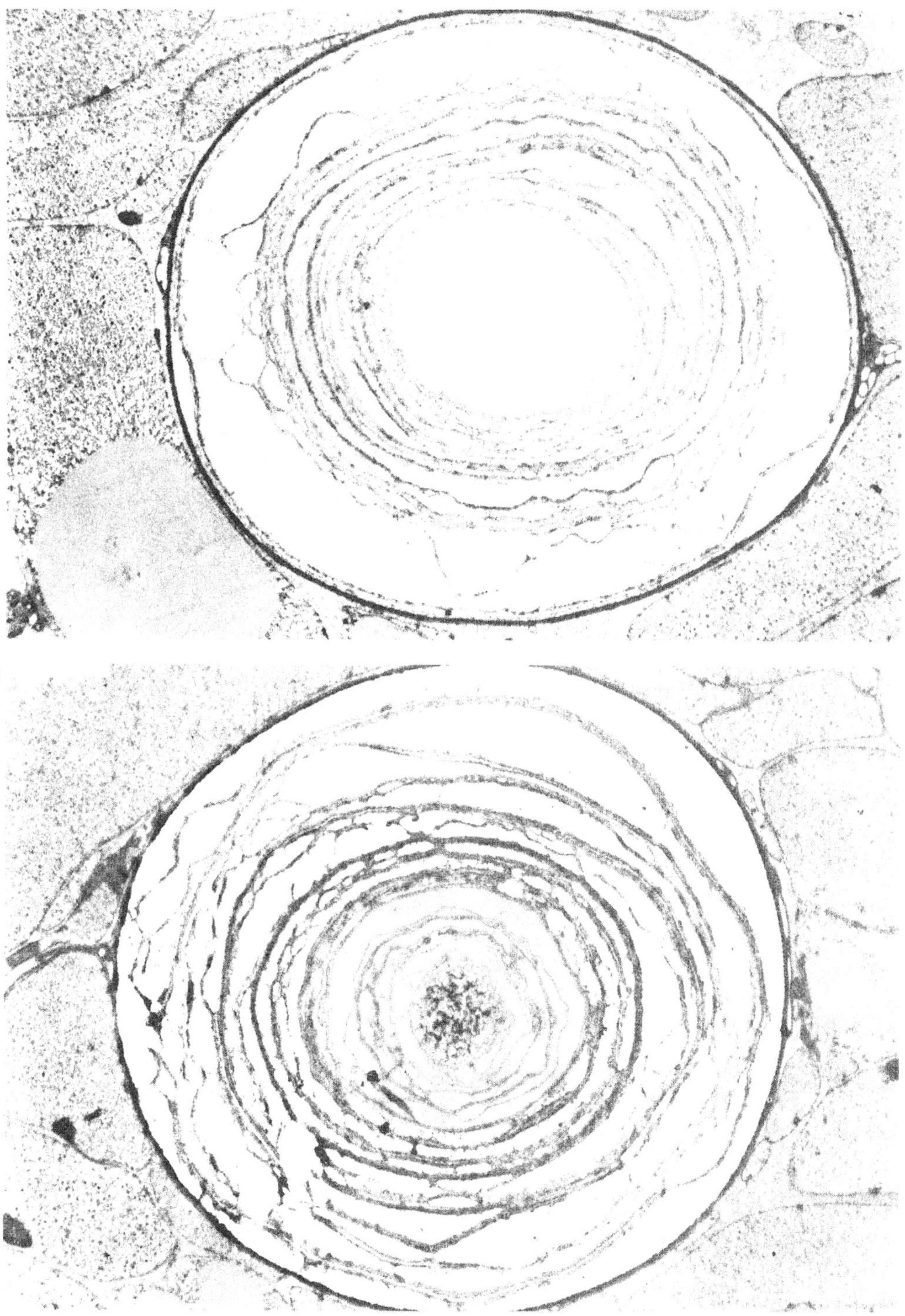

195

**196 Hydatid cysts in liver.** This patient had 3 cysts (marked by arrows) in the right lobe. Two were completely calcified and the large one in the background was partially calcified.

196

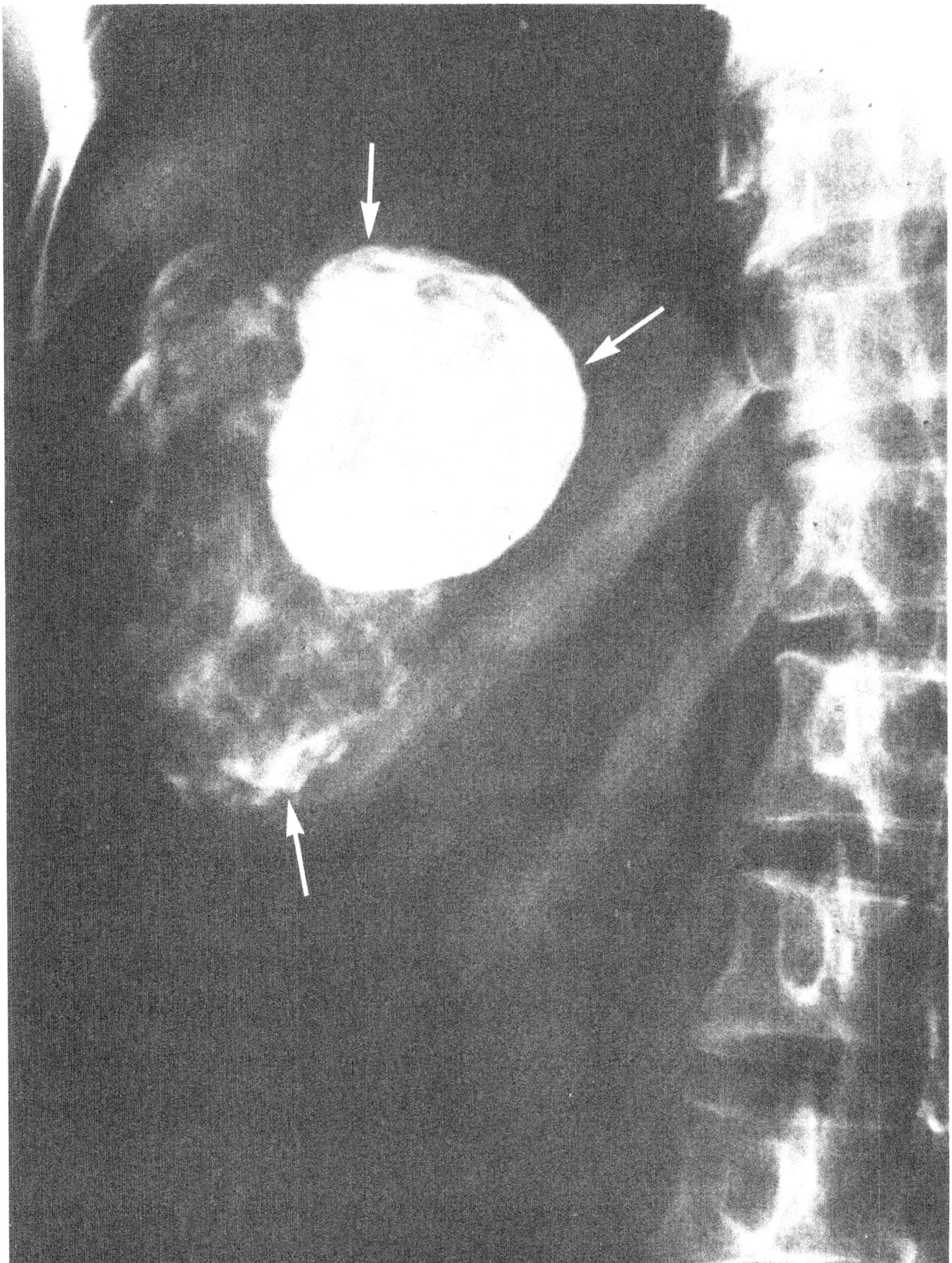

**197 Hydatid cyst in lung.** This patient had a single large cyst in the left lung. There was collapse and consolidation in part of the left lung due to pressure on the bronchus. Partial pneumonectomy was required to remove the cyst.

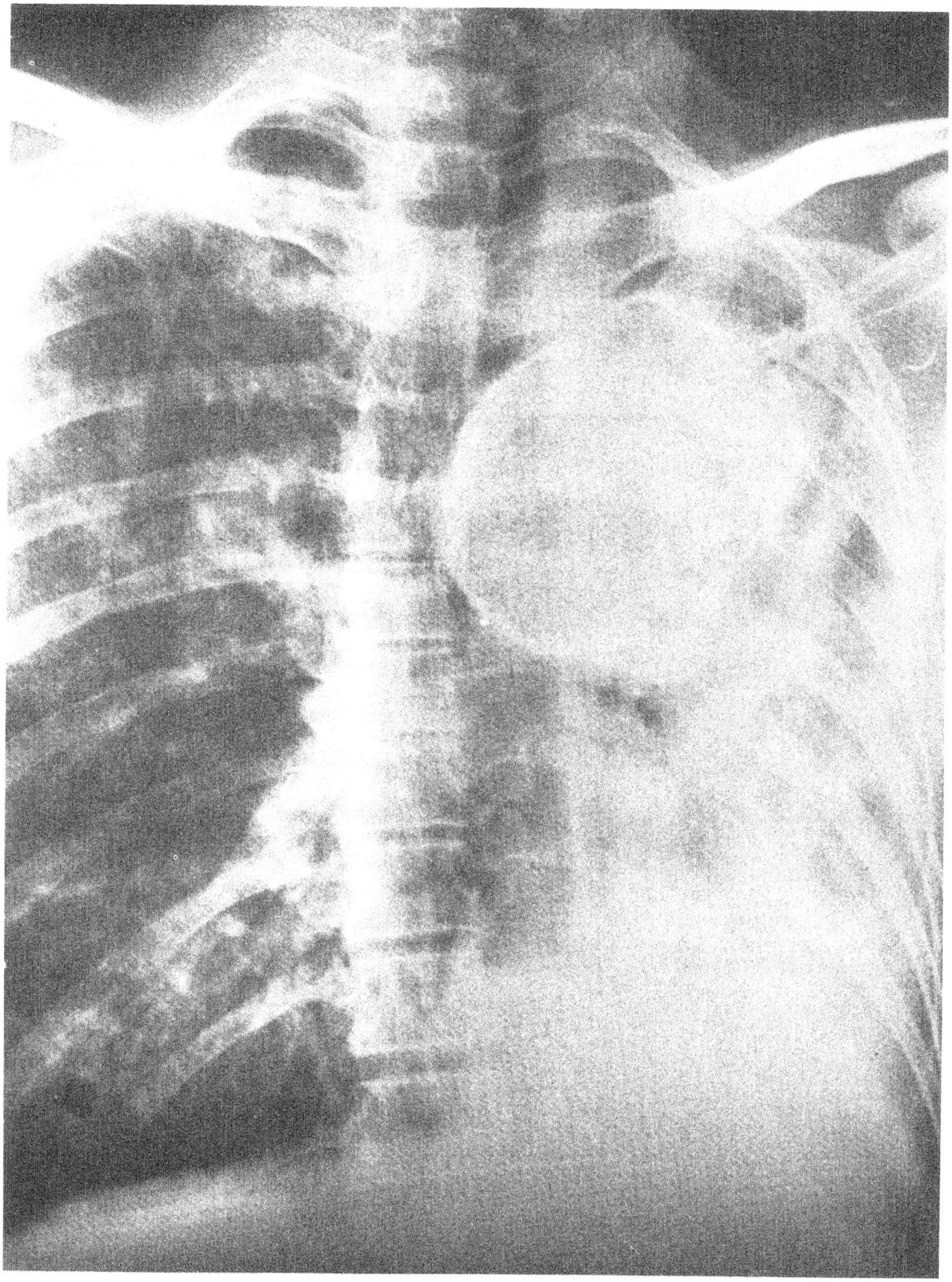

197

**198 Calcareous corpuscles from a proglottid of Taenia saginata.** These are found in large numbers in all cestode tissues and appear fluorescent green when exposed to ultraviolet light. Fresh preparation. X100. Enlarged by 5.4.

**199 Scolex of Diphyllobothrium latum showing the elliptical shape.** One of the grooves is visible. X100. Enlarged by 5.4

**200 Scolex of Taenia solium showing the armed scolex and 2 suckers.** The scolex is partly evaginated giving it a cup shaped appearance. Interference contrast. X100. Enlarged by 5.4

**201 Scolex of Hymenolepis nana showing 4 suckers and an armed scolex.** Interference contrast. X100. Enlarged by 5.4

**202 Scolex of Hymenolepis nana at higher magnification** (X400) to show the structure of hooks arranged in a single row. Enlarged by 5.4.

**203 Scolex of Taenia pisiformis.** Head on view showing the arrangement of hooks in a circular pattern. Interference contrast. X100. Enlarged by 23.4.

**204 Scolex of Dipyllidium caninum showing 2 suckers** (from the side) and an elongated scolex covered with spines. Interference contrast. X100. Enlarged by 5.4.

**205 Scolex of Dipyllidium caninum at higher magnification** (X400) showing the spines and their radial arrangement. Enlarged by 5.4.

198

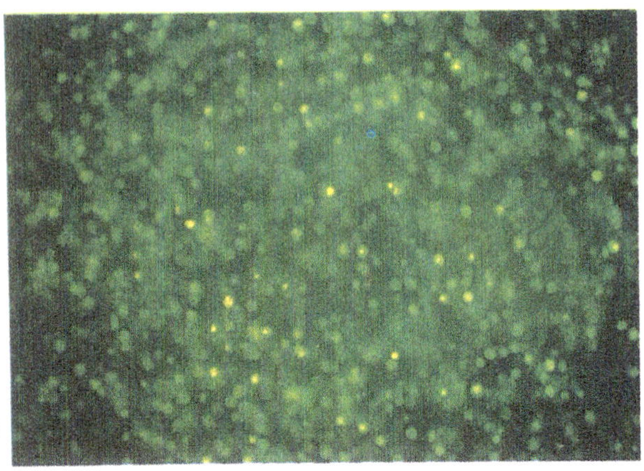

199

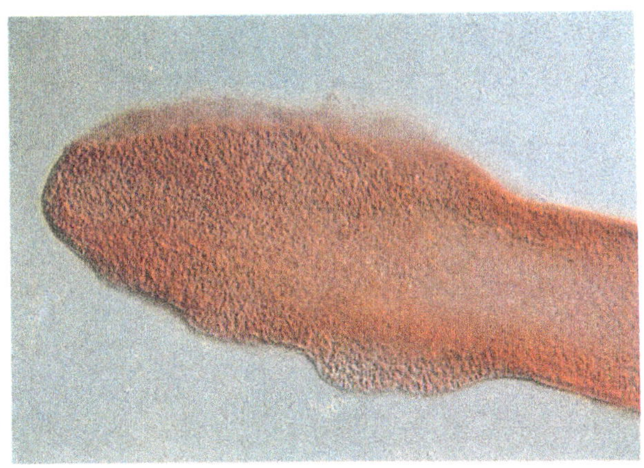

200

201

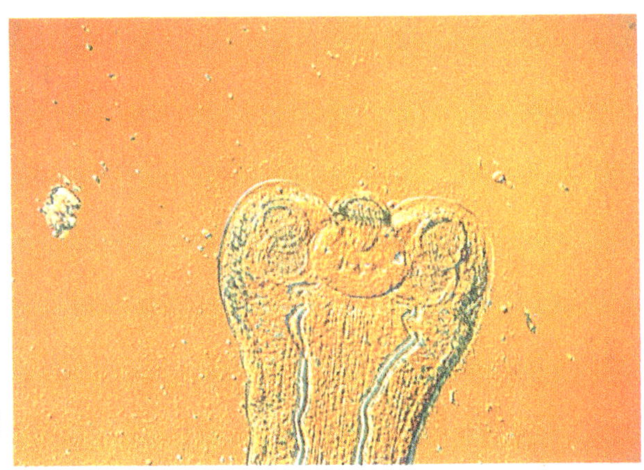

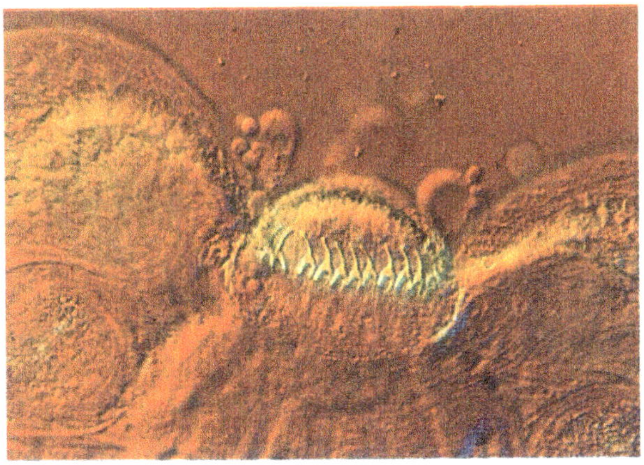
202

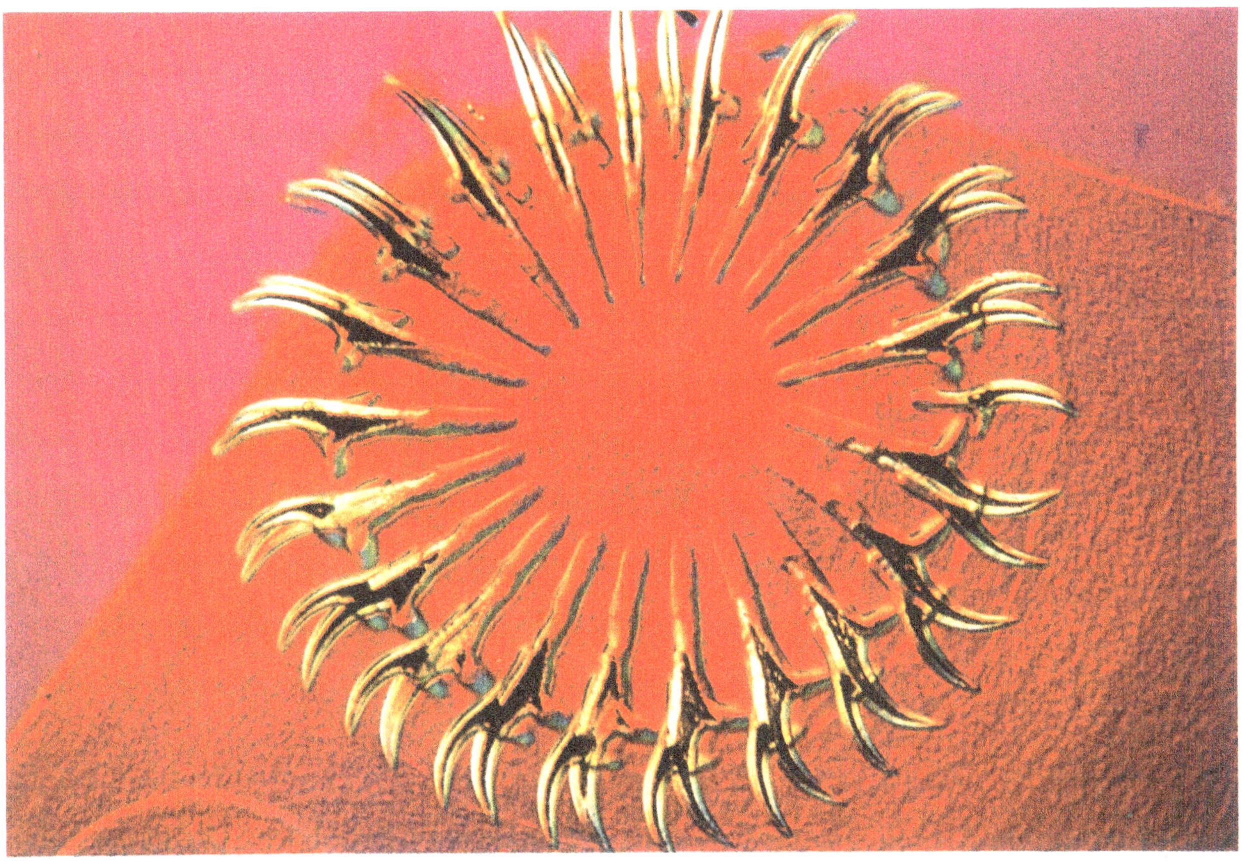
203

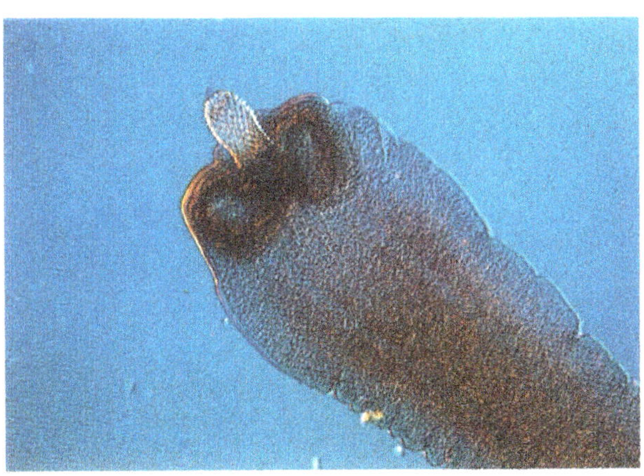
204

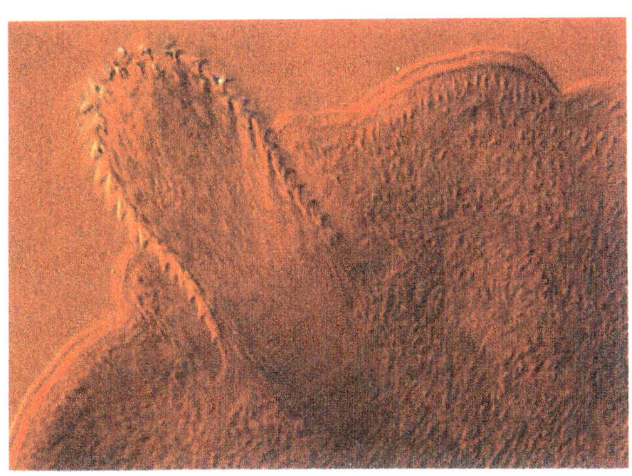
205

**206 Cysticercus bovis showing the bladder** and the scolex protruding into it. PAS. X50. Enlarged by 5.4.

**207 Cysticercoid larva of Hymenolepis nana.** There is no bladder and the scolex is embedded in solid tissue. Interference contrast. X100. Enlarged by 5.4.

**208 Hydatid cyst showing a row of brood capsules** attached to the germinal layer. H and E. X50. Enlarged by 5.4.

**209 A brood capsule with 4 inverted scolices,** a fully everted scolex and a partially everted scolex. X100. Enlarged by 5.4.

**210 A single inverted scolex from a brood capsule.** The hooks are visible in the centre. Interference contrast. X400. Enlarged by 5.4.

**211 Cysticercus cellulosae** in a heavily infected pig liver. (Courtesy of Dr Patrick Ko).

**212 Cysticercus cellulosae dissected out of pig muscle.** The opaque body inside the bladder is the scolex.

**213 Coenurus of Multiceps multiceps from sheep brain.** A large number of scolices are seen as white clusters attached to the internal layer of the cyst.

**214 Hymenolepis nana egg.** The 2 shells and the oncosphere can be seen. The polar filaments are clearly visible on one side. X100. Enlarged by 5.4.

**215 Hymenolepis diminuta egg.** The polar filaments are not present. X100. Enlarged by 5.4.

**216 Taenia saginata egg enclosed in an outer membrane.** This membrane disintegrates in the lumen of the bowel and is not often seen. X100. Enlarged by 5.4.

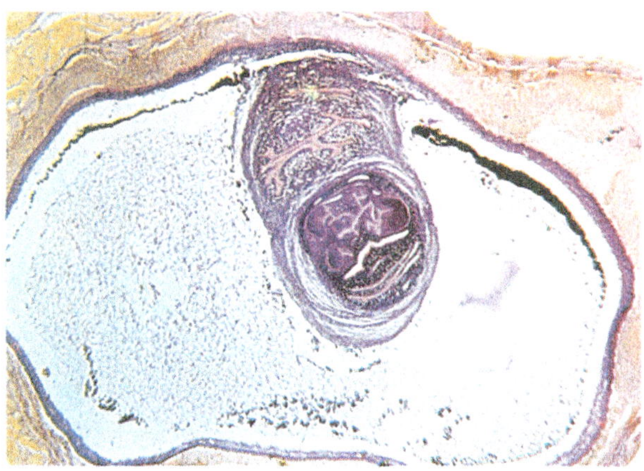

206

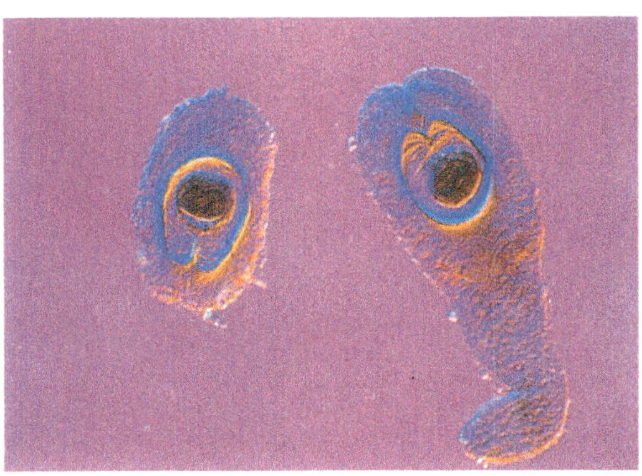

207

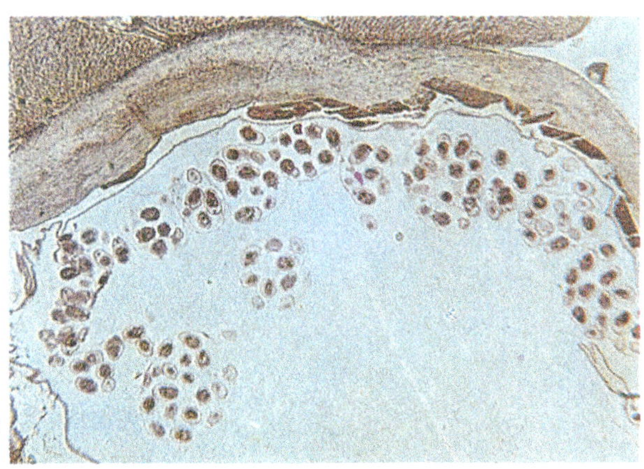

208

209

210
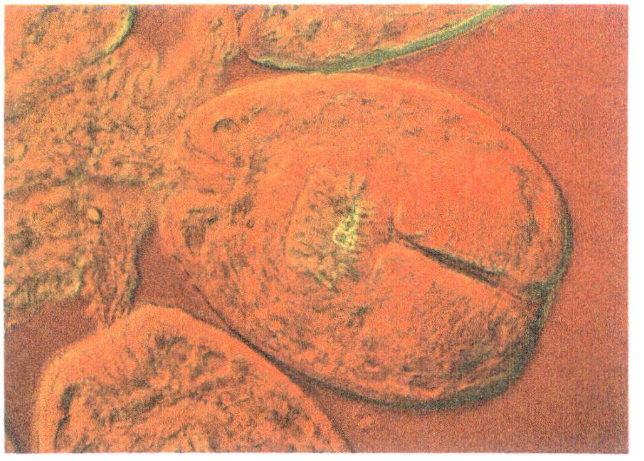

211
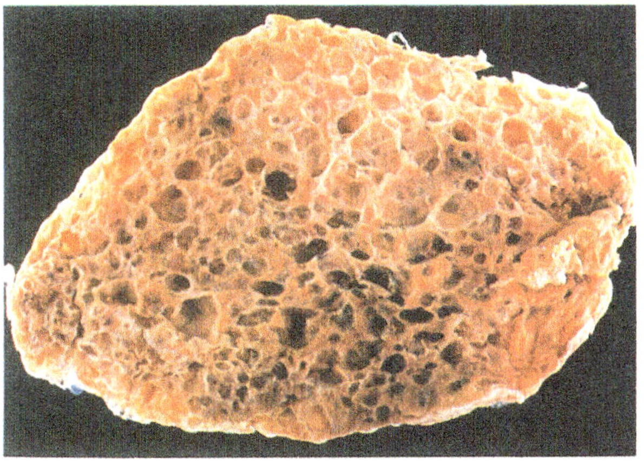

212

213

214

215
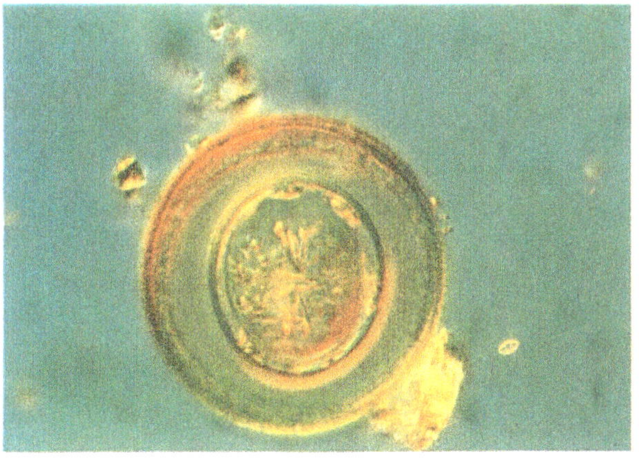

216
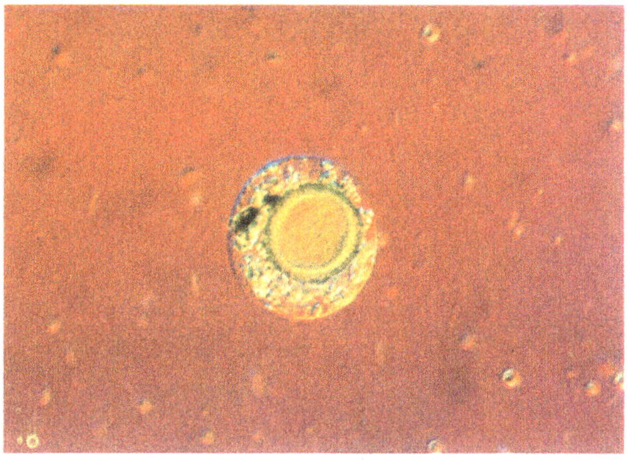

**217 Taenia saginata egg without the outer membrane,** showing the striated embryophore. X100. Enlarged by 5.4.

**218 Section of appendix** showing *Taenia* eggs in the lumen. H and E. X100. Enlarged by 5.4.

**219 An egg capsule of Dipyllidium caninum.** This specimen contains 9 eggs. Interference contrast. X50. Enlarged by 5.4.

**220 An egg of a Pseudophyllidian tapeworm** to which *Diphyllobothrium latum* belongs. It contains a ciliated coracidium. Interference contrast. X100. Enlarged by 5.4.

**221 An egg of a Pseudophyllidian tapeworm** showing hooks on coracidium. Interference contrast. X400. Enlarged by 9.6.

**222 An empty egg after the coracidium has escaped**. The detached operculum is now clearly seen. X100. Enlarged by 5.4.

**223 Coracidium with faintly visible cilia around it.** X100. Enlarged by 5.4.

217

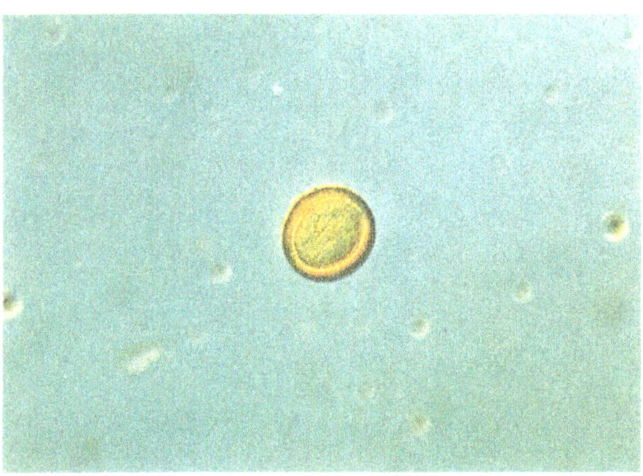

218

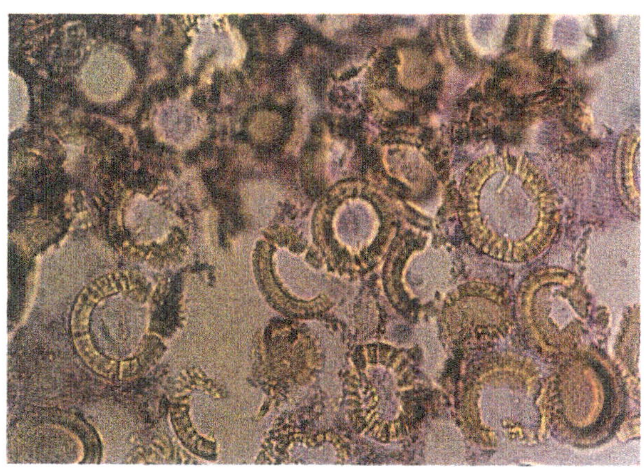

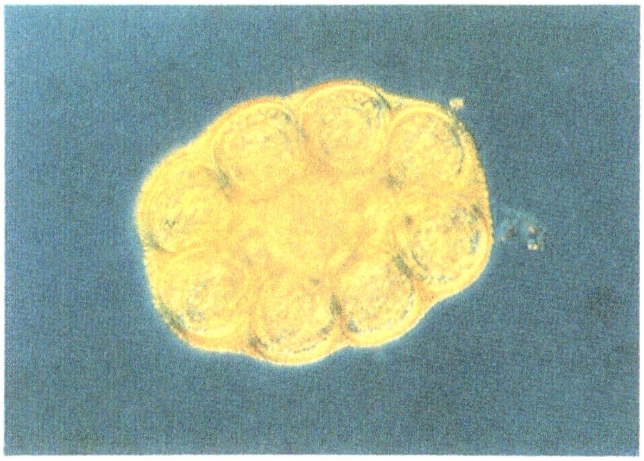
219

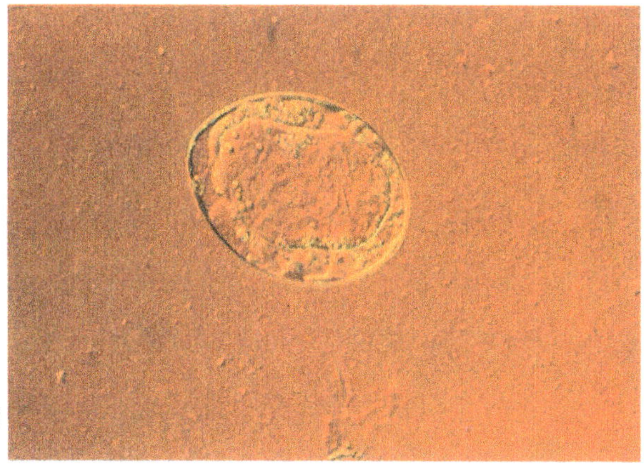
220

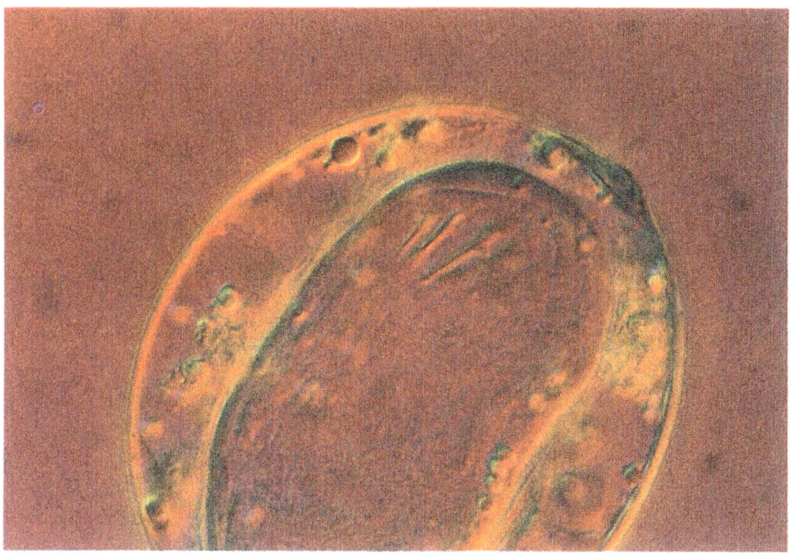
221

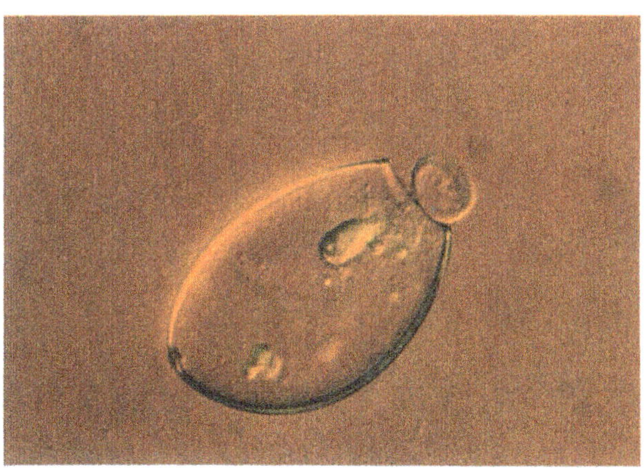
222

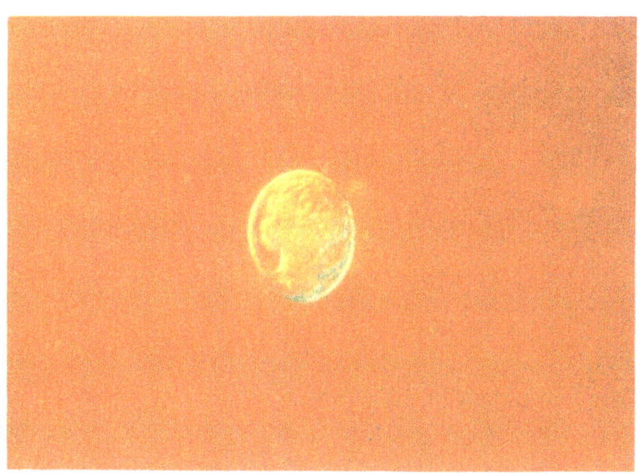
223

# Section 11
# Trematodes

## Classification of Medically Important Trematodes

| | |
|---|---|
| Phylum | Platyhelminthes |
| Class | Trematoda |
| Subclass | Digenea |
| Order | Prosostomata |
| Sub order | Strigeata |
| | Amphistomata |
| | Distomata |

| Suborder | Family | Genus | Species |
|---|---|---|---|
| Strigeata | Schistosomatidae | *Schistosoma* | *S. haemotobium* |
| | | | *S. mansoni* |
| | | | *S. japonicum* |
| | | | *S. intercalatum* |
| Amphistomata | Paramphistomatidae | *Gastrodiscoides* | *G. hominis* |
| | | *Watsonius* | *W. watsoni* |
| Distomata | Fasciolidae | *Fasciola* | *F. hepatica* |
| | | | *F. gigantica* |
| | | *Fasciolopsis* | *F. buski* |
| | Opisthorchidae | *Clonorchis* | *C. sinensis* |
| | | *Opisthorchis* | *O. felineus* |
| | | | *O. viverrini* |
| | Heterophyidae | *Heterophyes* | *H. heterophyes* |
| | | *Metagonimus* | *M. yokogawai* |
| | Troglotrematidae | *Paragonimus* | *P. westermani* |

## Terms used in Relation to Trematodes

**Monoecious** — individuals containing gonads of both sexes, i.e. hermaphroditic.

**Dioecious** — condition where male and female gonads are found in different individuals.

**Operculum** — a lid-like structure covering certain cestode and most trematode eggs.

**Miracidium** — the ciliated larva which hatches out from the trematode egg.

**Cercaria** — young flukes which develop from germ cells in sporocyst and redia.

**Metacercaria** — a stage in which the cercaria becomes encysted.

**Sporocyst** — a larval stage which is sac like and containing germinal cells which give rise either to a second generation of sporocyst or to redia.

**Redia** — a larval stage which is sac like but with a pharynx and a rudimentary intestine. It contains germinal cells which give rise either to a second generation of redia or to cercaria.

**Suckers** — adhesive organs. There are generally 2 suckers, oral and ventral.

**Acetabulum** — ventral sucker with well developed muscles.

**Cirrus** — the male copulatory organ.

**Cirrus pore** — the opening through which the cirrus is protruded.

**Cirrus pouch** — a hollow organ surrounding the introverted cirrus.

**Vas deferens** — a canal connecting vas efferens to cirrus.

**Vas efferens** — a canal extending from a testis to the vas deferens.

**Seminal receptacle** — a dilated organ in the female genital tract which stores sperms.

**Seminal vesicle** — the dilated lower part of the vas deferens which opens into cirrus.

**Ootype** — is the fertilising chamber where the ovum is fertilised by the spermatozoon.

**Laurer's canal** — a tubular structure connecting the oviduct with the exterior. Also known as the copulatory canal.

**Mehlis' gland** — a unicellular gland which encircles the ootype. Its function is not known.

**Vitelline glands** — the glands which provide substances for the development of the egg and the formation of the shell.

**Common genital pore** — the common opening formed by the fusion of the male genital duct and the uterus. This is usually enclosed in a small chamber.

**Flame cell** — an excretory cell with a tuft of cilia, whose beating resembles the flickering of a flame. The flame cells open into a collecting tubule. Also known as solenocyte.

**Parenchyma** — a loose collection of cells and fibres.

## Introduction

This group of parasites is also known as flukes. Their body is generally flattened and leaf like with the exception of schistosomes which are elongated. One or more muscular suckers are always present on their ventral surface. The alimentary canal consists of a short median anterior portion which bifurcates to form 2 posterior portions. The anterior portion of the gut has a muscular pharynx with the exception of schistosomes in which the pharynx is absent. The posterior portions end blindly and are therefore known as caeca. The caeca are usually unbranched but in some cases they show extreme branching.

Trematodes are monoecious, with the exception of schistosomes. The female reproductive organs consist of an ovary, seminal receptacle, Laurer's canal, ootype and uterus. The Mehlis' gland surrounds the ootype. Two masses of glandular structures known as vitelline glands open via vitelline ducts near the ootype. The ovum liberated by the ovary passes through an oviduct to the ootype. In the ootype fertilisation occurs by the discharge of sperms from the seminal receptacle. The shell material produced by the vitelline glands then covers the fertilised ovum. From the ootype the eggs are pushed into the uterus. The function of Mehlis' gland which surrounds the ootype is not known.

The male reproductive system consists of testes, vas efferens, vas deferens, seminal vesicle, cirrus, prostate and cirrus pouch. The number, position and shape of testes varies in different species and these features are used for the purpose of species identification. Spermatozoa are produced in the testes, pass through the tubular system of vasa efferentia and vas deferens and are temporarily stored in the seminal vesicle. The sperms are discharged into the female reproductive system by the cirrus, through an opening known as the common genital pore.

The excretory system consists of a bilateral array of flame cells which lead into tubules, collecting ducts and an excretory vesicle. The shape of the vesicle and the position and number of flame cells vary in different trematodes and is used for taxonomic purposes.

The nervous system is made up of 2 ganglia near the pharynx from which transverse and longitudinal nerve fibres extend anteriorly and posteriorly to cover the whole body.

The eggs of trematodes are operculated with the exception of schistosomes. In some cases they are passed with a miracidium or the ciliated larval stage. In others they are passed with an immature ovum, which requires to develop into a miracidium, in the external environment. The size of eggs vary from 140µm *(Fasciolopsis buski)* to 28µm *(Heterophyes heterophyes)*.

All trematodes pass through a phase of asexual development in the snail host. This form of development is initiated by the entry of a miracidium into snail tissues. When the miracidium reaches a suitable site in the snail it loses its cilia and changes into an elongated sporocyst. The sporocyst has a body wall, a body cavity and germinal cells. The germinal cells produce either daughter sporocysts or rediae. Both these stages lead to the formation of cercariae.

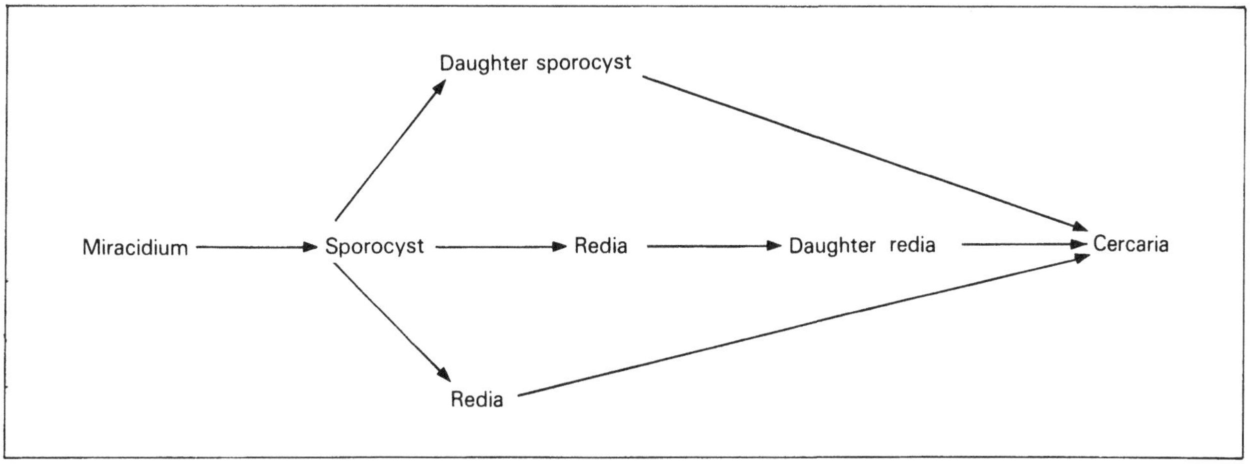

**224 Reproductive system in a monoecious trematode.** A simplified outline of genital structures is shown. The ovary and testes are drawn as circular bodies but their actual shape and size vary in different species. Similarly the size and shape of the vitelline glands also differ in different species. The ventral sucker which often overlaps the common genital atrium is not shown.

**225 Alimentary, excretory and nervous system in a trematode.**
a) *Opisthorcis* sp, an example of simple caeca which extend to the posterior end of the body and do not form any branches.
b) *Fasciola* sp, an example of highly branched caeca covering major part of the parasite body.
c) *Opisthorcis* sp, showing the arrangement of flame cells, the collecting tubules and a triangular excretory vesicle which opens posteriorly.
d) *Opisthorcis* sp, showing the 2 nerve ganglia lying in between the oral and ventral sucker and the longitudinal and transverse cords extending from the ganglia.

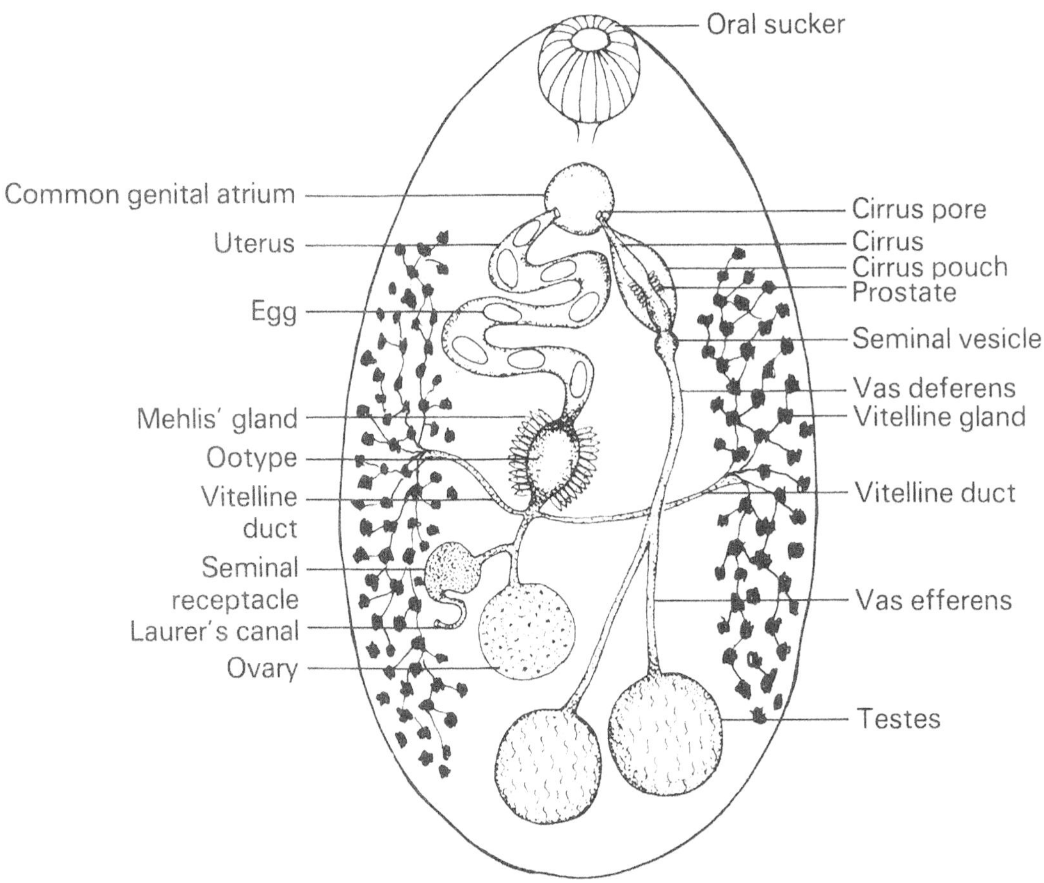


224

a b c d

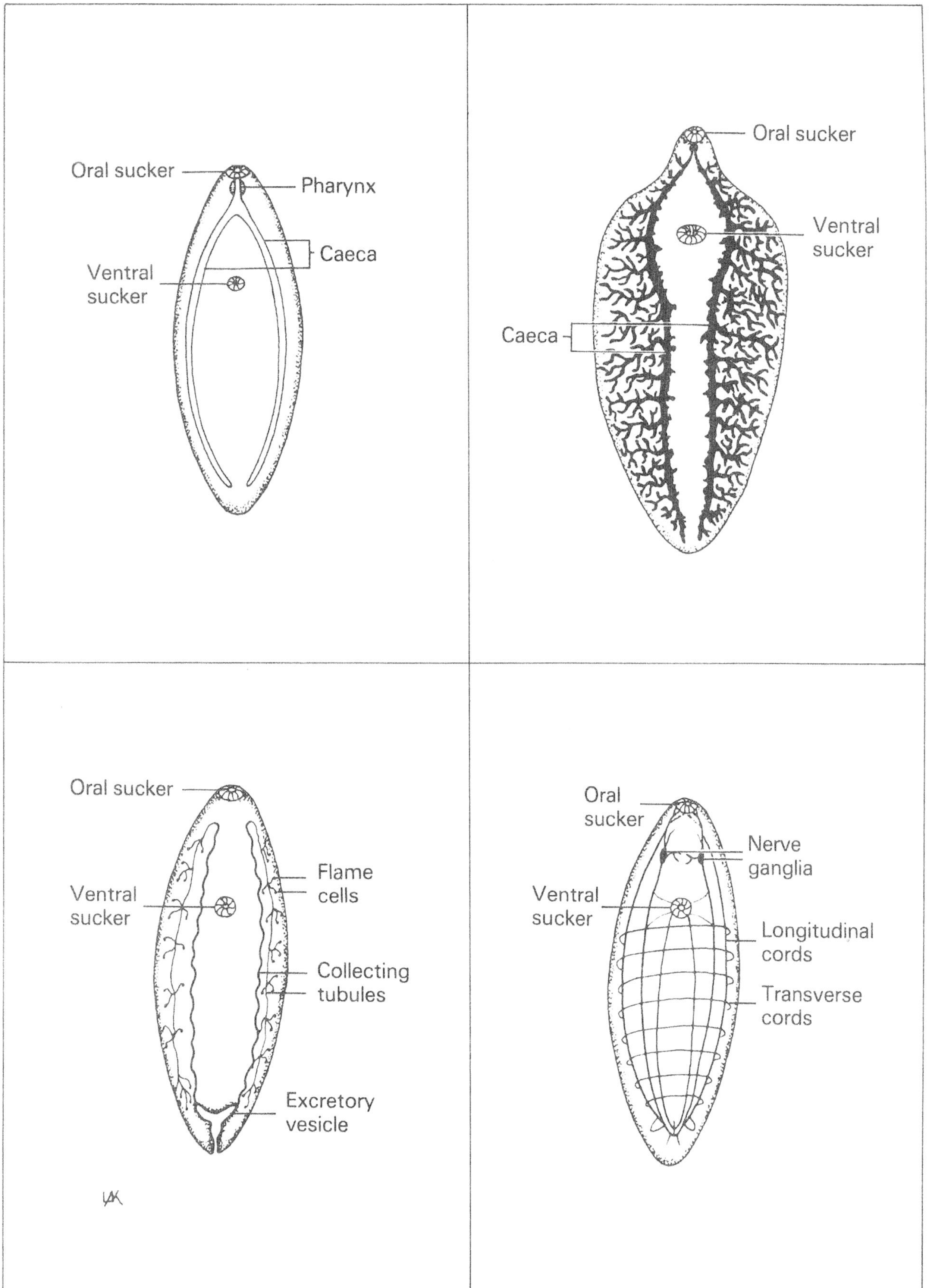
Oral sucker
Pharynx
Caeca
Ventral sucker
Oral sucker
Ventral sucker
Caeca
Oral sucker
Flame cells
Ventral sucker
Collecting tubules
Excretory vesicle
Oral sucker
Nerve ganglia
Ventral sucker
Longitudinal cords
Transverse cords

225

**226 Larval stages of trematode developing in snail.** Daughter sporocyst — Is sac like and contains germinal cells. The germinal cells produce cercarial embryos, which give rise to cercariae.
Mother redia — Is also sac like but has a distinct pharynx and a rudimentary intestine. The germinal cells produce redial embryos, which give rise to daughter rediae.
Daughter redia — In morphology it is like mother redia, excepting that it produces cercaria. In some trematodes cercariae are produced directly from the first generation redia without the formation of daughter redia.
Cercaria — The shape and size of cercaria vary amongst different groups of trematodes. In this case cercaria with a pair of fin-folds (Genus *Opisthorcis*) is shown. The eye spots are regarded as sense organs.

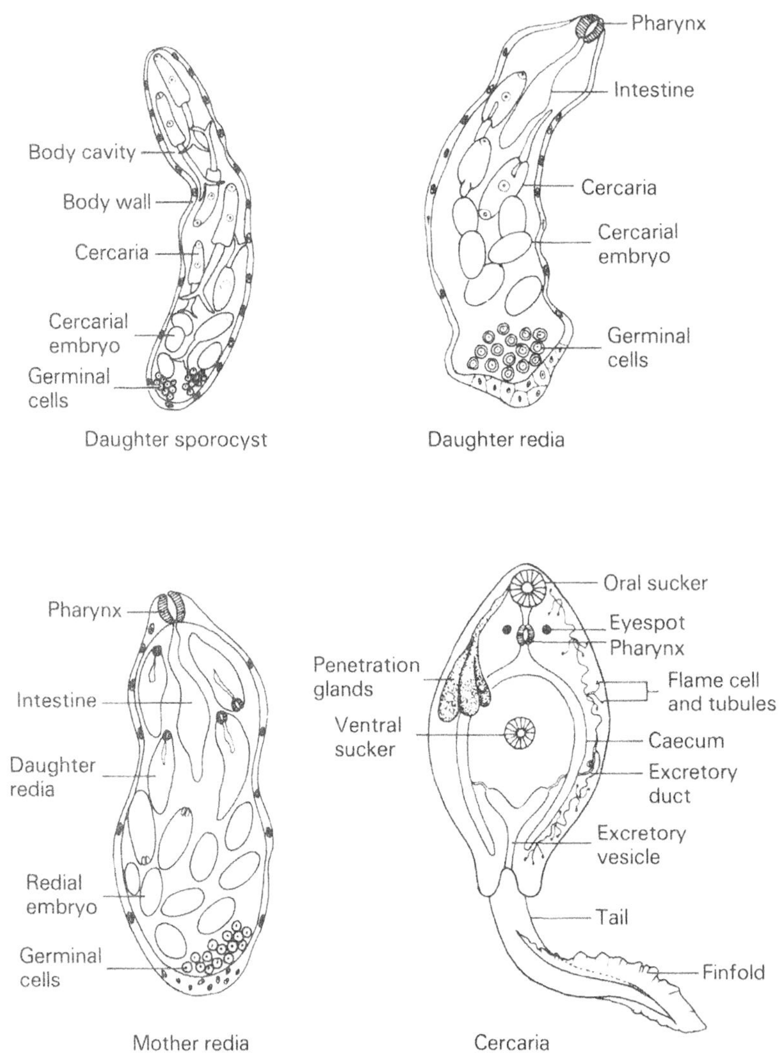


226

**227 Comparative size of adult trematodes.** These parasites were photographed at the same magnification (X6) to show the size differences.
1) *Fasciola gigantica*
2) *Paragonimus westermani*
3) *Gastrodiscoides hominis*
4) *Clonorchis sinensis*
5) *Opisthorcis viverrini*
6) *Dicrocoelum dendriticum*
7) *Metagonimus yokogawai*
8) *Schistosoma japonicum* (male and female in copulation)

227

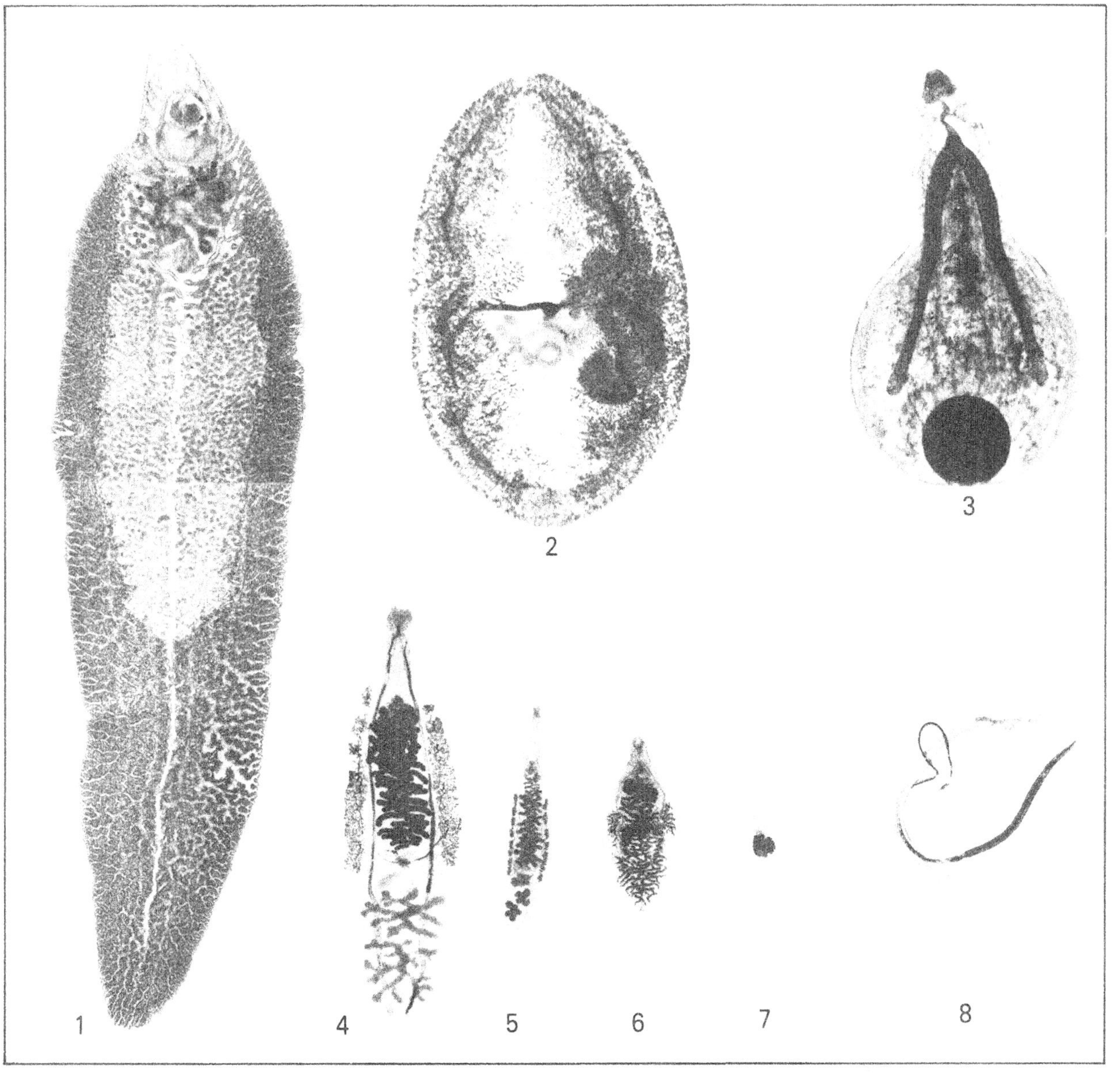

**228 Trematode eggs.**

1) *Fasciolopsis buski* are hen's egg-shaped with a small indistinct operculum. The size is 130 to 140μm x 80 to 85μm. They are unembryonated when passed out. They closely resemble eggs of *Fasciola hepatica* and *Fasciola gigantica.*

2) *Schistosoma mansoni* are elongated with a distinct lateral spine near 1 end. The size is 114 to 175μm x 45 to 68μm. They contain a miracidium when passed out.

3) *Schistosoma haematobium* are elongated with a small terminal spine. The size is 112 to 170μm x 40 to 70μm. They contain a miracidium when passed out.

4) *Schistosoma japonicum* are roundish with an indistinct blunt projection. The size is 70 to 100μm x 50 to 65μm. They contain a miracidium when passed out.

5) *Schistosoma intercalatum* are elongated with a terminal spine, which is sharper and longer than that of *S. haematobium.* The size is 140 to 240μ x 50 to 85μm. They contain a miracidium when passed out.

6) *Gastrodiscoides hominis* are elongated and spindle-shaped with an indistinct operculum at 1 end. The size is 150 to 152μm x 60 to 72μm. They are unembryonated when passed out.

7) *Paragonimus westermani* are ovoidal with a flattened operculum. The size is 80 to 118μm x 48 to 60μm. They are unembryonated when passed out.

8) *Clonorchis sinensis* are ovoidal with a distinct operculum. The operculum rests on a shoulder. There is usually a small knob at the posterior end. The size is 27 to 35μm x 12 to 70μm. They contain a miracidium when passed out. They closely resemble eggs of *Opisthorcis* spp.

9) *Heterophyes heterophyes* are ovoidal with a conical operculum. The size is 28 to 30μm x 15 to 17μm. They contain a miracidium when passed out. They closely resemble eggs of *Metagonimus yokogawai.*

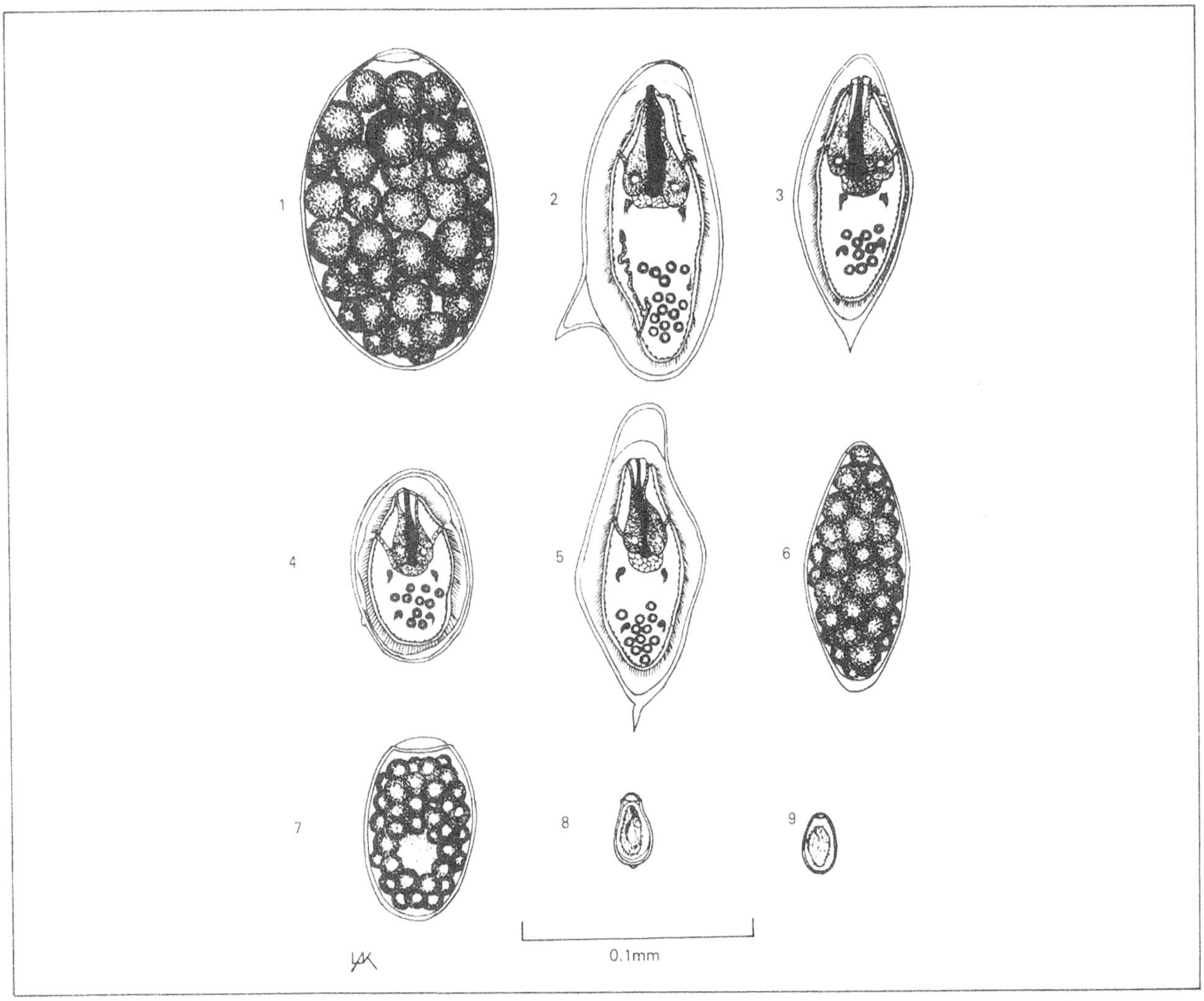


228

## Blood Flukes

*Schistosomes* — The 3 main species infecting man are *Schistosoma japonicum, Schistosoma mansoni* and *Schistosoma haematobium.* An additional species of schistosome, *Schistosoma intercalatum* is found in some parts of Africa (Wright et al., 1972).

Schistosomes are dioecious. The male has a deep ventral groove known as the gynaecophoric canal in which the female lies during copulation. Both sexes have 2 suckers, an anterior and a ventral sucker.

### *Life cycle* [229]

The egg contains a fully formed miracidium which hatches out when the egg is immersed in water, and swims actively to penetrate an appropriate snail host. After entry into the snail it changes into a sporocyst which produces daughter sporocysts and finally cercariae. In schistosomes redia are not produced. The cercariae emerge from the snail, mainly on exposure to light and infect the animal host by penetrating the skin. All schistosome cercariae have a bifid tail and do not possess a pharynx. The tail of the cercariae is shed during penetration and the parasite is converted into a schistosomula inside the host tissues. The schistosomula first enters the systemic circulation and then finds its way into the portal circulation. *S. mansoni* and *S. japonicum* worms mature in the mesenteric veins of the portal circulation. However, in the case of *S. haematobium,* worms generally remain in the systemic circulation and mature in the blood vessels of the vesical plexus. The eggs produced by *S. mansoni* and *S. japonicum* are discharged mainly in the faeces and by *S. haematobium* mainly in the urine.

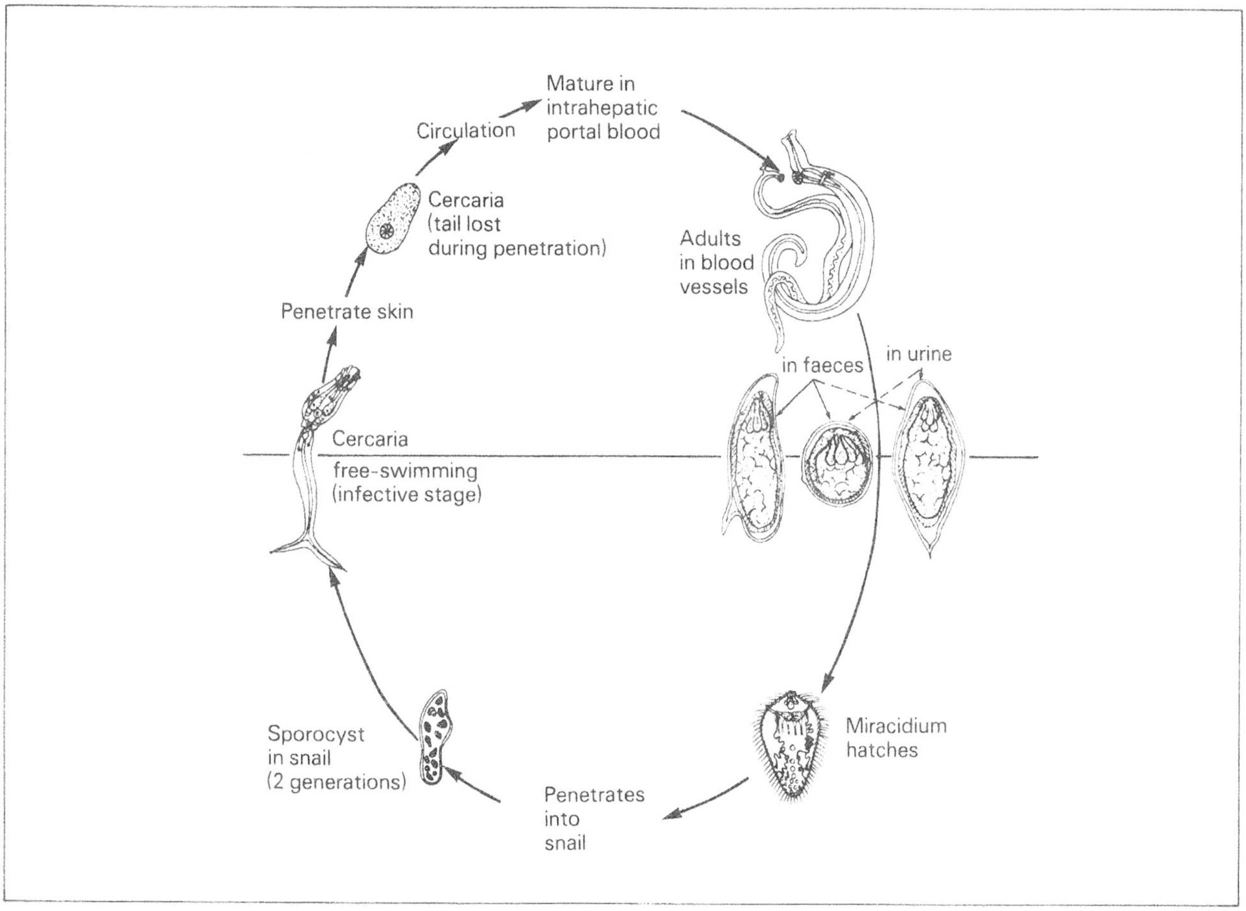


229

**230 Differential characteristics of schistosomes**

| | | |
|---|---|---|
| T = testes | VS = ventral sucker | OS = oral sucker |
| CJ = caecal junction | U = uterus | O = ovary |
| V = vitellaria | | |

| | *S. haematobium* | *S. mansoni* | *S. japonicum* |
|---|---|---|---|
| *Male* | | | |
| Tuberculations on cuticle | Fine | Coarse | None |
| Caecal junction | Middle | Anterior third of the body | Posterior third of the body |
| No. of testes | 4 to 5 | 6 to 9 | 6 to 7 |
| *Female* | | | |
| Position of ovary | Posterior half of the body | Anterior half of the body | Middle of the body |
| Uterus | Long contains 20 to 30 ova | Short contains 1 to 4 ova | Long contains 50 to 100 ova |

**231 Location of S. japonicum and S. mansoni in mesenteric veins.** *S. japonicum* is located mainly in the superior mesenteric vein and its branches. *S. mansoni* is located mainly in the inferior mesenteric vein and its branches. It is generally believed that *S. japonicum* discharges more eggs per worm than *S. mansoni.* The pathology produced by *S. japonicum* is, therefore, greater. The eggs travel in 2 directions as shown by the larger arrows. Some are able to find their way into the lumen of the bowel and appear in the faeces. Others flow along the direction of the blood, in the portal circulation, to enter the liver. Most of them are trapped in the liver and give rise to pathology in that organ. Some eggs are able to find their way through the liver tissue to enter the systemic circulation. These may circulate with the blood and settle down in any other organ of the body. Fibrosis of the liver may produce portal hypertension leading to the flow of blood along the dotted arrows. This may lead to splenic enlargement, oesophageal varices, haemorrhoids and ascites.

**232 Location of S. haematobium in vesical venous plexus.** *S. haematobium* is located mainly in the vesical venous plexus surrounding the urinary bladder. The eggs as in *S. japonicum* and *S. mansoni* travel in 2 directions. Some are able to find their way into the cavity of the bladder and appear in urine. Others are swept along with the flow of blood into the internal iliac vein and enter the systemic circulation. A large proportion of eggs are trapped in the wall of the bladder, where they may give rise to calcification. Constriction of the orifice of the ureter may produce kidney damage and hydronephrosis.

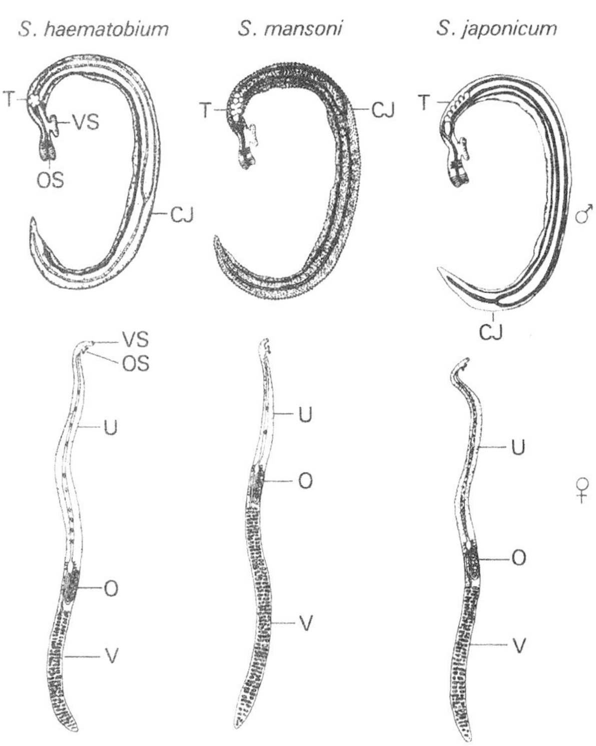
S. haematobium
S. mansoni
S. japonicum
T
VS
OS
CJ
♂
U
O
V
♀

230

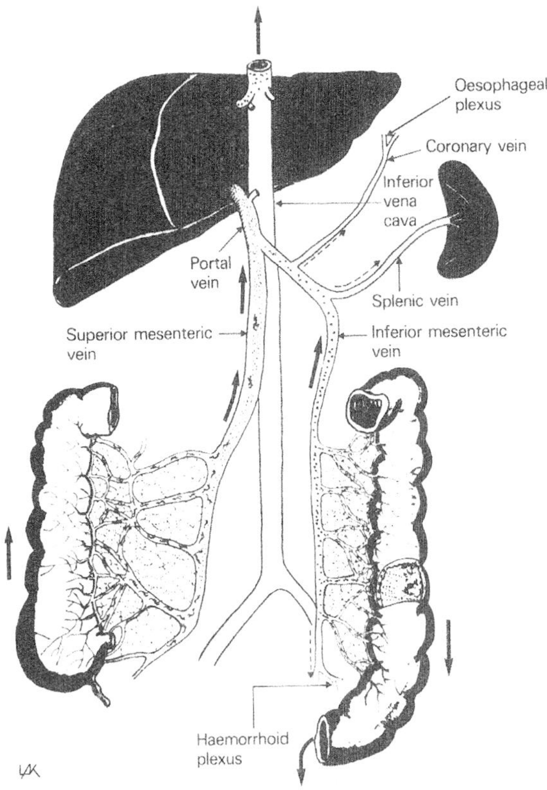
Oesophageal plexus
Coronary vein
Inferior vena cava
Portal vein
Splenic vein
Superior mesenteric vein
Inferior mesenteric vein
Haemorrhoid plexus

231

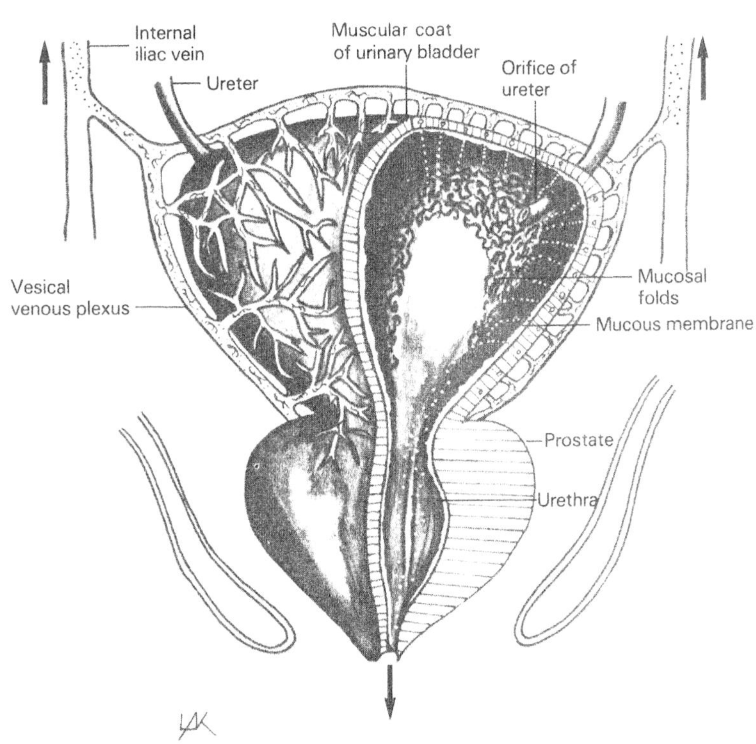
Internal iliac vein
Ureter
Muscular coat of urinary bladder
Orifice of ureter
Vesical venous plexus
Mucosal folds
Mucous membrane
Prostate
Urethra

232

**233 Calcification of bladder due to S. haematobium infection.** In chronic infection the walls of the bladder become calcified and are revealed on X-ray examination. (Courtesy Dr M.M. El-Mehairy).

**234** *Schistosoma mansoni* worms in copulation. The male has distinct tubercles on its cuticle, differentiating it from other species of schistosomes. The female appears as a thin tube like structure emerging from the gynaecophoric canal. Scanning electron micrograph (Courtesy of Dr M.M. Wong).

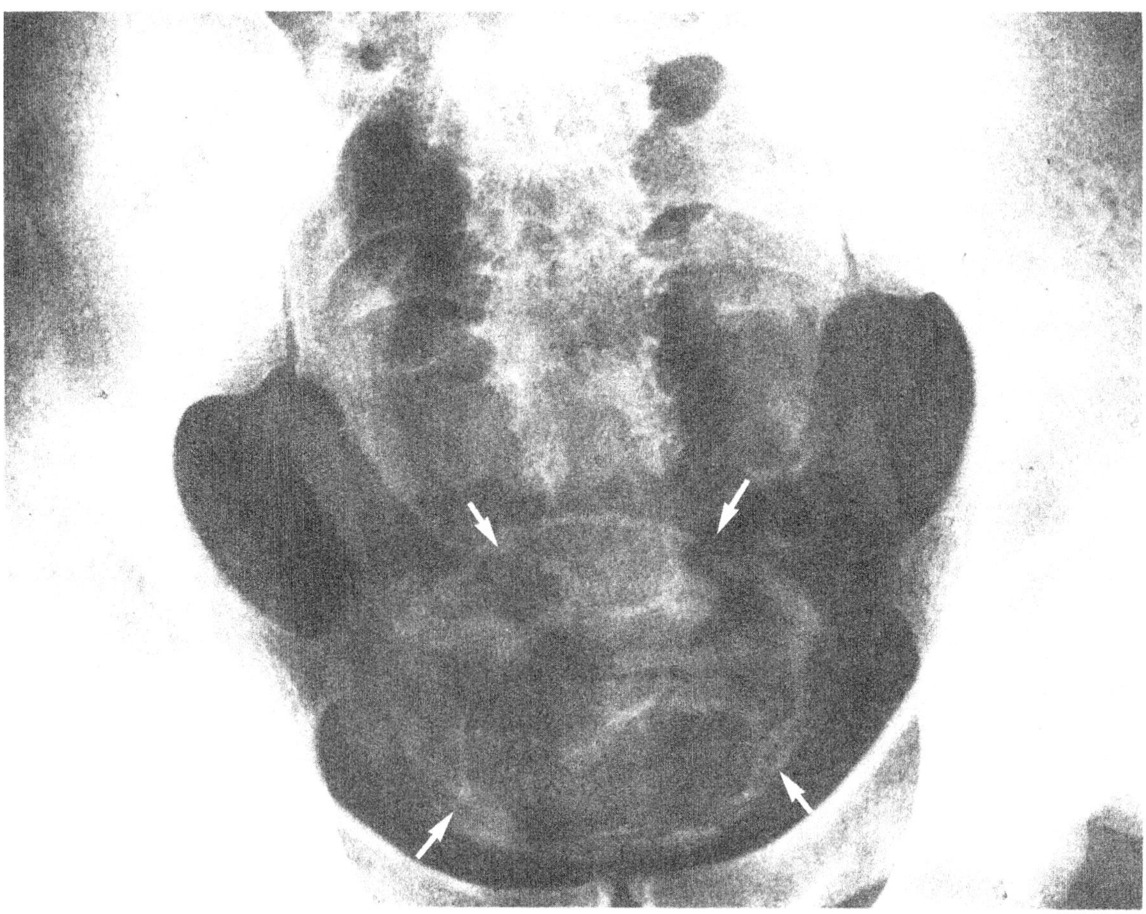

233

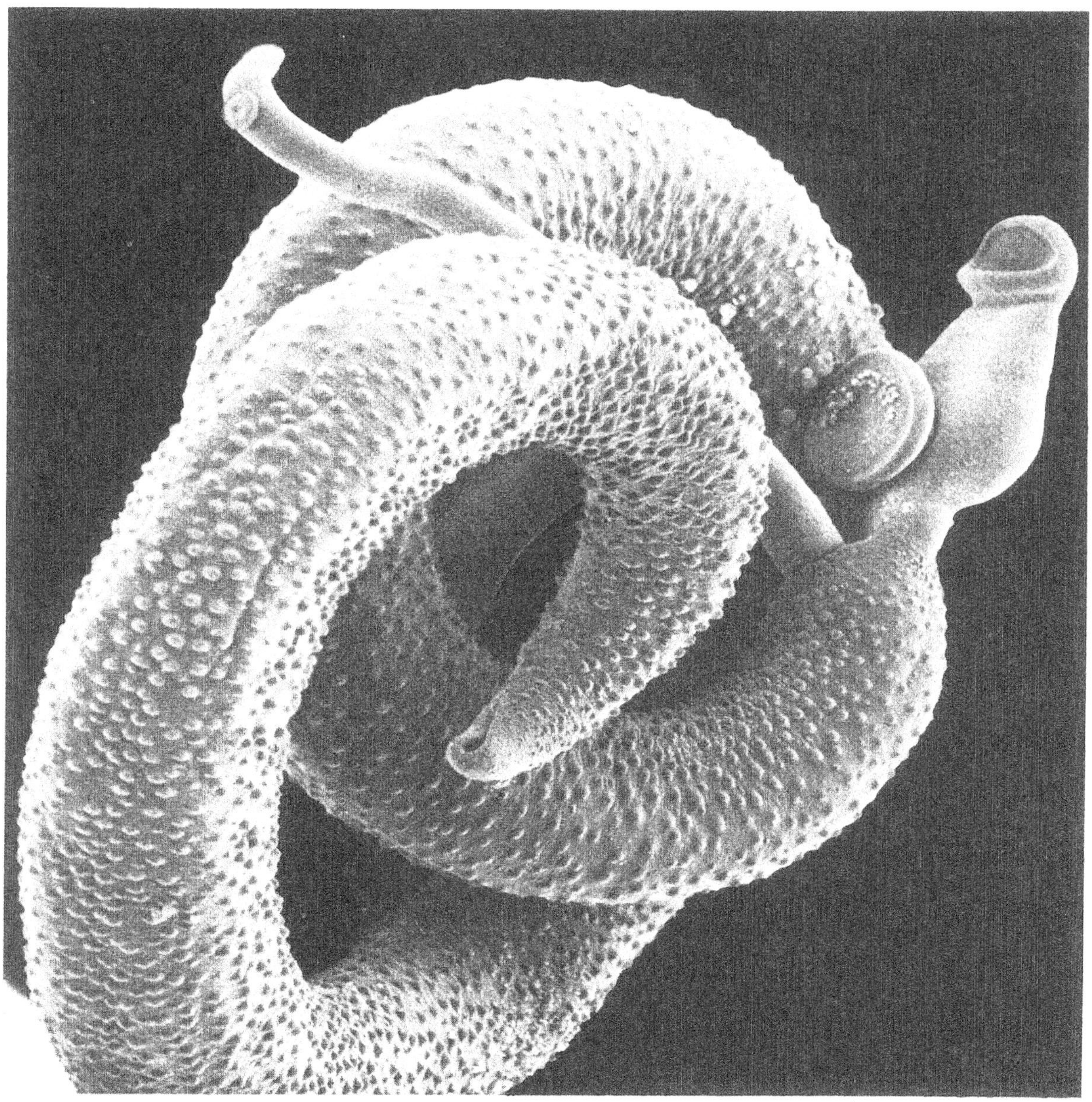

234

*Clinical aspects*

It is the egg which is the main cause of pathology in schistosomiasis. The eggs penetrate the blood vessels and the host tissues by discharging proteolytic enzymes, through ultramicroscopic pores in their shell. However, many eggs become stranded in the tissues or are carried by the blood stream to other organs of the body. The host reaction to the eggs may vary from small granulomas to extensive fibrosis. The extent of damage is generally related to the number of eggs present in the tissues.

Heavy infection of *S. haematobium* may produce chronic cystitis and urethritis. Fibrosis and eventually obstruction of the urethral orifice may lead to hydronephrosis and renal failure. The infection may also predispose to bladder stones and carcinoma. *S. haematobium* eggs may be dispersed into various other organs of the body via the circulatory system. In *S. mansonia* and *S. japonicum* eggs are first carried to the liver and from there they may enter the systemic circulation resulting in lesions in various organs of the body. In the liver they may give rise to fibrosis and portal hypertension. This in turn will give rise to splenomegaly, ascites, oesophageal varices and haemorrhoids. In experimental animals, the granuloma around the eggs in *S. japonicum* appears to be mainly antibody mediated in contrast to *S. mansoni* where it appears to be cell mediated (Warren and Domingo, 1970; Williams et al., 1972). In *S. haematobium* de-sensitisation often occurs after an acute inflammatory response with amelioration of disease and clinical improvement (von Lichtenburg et al., 1971).

*Diagnosis*

Detection of eggs in the faeces *(S. japonicum* and *S. mansoni)* and urine *(S. haematobium)* is the most important method of diagnosis. In light infection rectal biopsy is useful for *S. japonicum* and *S. mansoni*. Cystoscopy is similarly of value in *S. haematobium* infection. Various immunodiagnostic tests are available. These include intradermal, Cercarian Hullen reaction (CHR), the fluorescent antibody test using cercaria (FAT) and the circumoval precipitin reaction (COP).

## Intestinal Flukes

*Fasciolopsis buski* — This is a large trematode measuring 20 to 75mm by 8 to 20mm. It is located in the small intestine. The main morphological features are extensive, highly branched testes which occupy almost the posterior two-thirds of the body. The oral sucker is smaller than the ventral sucker. There is a small branched ovary and a short convoluted uterus.

*Life cycle of Fasciolopsis buski* [235]

Man is infected by ingesting metacercaria from the skin of various aquatic plants. The larva excysts in the duodenum and matures in about 3 months' time.

*Clinical aspects*

Light infections are generally asymptomatic. In heavy infections various gastro-intestinal symptoms such as colic, diarrhoea and vomiting may occur. There is generally a marked eosinophilia and in some cases oedema of the face and urticarial lesions on the body may occur.

*Diagnosis*

This is made by the presence of eggs in the faeces.

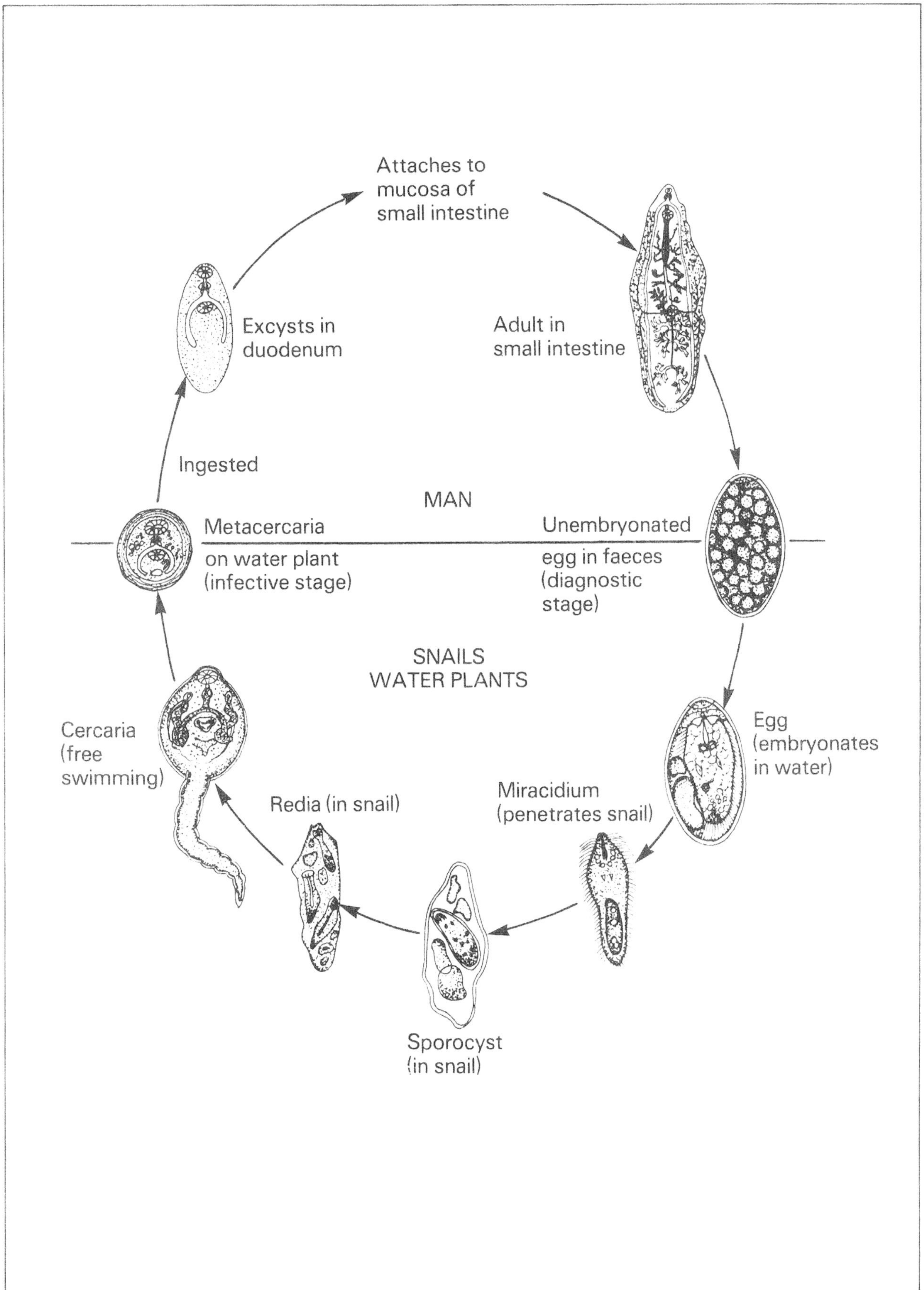
Attaches to
mucosa of
small intestine
Adult in
small intestine
Excysts in
duodenum
Ingested
MAN
Metacercaria
on water plant
(infective stage)
Unembryonated
egg in faeces
(diagnostic
stage)
SNAILS
WATER PLANTS
Egg
(embryonates
in water)
Miracidium
(penetrates snail)
Sporocyst
(in snail)
Redia (in snail)
Cercaria
(free
swimming)

235

*Gastrodiscoides hominis* — The main morphological feature is a large acetabulum situated at the posterior end of the worm. Man is infected by ingesting metacercaria from the skin of various aquatic plants. Pigs are the main animal reservoir in the endemic areas. The worm is located in the large intestine. In heavy infections patients have diarrhoea and colic. Diagnosis is made by the presence of eggs in the faeces.

*Echinostome spp* — A number of species of *Echinostome* have been reported from man. These include *E. ilocanum, E. lindoense, E. malayanum, E. revolutum* and *E. cinetorchis.* All the *Echinostomes* are characterised by the possession of a colar of spines at the anterior end.

Man is infected by the ingestion of metacercaria located in the tissues of edible snails. The adults are located in the small intestine and may give rise to diarrhoea and colic. Diagnosis is made by the presence of eggs in the faeces.

*Heterophyes heterophyes* — This is a minute parasite measuring 1 to 1.7mm by 0.3 to 0.4mm. The main morphological feature is the genital sucker situated posterior to the acetabulum.

Man is infected by eating raw fish and the worms are located in the small intestine. In heavy infections they may produce diarrhoea and colic. Sometimes the worms penetrate the mucous membrane and release the eggs into the circulatory system. In such cases the eggs may produce lesions in various organs of the body including the heart (Kean and Breslau, 1964). Diagnosis is made by the presence of the eggs in the faeces.

*Metagonimus yokogawai* — This is also a small parasite measuring 1 to 2.5mm by 0.4 to 0.75mm. The main morphological feature is that the ventral sucker is not in the centre but lies lateral to the genital opening.

Man is infected by eating raw fish. The worms are located in the small intestine. It can produce diarrhoea and colic. Occasionally the worms invade the intestinal tissues. Diagnosis is made by the presence of eggs in the faeces.

## Liver Flukes

*Fasciola hepatica* — This is a large trematode measuring 30 by 13mm. The main morphological feature is the extensive branching of the vitelline glands, testes and intestinal caeca.

*Life cycle of Fasciola hepatica* [236]

Man is occasionally infected by ingestion of metacercaria found on vegetation. The worms are located in the biliary tract and the eggs are passed in the stools. It can produce biliary colic, jaundice, generalised abdominal pain, cholecystitis and cholelithiasis. Diagnosis is made by the presence of eggs in the faeces.

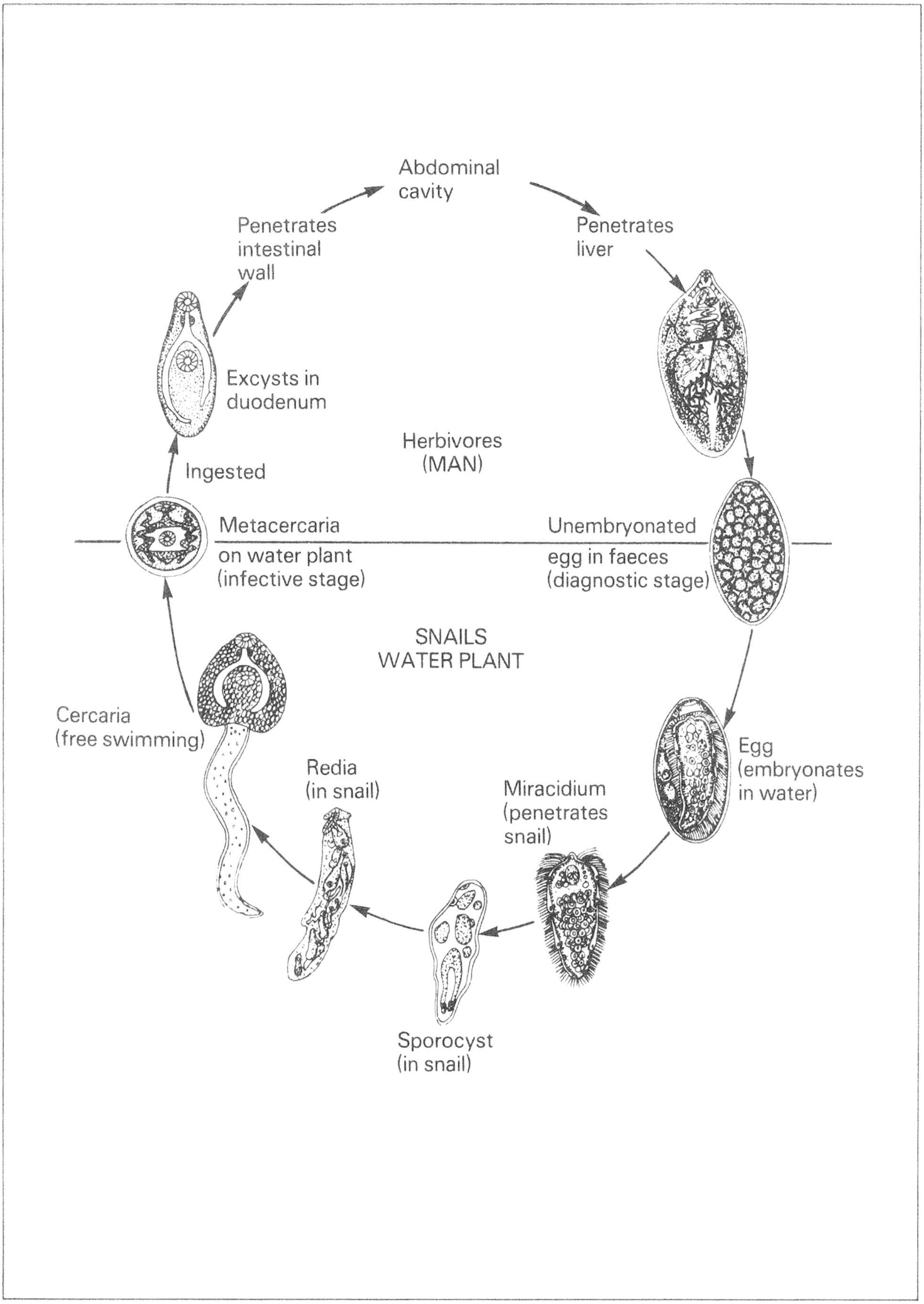
Abdominal cavity
Penetrates intestinal wall
Penetrates liver
Excysts in duodenum
Herbivores (MAN)
Ingested
Metacercaria on water plant (infective stage)
Unembryonated egg in faeces (diagnostic stage)
SNAILS WATER PLANT
Cercaria (free swimming)
Egg (embryonates in water)
Redia (in snail)
Miracidium (penetrates snail)
Sporocyst (in snail)

236

*Fasciola gigantica* — The worm resembles *F. hepatica* in its morphology and life cycle, excepting that it is larger and has a more attenuated shape.

*Dicrocoelium dendriticum* — This is a small trematode measuring 5 to 15mm by 1.5 to 2.5mm. The main morphological feature is that the 2 testes lie anterior to the ovary.

Man is occasionally infected by ingesting ants *(Formica fusca)* containing metacercaria (Krull and Mapes, 1952). The worms are located in the biliary tract. It can produce biliary colic, jaundice and hepatitis. Diagnosis is made by the presence of eggs in the faeces.

*Clonorchis sinensis* — This is a moderate size trematode measuring 10 to 25mm by 3 to 5mm. The main morphological feature is the elongated and coiled uterus which lies in the centre of the worm. As the worm is transparent the internal structures are visible even in unstained preparation.

*Life cycle of Clonorchis sinensis* [237]

Man is infected by eating uncooked fish and the worms are located in the biliary tract and the pancreatic ducts. In the majority of cases the infection is asymptomatic. In some patients, with heavy infection, jaundice and hepatomegaly may occur. The pathology in these cases consists of biliary cirrhosis and cholangitis. Malignant changes of the liver and pancreas have been attributed to this parasite. Diagnosis is made by the presence of eggs in the faeces or by examination of the duodenal aspirate.

*Opisthorchis viverrini* — The adult worm is generally smaller than *C. sinensis* and the testes are lobed and not dendritic as in *Clonorchis.* In all other respects the parasite resembles *C. sinensis.* Morphologically, adult *O. viverrini* is identical to *O. felineus* (Wykoff et al., 1965).

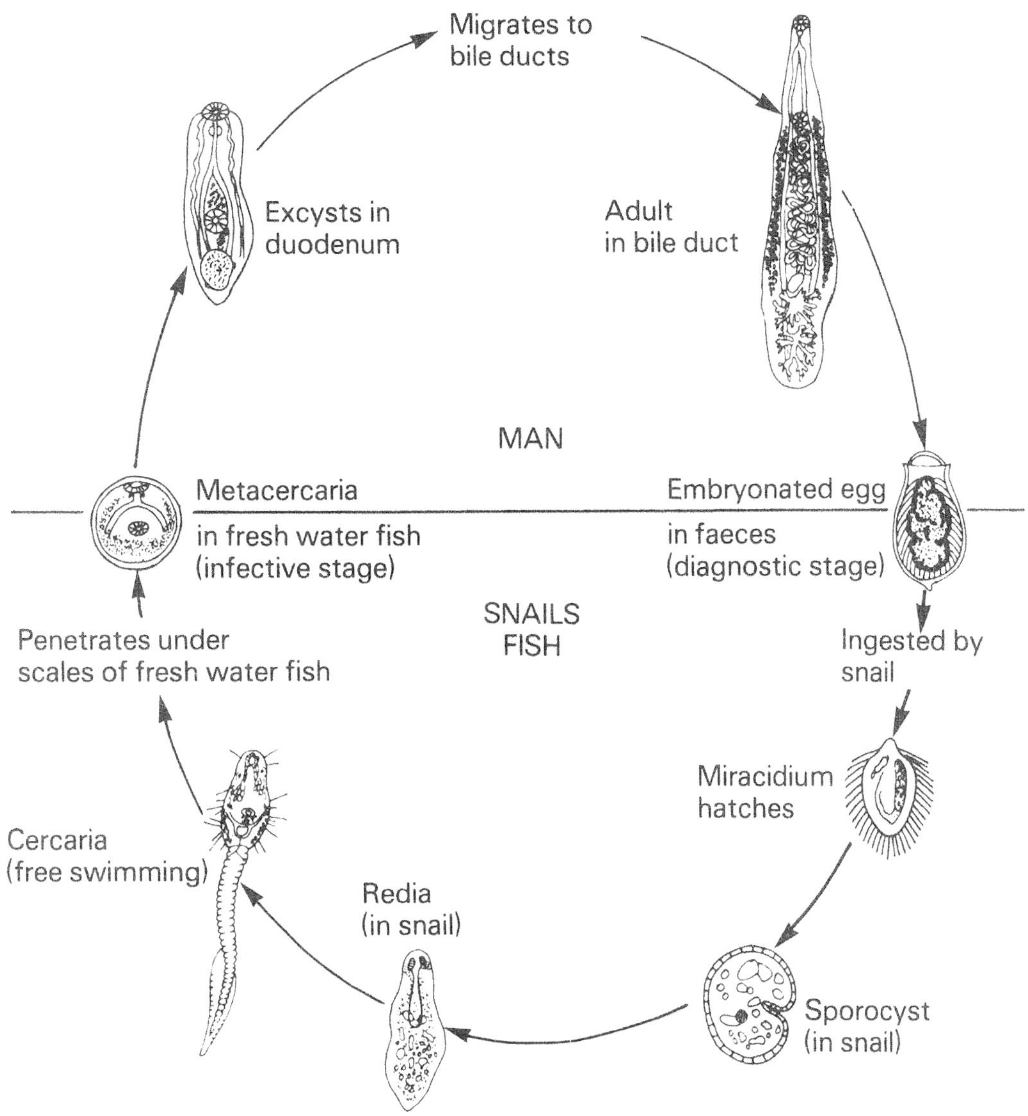
Migrates to
bile ducts
Excysts in
duodenum
Adult
in bile duct
MAN
Metacercaria
in fresh water fish
(infective stage)
Embryonated egg
in faeces
(diagnostic stage)
SNAILS
FISH
Penetrates under
scales of fresh water fish
Ingested by
snail
Miracidium
hatches
Cercaria
(free swimming)
Redia
(in snail)
Sporocyst
(in snail)

237

**238 Location of Clonorchis sinensis.** Adult *Clonorchis* are mainly located in the distal biliary passages of the liver and gradually increase in number with repeated infections. The worms may also be found in the gall bladder and occasionally in the pancreatic ducts. Pathology is mostly produced by the irritation and obstruction of the biliary passages by the adult worms. The eggs are discharged into the lumen of the duodenum through the common bile duct.

238

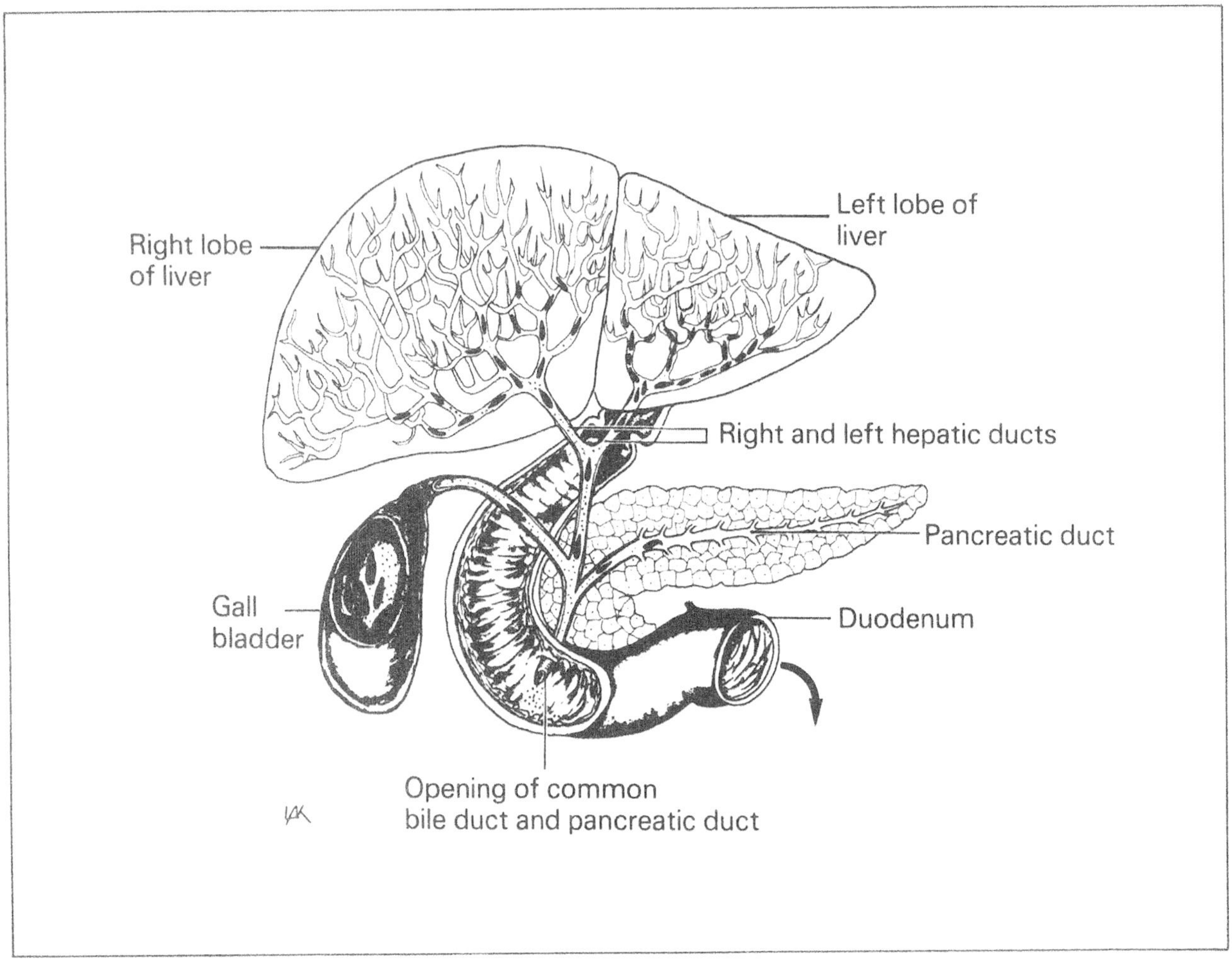


## Pulmonary Fluke

*Paragonimus westermani* — The worms are moderately large measuring 7.5 to 12mm by 4 to 6mm. The main morphological feature is the finely branched vitelline glands extending from the anterior to the posterior end. The ovary and testes are located in the centre.

*Life cycle* [239]

Man is infected by eating fresh water crabs or crayfish containing metacercaria. The worms are usually located in the pulmonary parenchyma and after a period of time they are surrounded by a capsule of fibrous tissue. The patient may have dispnoea, fever and pain in the chest. Rusty sputum and haemoptysis are common. Worms may also be found in ectopic sites such as brain, liver, testes and other organs of the body. In these abnormal sites the parasite has a tendency to form an abscess.

Diagnosis is made by the presence of eggs in the sputum and faeces.

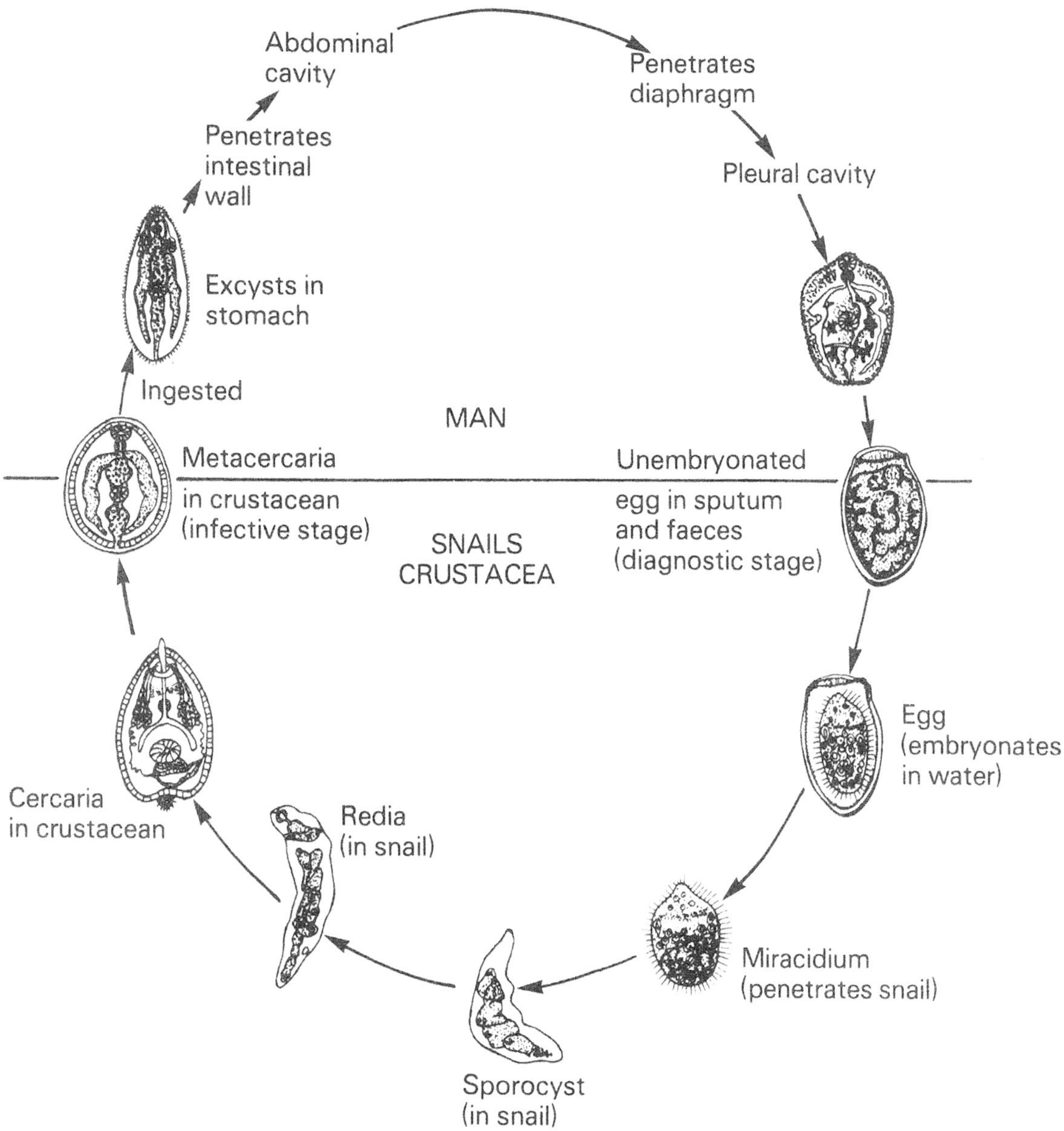
Abdominal
cavity
Penetrates
diaphragm
Penetrates
intestinal
wall
Pleural cavity
Excysts in
stomach
Ingested
MAN
Metacercaria
in crustacean
(infective stage)
Unembryonated
egg in sputum
and faeces
(diagnostic stage)
SNAILS
CRUSTACEA
Egg
(embryonates
in water)
Cercaria
in crustacean
Redia
(in snail)
Miracidium
(penetrates snail)
Sporocyst
(in snail)

239

**240 Location of Paragonimus westermani.** Adult worms are normally located in the lungs. The eggs discharged by the worms enter the bronchi and appear in the sputum. If the sputum is swallowed the eggs pass through the intestinal tract and are discharged with the faeces. Occasionally the eggs may enter the circulation and be distributed to various organs of the body. Adult worms may also be located in ectopic sites, such as the brain. In the brain, this may lead to abscess formation, with signs and symptoms of a space occupying lesion.

240

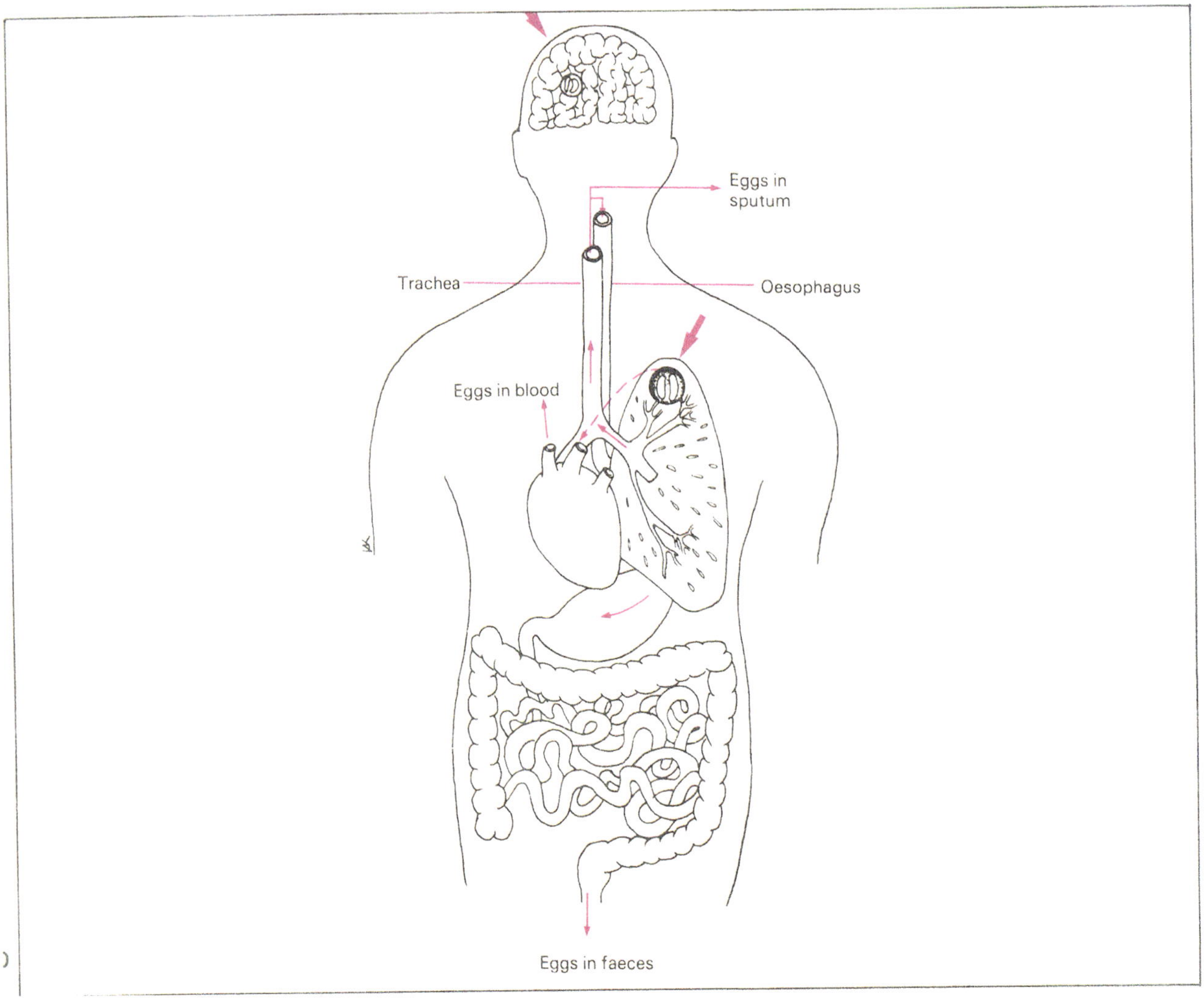


## References


Kean, B.H. and Breslau, R.C.: Parasites of the human heart: in Cardiac Heterophyidiasis, chapter X. pp95-103 (Grune and Stratton, New York 1964).

Krull, W.H. and Mapes, C.R.: Studies on the biology of *Dicrocoelium dendriticum* (Rudolphi, 1819) Looss, 1899 (Trematoda: Dicrocoeliidae), including its relationship to the intermediate host Cionella lubrica (Muller). VII. The second intermediate host of *Dicrocoelium dendriticum.* Cornell Veterinarian 42: 603 (1952).

von Lichtenberg, F., Edington, G.M., Nwabuebo, I., Taylor, J.R. and Smith, J.H.: Pathological effects of schistosomiasis in Ibadan, Western State of Nigeria. II. Pathogenesis of lesions of the bladder and ureters. American Journal of Tropical Medicine and Hygiene 20: 244 (1971).

Warren, K.S. and Domingo, E.O.: Granuloma formation around *Schistosoma mansoni, S. haematobium,* and *S. Japonicum* eggs. Size and rate of development, cellular composition, cross-sensitivity, and rate of egg destruction. American Journal of Tropical Medicine and Hygiene 19: 292 (1970).

Williams, J.S., Sadun, E.H. and Gore, R.W.: Immunological reactions in chimpanzees experimentally infected with *Schistosoma japonicum*. Experimental Parasitology 32: 217 (1972).

Wright, C.A., Southgate, V.R. and Knowles, R.J.: What is *Schistosoma intercalatum* Fisher 1934? Transactions of the Royal Society of Tropical Medicine and Hygiene 66: 28 (1972).

Wykoff, D.E., Harinasuta, C., Juttijudata, P. and Winn, M.M.: *Opisthorchis viverrini* in Thailand — the life cycle and comparison with *O. fileneus.* Journal of Parasitology 51: 207 (1945).

**241 Schistosoma japonicum male and female.** The female worm is lying in the gynaecophoric canal of the male and appears dark in colour. Ovary appears as a depigmented ovoid body in the centre of the female. X50. Enlarged by 5.4.

**242 Section of a vein showing schistosomes in cross-section.** The males appear curved with the pigmented females lying in the concavity. Note the absence of worms in the artery which is lying adjacent to the vein. H and E. X50. Enlarged by 5.4.

**243 S. mansoni egg. The lateral spine is visible but not in focus.** The anterior part of the miracidium is directed towards the broad end of the egg. X400. Interference contrast. Enlarged by 5.4.

241

242
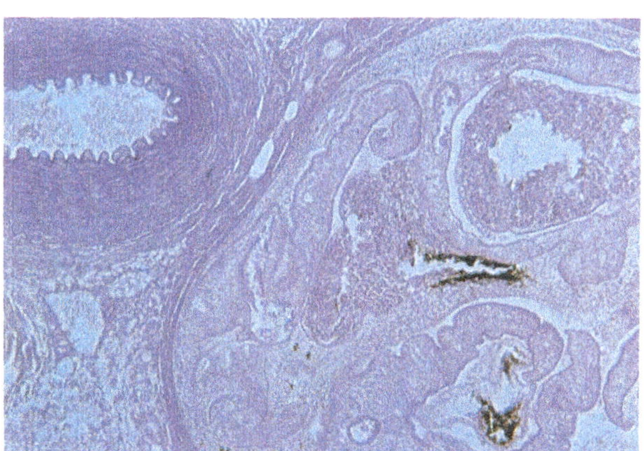

243

**244 The same egg with the lateral spine in focus.** The spine is long and sharply pointed. X400. Interference contrast. Enlarged by 5.4.

**245 S. haematobium egg.** A small terminal spine is visible. X400. Interference contrast. Enlarged by 9.6.

**246 S. japonicum egg from faeces.** The small spine is generally not visible as the egg surface is often covered with faecal debris. The miracidium surrounded by a membranous sac can be seen inside the egg. X400. Interference contrast. Enlarged by 5.4.

**247 S. japonicum eggs dissected from the uterus.** As the eggs are free from faecal debris, in one of them a small curved spine can be seen. X400. Interference contrast. Enlarged by 9.6.

**248 Three important snail hosts of schistosomes.** *Australorbis sp (S. mansoni)* is on the left and *Onchomelania sp (S. japonicum)* is on the right. In the centre is the *Bulinus sp (S. haematobium).*

**249 S. japonicum cercaria.** Live preparation showing the bifid tail. X100. Interference contrast. Enlarged by 5.4.

**250 S. mansoni cercaria.** Stained preparation. The anterior sucker is visible as a dark circular structure. X100. Enlarged by 5.4.

**251 S. mansoni egg in the intestine.** The lateral spine is clearly seen. There is a strong polymorphonuclear reaction. The nuclei of polymorphs appear as small granulated bodies. H and E. X800. Enlarged by 5.4.

**252 S. mansoni eggs in the tissues of the intestine.** The eggs are distributed both in the mucosa and submucosa. H and E. X400. Enlarged by 5.4.

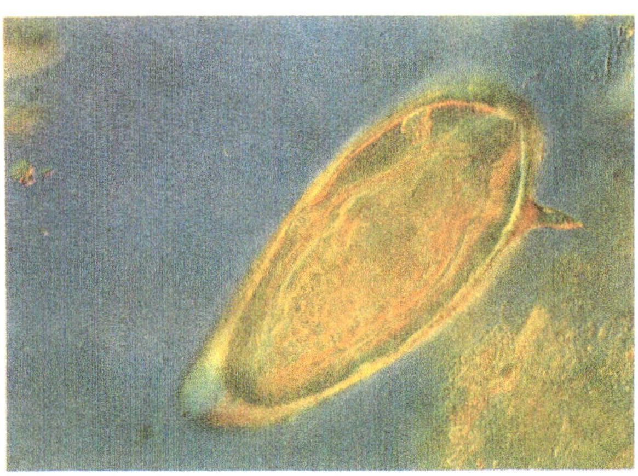

244

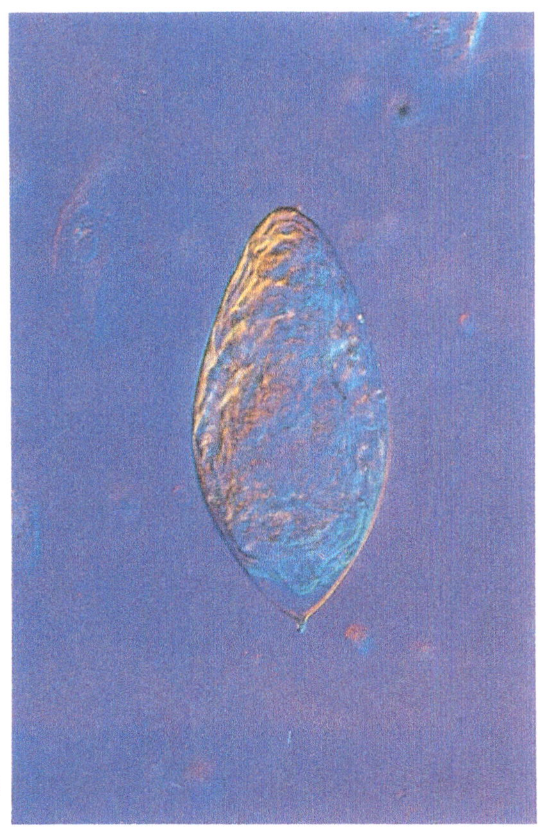

245

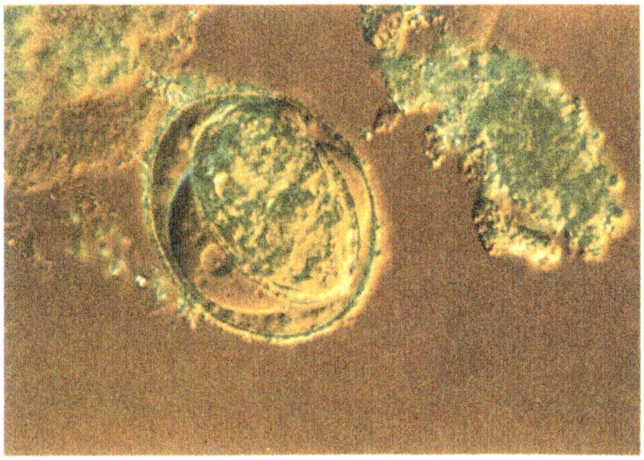

246

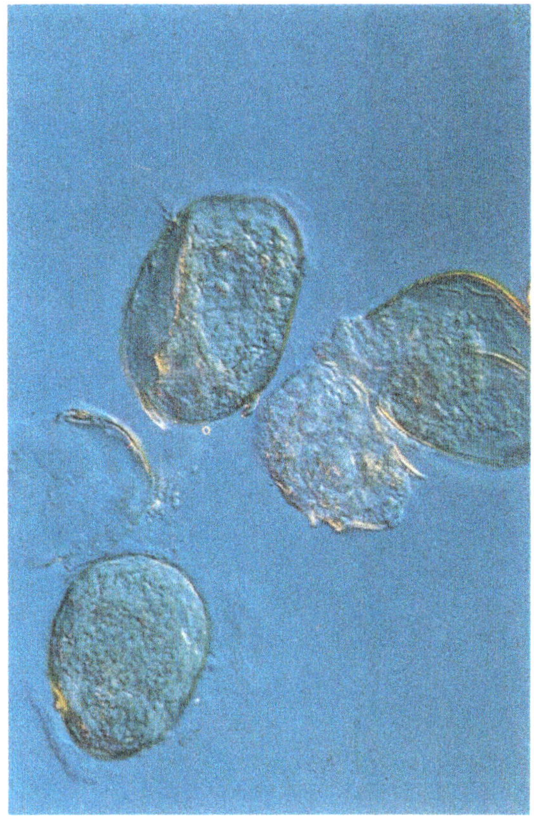

247

248

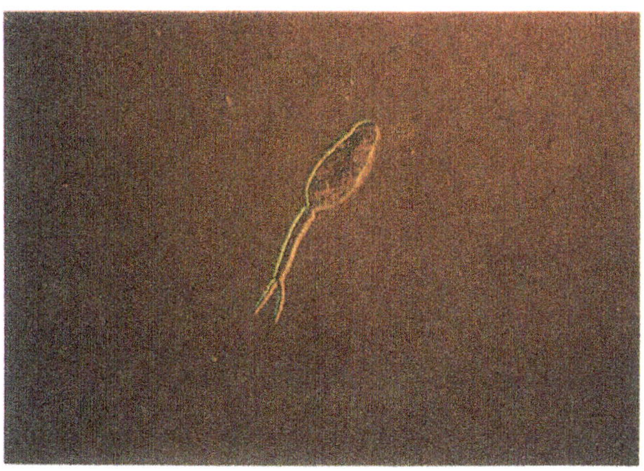

249

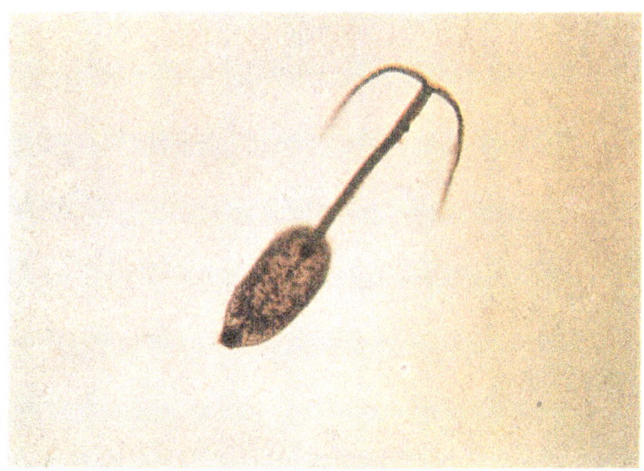

250

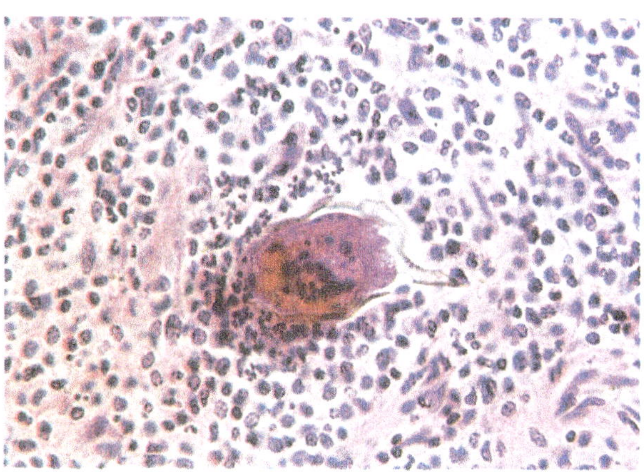

251

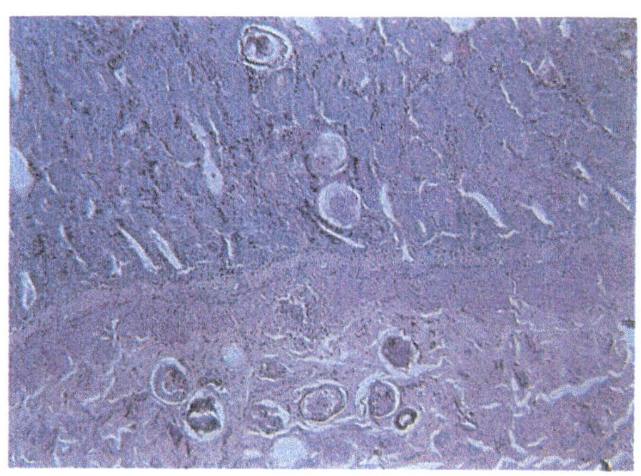

252

**253 S. haematobium eggs in the tissues of the urinary bladder.** The calcified eggs appear dark in colour. H and E. X400. Enlarged by 5.4.

**254 S. haematobium eggs in the tissues of the urinary bladder.** Terminal spine and miracidium can be seen in some eggs. H and E. X800. Enlarged by 5.4.

**255 S. japonicum egg in the liver showing a granulomatous reaction.** Consisting predominantly of plasma cells. H and E. Enlarged by 5.4.

**256 S. japonicum infection of the liver.** In chronic infection there is extensive infiltration with fibrous tissue. The eggs appear as dark bodies scattered in the portal tracts. H and E. X400. Enlarged by 5.4.

**257 S. japonicum eggs.** From the same tissue as in illustration 256. H and E. X800. Enlarged by 5.4.

**258 S. japonicum egg in the liver** being ingested by a giant cell. H and E. X800. Enlarged by 5.4.

**259 S. japonicum eggs in the liver after staining with PAS method.** The eggs are strongly PAS positive and are surrounded by fibrous tissue. The adjacent liver tissue shows kupfer cells containing pigment. Presence of pigment in the kupfer cells is characteristic of chronic schistosomiasis. X400. Enlarged by 9.6.

**260 Fluorescent antibody test using cercarial antigen.** In this case *S. mansoni* cercaria are used. The serum is from a *S. japonicum* case. The test is not species specific and cross reaction is seen between the 3 species. During antigen preparation the cercarial tail generally breaks off. Both the head and the tail show a positive reaction. X400. Enlarged by 5.4.

**261 Fluorescent antibody test using cercarial antigen** showing the reaction around the head, which appears as a zone of bright green fluorescence. X800. Enlarged by 5.4.

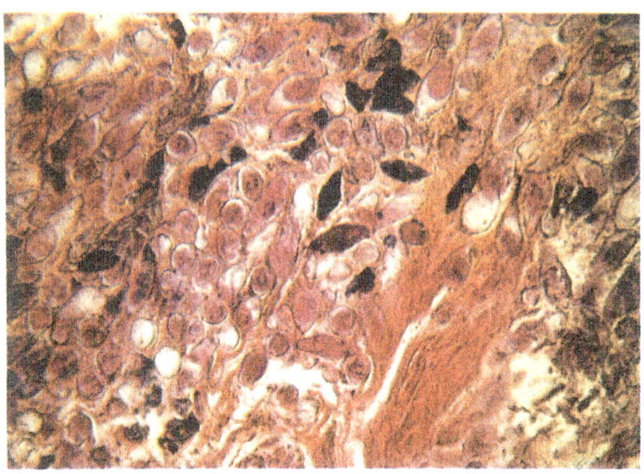

253

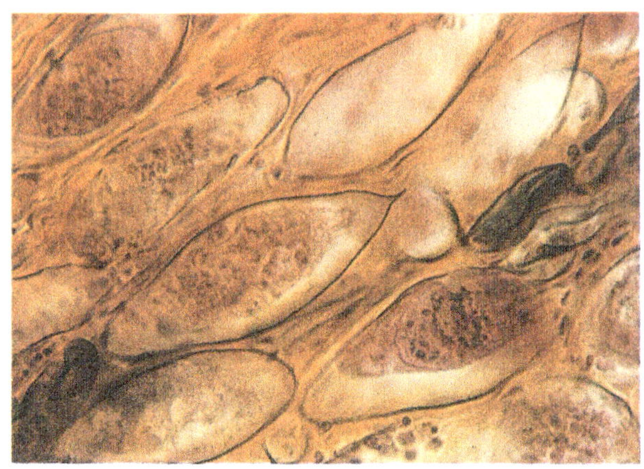

254

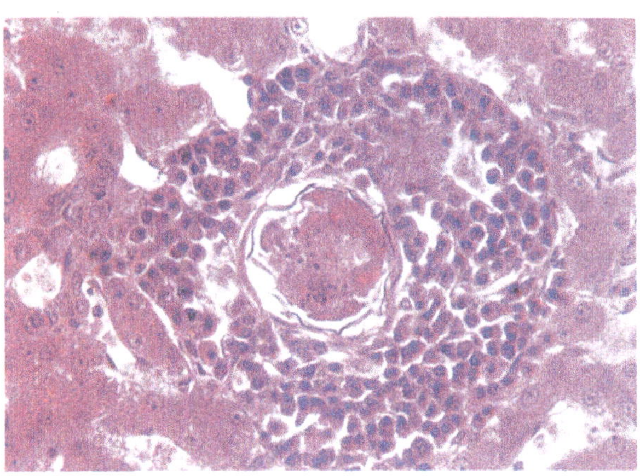

255

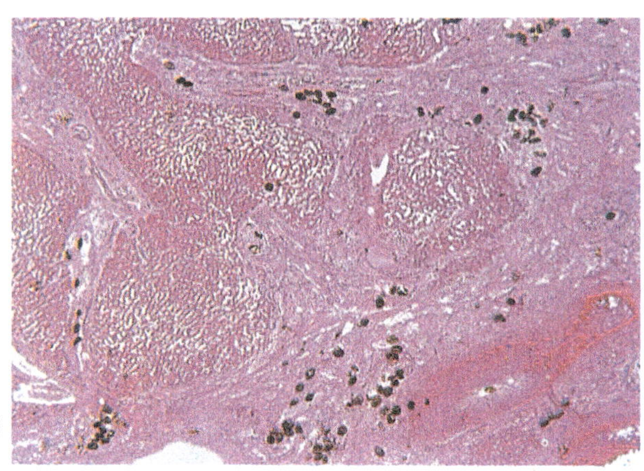

256

257
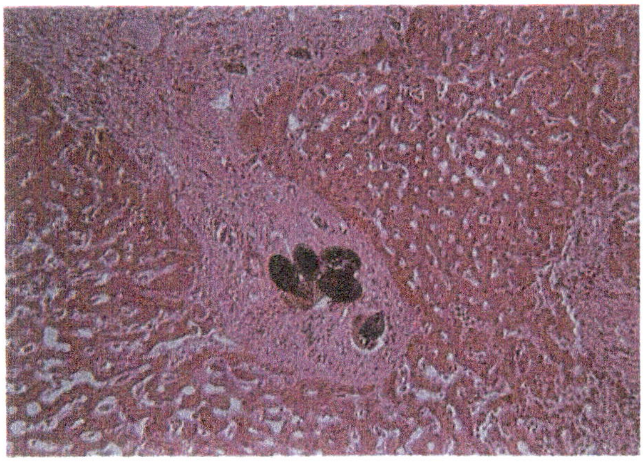

258
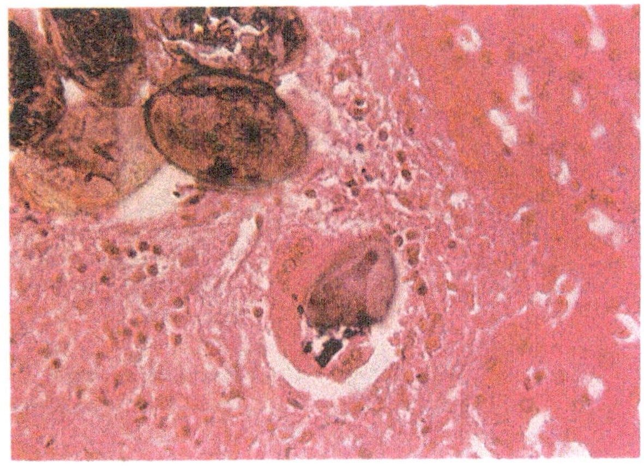

259
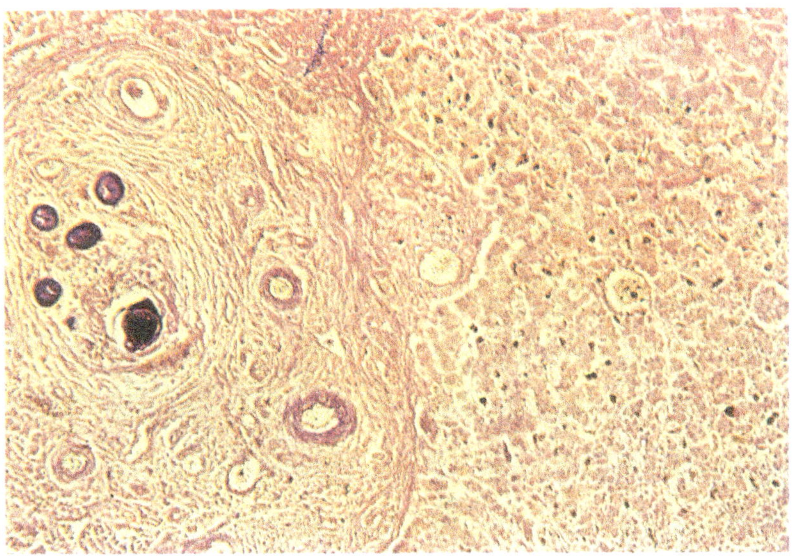

260
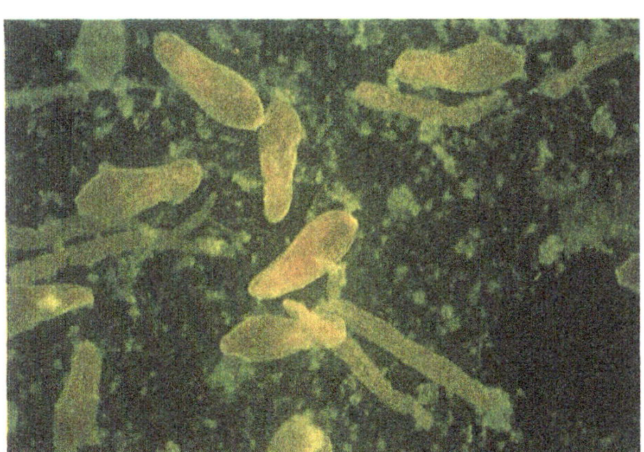

261
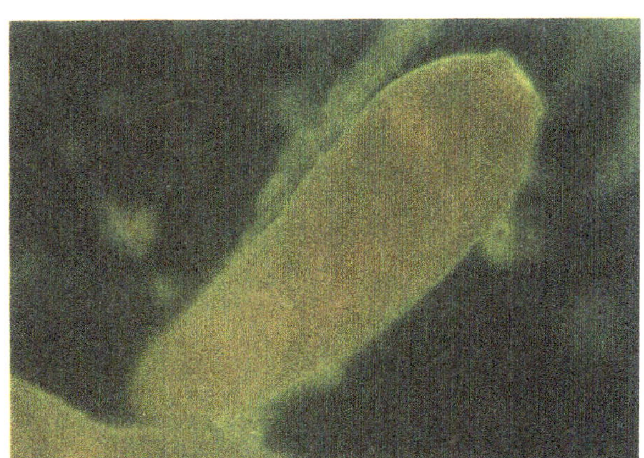

**262 Paragonimus sp from the lungs of an infected animal.** The parasites are fleshy worms with convex surfaces. The sides appear dark because of the presence of vitelline glands.

**263 Cross-section of Paragonimus westermani lying inside a cavity in the lung.** The cavity is lined with a thick fibrous capsule and contains 2 parasites., They are often found in pairs as shown here. H and E. Enlarged by 5.4.

**264 A cross-section of Paragonimus sp in lungs.** The tegument is covered with spines which appear reddish in colour. The vitellaria are dark bodies scattered along the periphery. One of the intestinal caeca is also visible. X100. Enlarged by 5.4.

**265 Potamon sp, a crab host of Paragonimus.** This is a fresh water crab and is sometimes eaten raw when it could transmit infection.

**266 A young Paragonimus hatching out of a metacercaria.** This usually takes place in the duodenum of the animal host. X100. Interference contrast. Enlarged by 5.4.

**267 A freshly passed egg of Paragonimus sp** showing the distinct operculum and a thick shell. It is unembryonated at this stage. X100. Interference contrast. Enlarged by 5.4.

**268 A miracidium hatching out of a Paragonimus egg.** The egg takes about 3 weeks in water to embryonate. X100. Interference contrast. Enlarged by 9.6.

**269 Cross-setion of Clonorchis sinensis adults in a bile duct.** There is marked fibrosis of the bile duct which is characteristic of chronic clonorchiasis. H and E. X5. Enlarged by 5.4.

**270 Testes of Clonorchis sinensis** are deeply branched or fimbriated and lie in a tandem. The unbranched caeca can be seen on the sides. X100. Enlarged by 5.4.

262

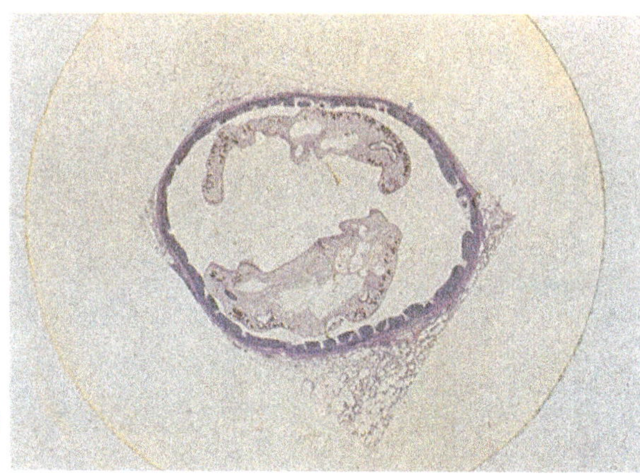

263

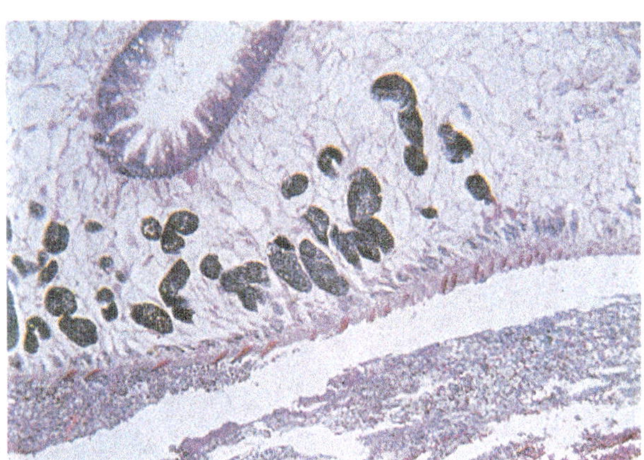

264

265

266

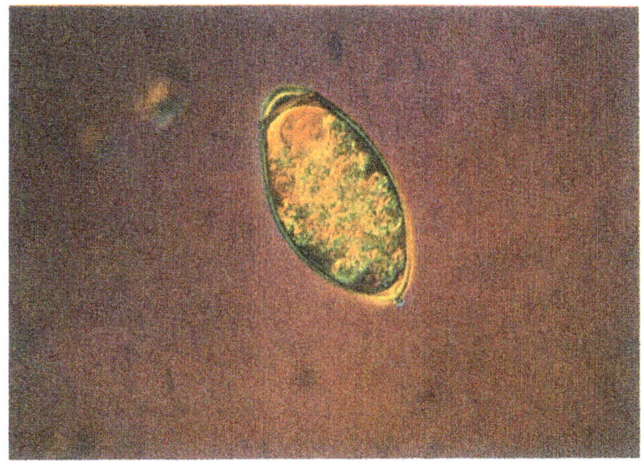
267

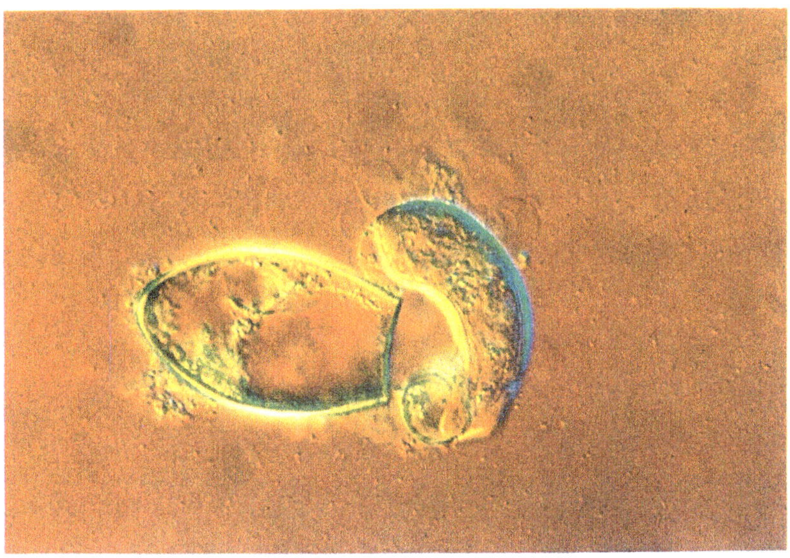
268

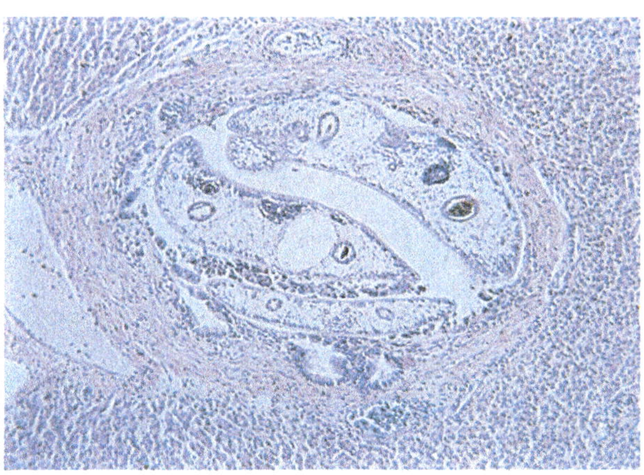
269

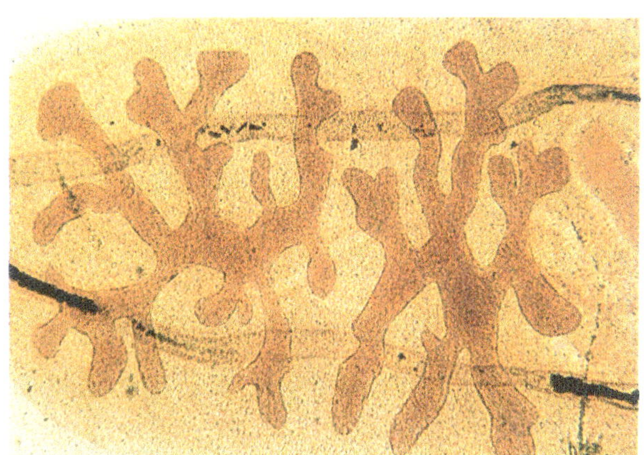
270

**271 Testes of Opisthorcis felineus** are lobed and lie at a slight angle to each other. X100. Enlarged by 5.4.

**272 Clonorchis sinensis eggs in faeces.** The eggs have a distinct operculum and contain a miracidium. X100. Interference contrast. Enlarged by 5.4.

**273 Clonorchis sinensis eggs in faeces** at a higher magnification (X400). The operculum, miracidium and the posterior knob can be seen. Interference contrast. Enlarged by 5.4.

**274 Clonorchis sinensis metacercaria** are usually found in the fish of the carp family. In this case grass carp (*Ctenopharyngodon idellus*) is shown. (Courtesy of Dr Ronald C. Ko).

**275 Metagonimus yokogawai** is one of the smallest trematodes infecting man. The 2 round bodies at the posterior end are the testes. X100. Enlarged by 5.4.

**276 Gastrodiscoides hominis showing the pyriform shape.** It has a deep concavity on the ventral surface, which contains a large sucker.

**277 Echinostome sp showing the scaly integument** and the circumoral collar with spines. X100. Interference contrast. Enlarged by 5.4.

**278 Miracidia of Fasciola hepatica.** The cilia appear bluish in colour and surround the parasite. Each miracidium has two pigmented areas or eye spots at the anterior end. X100. Enlarged by 5.4.

**279 Fasciola hepatica in bile duct.** The tegument is covered with spines and the dark bodies inside the parasite are mostly vitelline glands. Eggs can be seen lying in the lumen of the bile duct. H and E. X100. Enlarged by 5.4.

**280 Water chestnut — Eliocharis tuberosa** is commonly eaten raw in the Far East and is an important source of infection of *Fasciolopsis buski,* as the cercaria encysts on their surface.

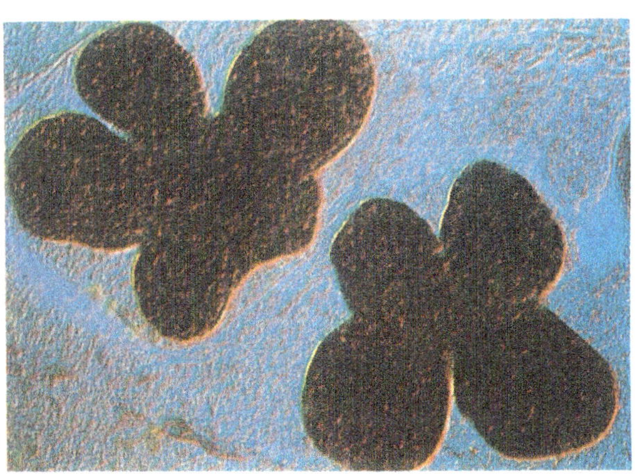

271

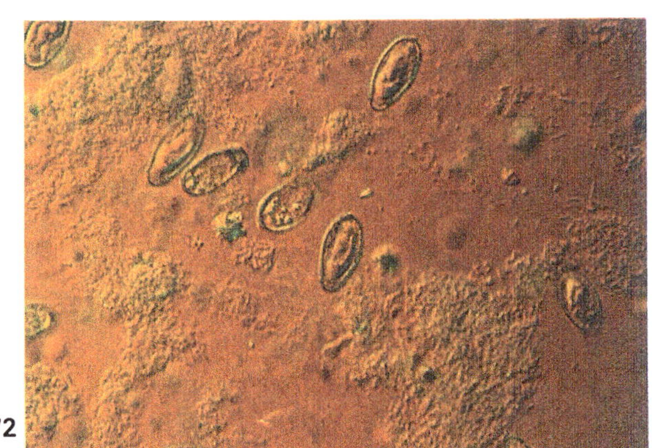

272

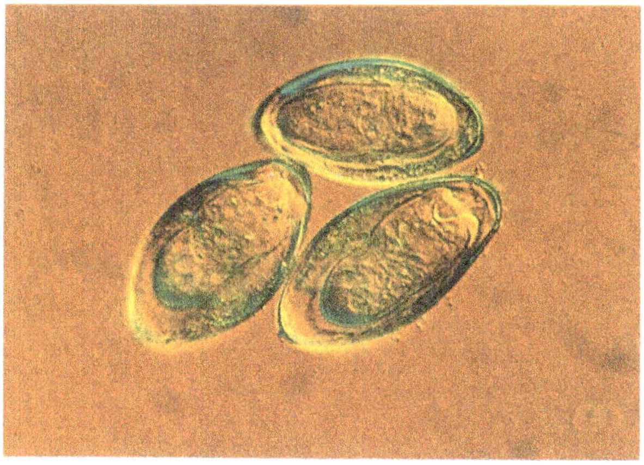
273

274

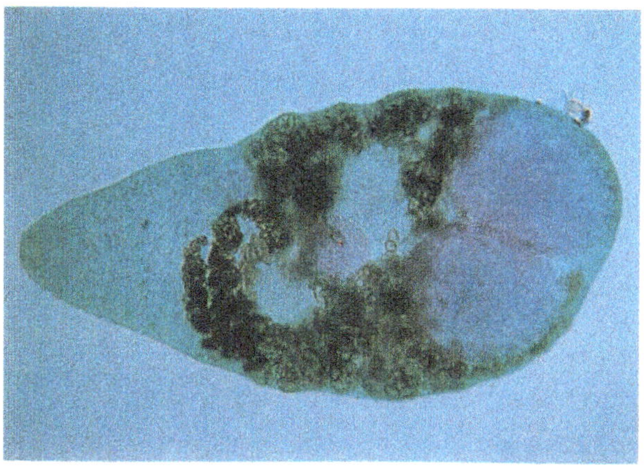
275

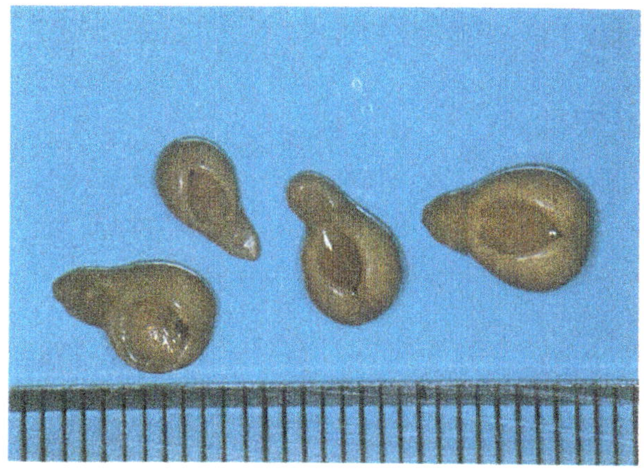
276

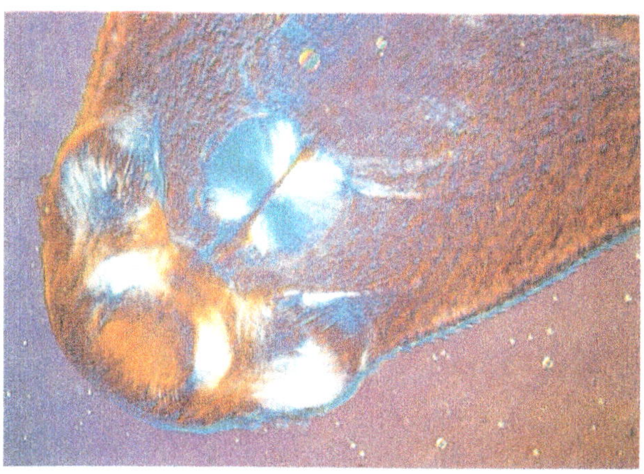
277

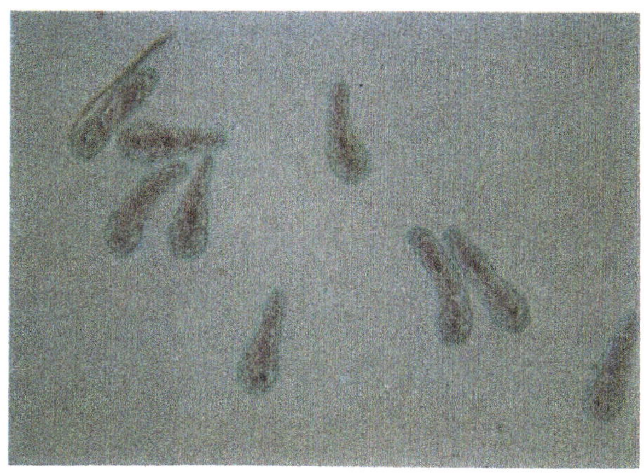
278

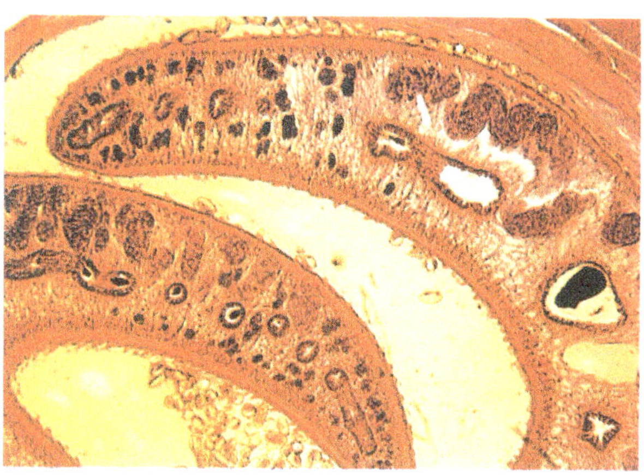
279

280

# Section 12
# Nematodes

## Classification of Medically Important Nematodes

Class Nematoda

| Subclass | Order | Superfamily | Genus |
|---|---|---|---|
| *Adenophorea* (formerly Aphasmidia) | Enoplida | Trichinelloidea | *Trichinella*<br>*Trichuris*<br>*Capillaria* |
| *Secernentea* (formerly Phasmidia) | Rhabditida | Rhabditoidea | *Strongyloides* |
| | Strongylida | Ancylostomatoidea | *Ancylostoma*<br>*Necator*<br>*Ternidens* |
| | | Metastrongyloidea | *Angiostrongylus*<br>*Metastrongylus* |
| | | Trichostrongyloidea | *Tricostrongylus* |
| | Ascaridida | Ascaridoidea | *Ascaris*<br>*Toxocara*<br>*Anisakis*<br>*Lagochilascaris* |
| | Oxyurida | Oxyuroidea | *Enterobius* |
| | Spirurida | Spiruroidea | *Gongylonema* |
| | | Thelazoidea | *Thelazia* |
| | | Gnathostomatoidea | *Gnathostoma* |
| | | Filarioidea | *Wuchereria*<br>*Brugia*<br>*Onchocerca*<br>*Loa loa*<br>*Dipetalonema*<br>*Mansonella*<br>*Dirofilaria* |
| | | Dracunculoidea | *Dracunculus* |

## Terms Used in Relation to Nematodes

**Cuticle** (Cuticula) — the external non-cellular hyaline layer covering the nematode.

**Alae** — ridge like extensions of the cuticle. In the anterior region they are known as cervical alae, in the posterior region as the caudal alae.

**Papillae** — protuberances on the cuticle. As in the case of alae, these may be cervical or caudal.

**Buccal capsule** — is the mouth cavity of the nematode.

**Hypodermis** (Epidermis) — a layer between the cuticle and the somatic musculature.

**Bacillary band** — a row of longitudinal cells formed by the hypodermis. Seen in Trichinelloidea.

**Coelomyarian** — somatic muscle cells in which the muscle fibres extend along the sides of the cell in addition to lying perpendicular to the hypodermis.

**Platymyarian** — somatic cells in which the muscle fibres lie only perpendicular to the hypodermis.

**Meromyarian** — a nematode in which few muscle cells are seen between chords e.g. *Enterobius* and hookworms.

**Polymyarian** — a nematode in which many muscle cells are seen between chords e.g. *Ascaris,* filarial worms and *Dracunculus.*

**Amphid** — sensory receptors situated near the anterior end of the nematode.

**Phasmid** — sensory receptors situated near the posterior end of the nematode.

**Spicule** — the male copulatory organ which is elongated and protrusible. Its length, shape and number vary in different species.

**Copulatory bursa** — a membranous expansion of cuticle at the posterior end of certain nematodes. It is supported by rib-like structures known as rays and is used for holding the female during copulation.

**Gubernaculum** — a protuberance on the wall of cloaca. It apparently guides the spicule during copulation.

**Cloaca** — a common opening for the rectum and the genital tract.

**Stichosome** — glandular cells (stichocytes) arranged in a row along the oesophagus. Seen in Trichinelloidea.

**Pseudocelom** (Pseudocoel) — body cavity of nematode which is filled with fluid and in which the internal organs are suspended.

**Oesophagus** (Pharynx) — the part connecting the buccal capsule to the intestinal tract.

**Corpus** — anterior portion of the oesophagus. In *Rhabditis* type it is divisible into procorpus and metacorpus.

**Isthmus** — middle portion of the oesophagus.

**Bulb** — posterior portion of the oesophagus.

**Didelphic** (bicornate) — having a double set of reproductive system.

**Monodelphic** — having a single set of reproductive system in the female nematode.

**Viviparous** (Larviparous) — a species which discharges larvae instead of eggs.

**Oviparous** — a species which discharges eggs containing larvae.

**Parthenogenesis** — reproduction without fertilisation of the ovum by the sperms.

**Autoinfection** — self infection with the parasite.

**Retroinfection** — a form of autoinfection in which the larvae hatch out near the anus and migrate back into the large intestine to develop into adults.

**Rhabditiform larva** — is a larva with a short oesophagus, which has a bulb at its posterior end.

**Filariform larva** — is a larva with an elongated oesophagus which does not have a bulb at its posterior end.

**Paratenic host** — is a transport host in which the particular parasitic stage does not grow, but remains viable.

**Periodicity** — term used to denote the presence or absence of microfilariae in blood.

**Nocturnal periodicity** — when microfilariae show a pronounced peak during night time. The rest of the period they are absent or scanty.

**Diurnal periodicity** — when microfilariae show a pronounced peak during daytime. The rest of the period they are absent or scanty.

**Subperiodicity** — when microfilariae are present in appreciable numbers throughout the 24-hour period.

**Innen-Korper** (Inner body) — an elongated structure seen in the mid region of microfilaria. This along with the pharyngeal thread is regarded as the modified intestine of the later developing larval stages.

**R cells** — embryonic rectal cells seen in microfilaria.

**G cell** (GI or RI cell) — a large cell close to innenkorper seen in microfilaria. A part of the future intestine.

## Introduction

Nematodes are elongated, cylindrical and unsegmented worms. With the exception of the parasitic form of *Strongyloides stercoralis,* all medically important nematodes are dioecious. They range in size from a few millimetres to many centimetres.

The body wall is divisible into the external cuticle, the hypodermis and the somatic musculature. The hypodermis projects into the body in the form of lateral, ventral and dorsal chords. The cuticle may show markings and projections of various types and these are useful in species identification especially in tissue sections.

The alimentary canal is an elongated tube consisting of an oral cavity, an oesophagus, a midgut, a hindgut or rectum which opens in a subterminal anus. The oesophagus is muscular with the exception of Trichinelloidea.

The excretory system consists of two tubes running inside the lateral chords. At the anterior end these tubes are interconnected and open in the mid ventral region as an excretory sinus.

The nervous system consists of a nerve ring encircling the oesophagus, from which trunks radiate anteriorly and posteriorly.

The male genital system is tubular and can be differentiated into a small ejaculatory duct, a seminal vesicle, a vas deferens and a testis. The ejaculatory duct opens along with the rectum at the cloaca.

The female genital system is also tubular and may have a didelphic or monodelphic arrangement. Each tube consists of an ovary, an oviduct, a seminal receptacle, a uterus, a vagina and a vulva.

Nematodes generally undergo 4 moults during their life. The first moult and occasionally the second moult may take place in the egg. The remaining moults occur in the definitive host. An external stimulus apparently triggers a response in the nematode which leads to production of enzymes which initiate moulting. Before moulting, the hypodermis thickens with accumulation of ribosome like granules. The new cuticle then forms below the old cuticle, which is either partially resorbed or shed intact (Bird, 1971).

**281** **General morphology of a hypothetical nematode.**

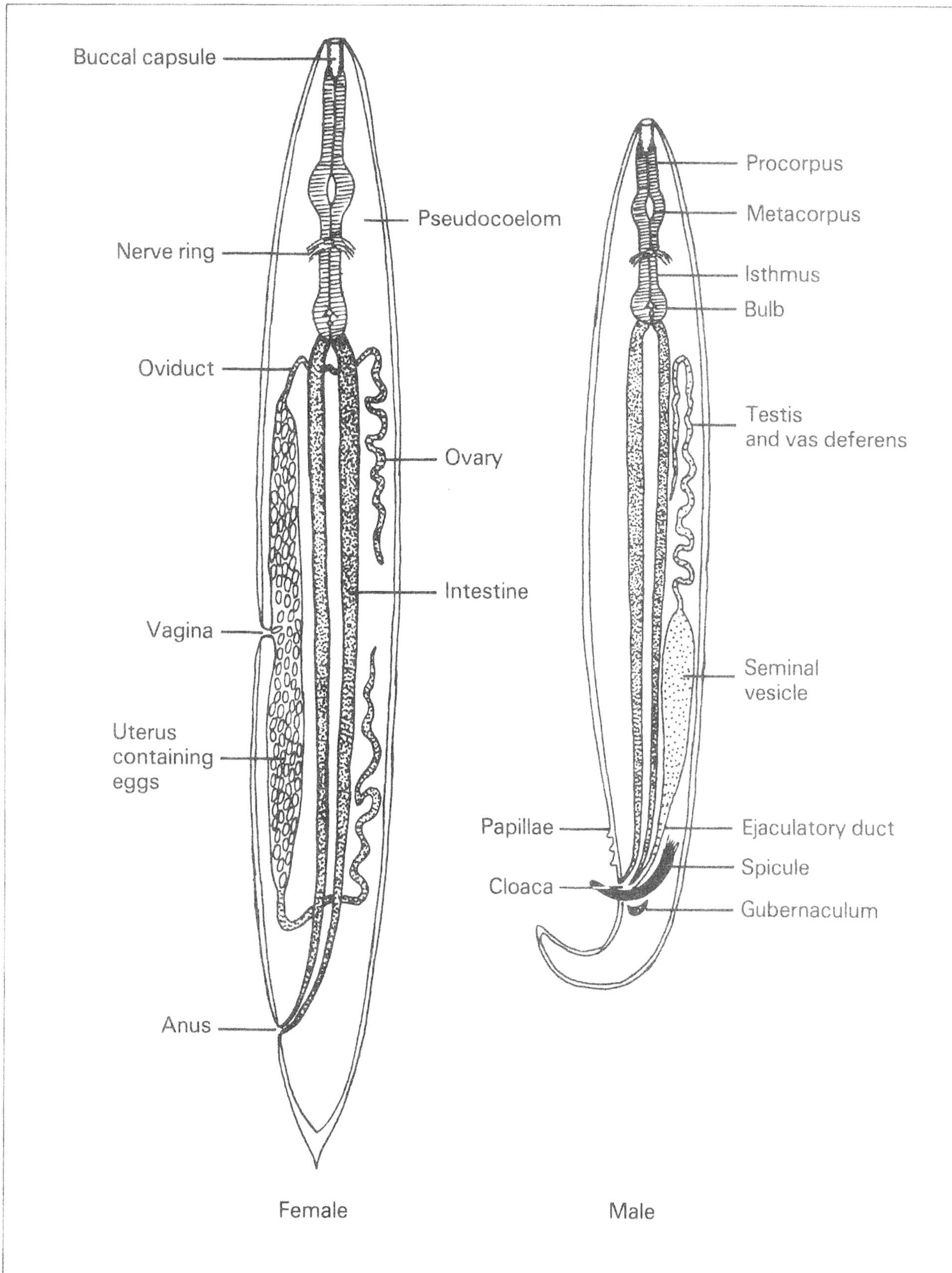


281

**282 Internal structure of a nematode.** In cross-section, a nematode (in this case an *Ascaris*) appears as a tube within a tube. The outer tube consists of the body wall with its appendages and the inner tube the oesophagus with its triradiate lumen. The body wall is made up of cuticle and the hypodermis. The hypodermis projects into the body cavity in the form of chords. The somatic muscles are unique in that each throws out an extension which attaches it to one of the chords. The oesophagus is muscular and the muscle fibres extend from the wall to its lumen. Contraction of these muscles opens the lumen and allows the entry of food.

DC = dorsal chord (contains the dorsal nerve)
MA = muscle arms
C = cuticle
H = hypodermis
SMC = somatic muscle cells
LC = lateral chord (contains the excretory canal)
LO = lumen of oesophagus
OMC = oesophageal muscle fibres
VC = ventral chord (contains the ventral nerve)

**283 a) Body walls of nematode.** A to F diagrams showing outer cuticle, longitudinal markings, lateral alae, muscle layer (msc) and hypodermal chords (ch). A showing alae absent. B to C showing sublateral alae. D showing alae typical of ascarids and some oxyurids e.g. *Enterobius*. E showing Trichostrongyloid type. F showing Chromadorid type (free-living).

**b) Muscle cells of nematode** showing basement membrane (m), contractile fibres (f), sarcoplasm (s) and nucleus (n).

A = Platymyarian. B = Transitional. CD = Coelomyarian.

Redrawn from Chitwood, M. and Lichtenfels, J.R.: Identification of parasitic metazoa. Experimental Parasitology 32: 407 (1972).

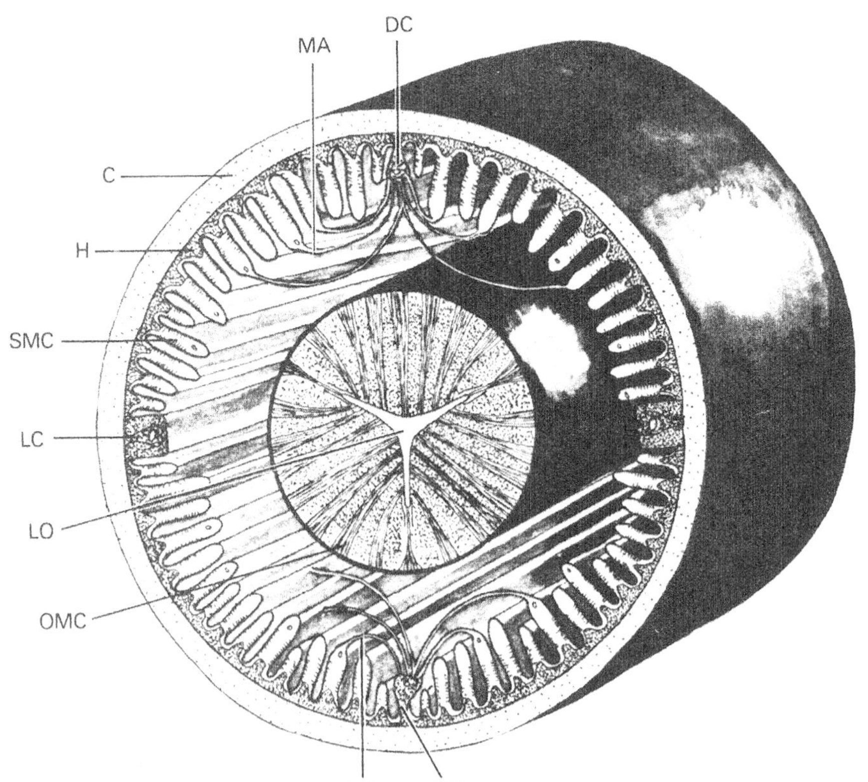


282

a

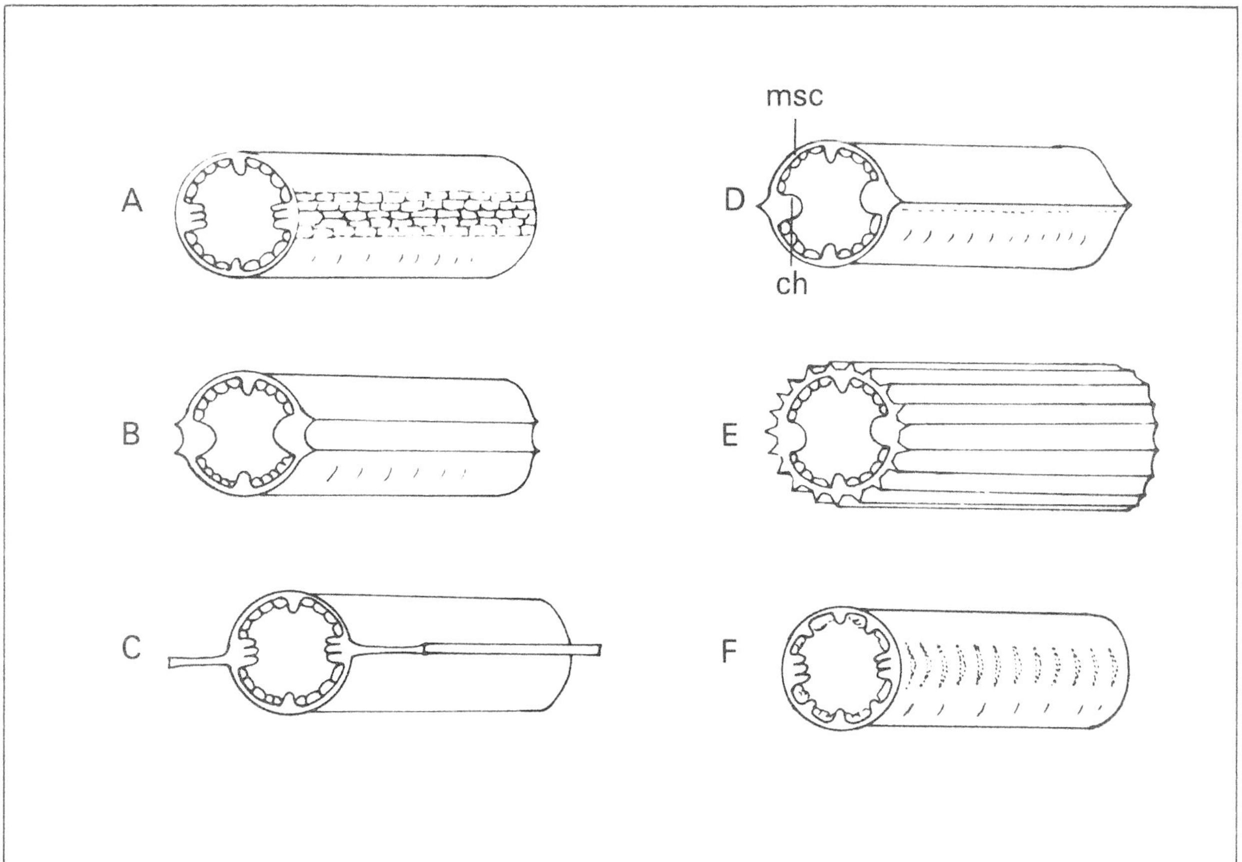


b

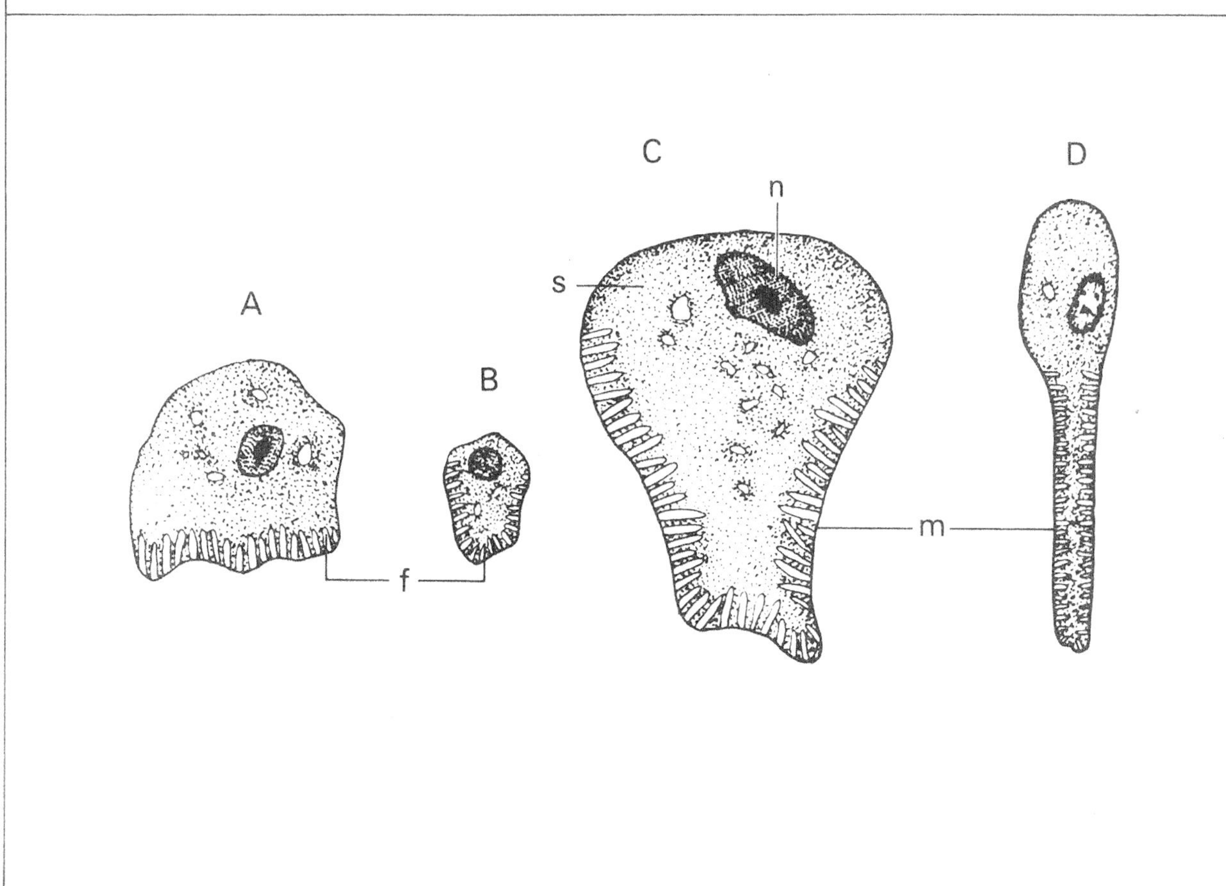


283

## Wuchereria bancrofti [284] and Brugia malayi

Adults are elongated thread like worms, which live in the lympatics of various parts of the body. The male measures approximately 4cm in length and 0.1mm in breadth. The adult female is 8 to 10cm in length and 0.2 to 0.3mm in breadth. The female is viviparous and liberates sheathed microfilariae into the lymph, from where they find their way into the bloodstream. The female is didelphic and the presence of paired uteri are useful in diagnosing it in tissue sections.

The microfilariae show nuclei in their body after staining. The arrangement, size and shape of these nuclei is used for differentiation of various species. Recent studies have demonstrated hooks and spines in the cephalic space and a pharyngeal thread extending from the anterior end to the innenkorper (Simpson and Laurence, 1972; Tongu, 1974). The microfilariae show nocturnal periodicity in many parts of the world. This occurs because of their concentration in the lungs during day time and their appearance in the peripheral circulation during the night. If the microfilariae are ingested by a suitable vector they penetrate the stomach wall and enter the thoracic musculature, after which they develop into infective larvae and enter the host tissues through the aperture made by the mosquito mouth parts.

### *Clinical aspects*

The infection passes through 2 stages:
a) The acute stage
b) The chronic stage

During the acute stage there is lymphadenitis and lymphangitis. The chronic stage may show elephantiasis, hydrocoele and chyluria. Lymphatic obstruction leading to lymph stasis and lymphoedema are believed to cause elephantiasis. Rupture of the lymphatic varices gives rise to chyluria. In *W. bancrofti* involvement of genitalia and hydrocoele is more common than in *B. malayi.*

### *Diagnosis*

Laboratory diagnosis is based on detection of microfilariae in the peripheral blood. Various methods are available for this purpose and include concentration techniques (Denham, 1975).

## Eosinophilic Lung

This condition is also known as tropical pulmonary eosinophilia. It is an allergic reaction to microfilariae which are trapped in the pulmonary tissues of a hypersensitive individual [284]. The peripheral blood does not show any parasites but there is a great increase in eosinophils, raised ESR and antibodies to filaria. The clinical picture may resemble asthma or bronchitis. Radiological examination of the chest often shows diffuse mottling.

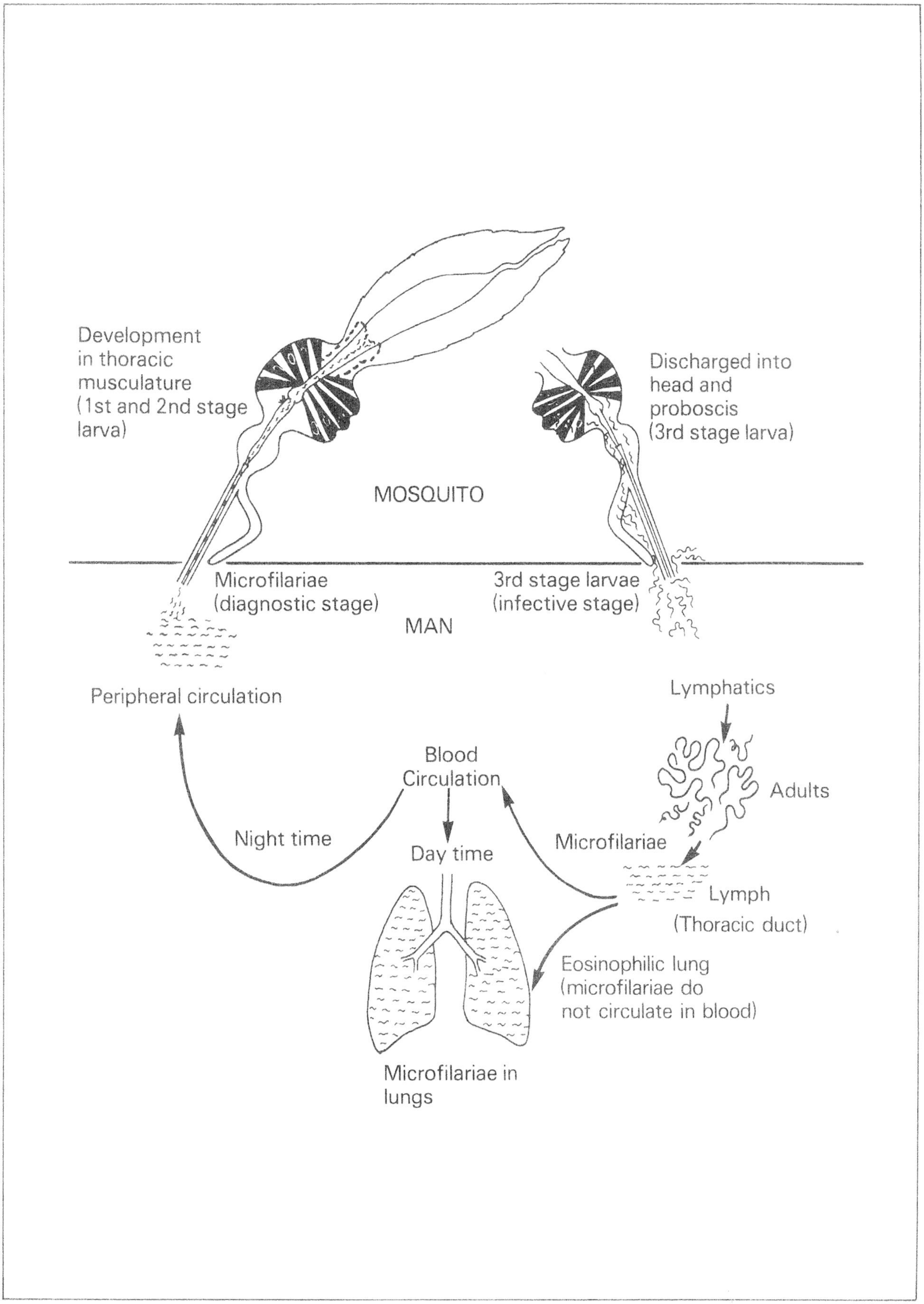
Development
in thoracic
musculature
(1st and 2nd stage
larva)
Discharged into
head and
proboscis
(3rd stage larva)
MOSQUITO
Microfilariae
(diagnostic stage)
3rd stage larvae
(infective stage)
MAN
Peripheral circulation
Lymphatics
Blood
Circulation
Adults
Night time
Day time
Microfilariae
Lymph
(Thoracic duct)
Eosinophilic lung
(microfilariae do
not circulate in blood)
Microfilariae in
lungs

284

**285 Differential characteristics of microfilariae.**

| | 1 *W. bancrofti* | 2 *B. malayi* | 3 *L. loa* | 4 *O. volvulus* | 5 *D. perstans* | 6 *D. streptocerca* | 7 *M. ozzardi* |
|---|---|---|---|---|---|---|---|
| Length | 200-300µ | 220-250µm | 250-300µm | 250-300µm | 150-200µm | 180-240µm | 150-200µm |
| Diameter | 8µm | 6µm | 8µm | 8µm | 4µm | 5µm | 4µm |
| Sheath | Present (stains slightly with Giemsa) | Present (stains deeply with Giemsa) | Present (almost colourless with Giemsa) | Absent | Absent | Absent | Absent |
| Body curves | Regular smoothly curved | Irregular and twisted | Irregular and twisted | Regular slightly twisted | Regular often forms loops | Tail usually curved | Regular slightly twisted |
| Cephalic space | Small | Large | Large | Large and bulbous | Large | Large | Large |
| Body nuclei | Coarse Well separated | Coarse Tend to overlap | Coarse Tend to overlap | Coarse Mostly separated | Medium sized Tend to overlap | Fine Mostly separated | Fine Mostly separated |
| Tail end | No nuclei Pointed tip | 2 widely spaced nuclei Blunt tip | Nuclei present Rounded tip | No nuclei Pointed tip | Nuclei present Rounded tip | Nuclei present Curved tip | No nuclei Pointed tip |

**Diagrammatic illustration of** ***B. malayi*** **microfilaria.** (Courtesy of Dr Y. Tongu).

H = hook
NR = nerve ring
EP = excretory pore
EC = excretory cell
I = inenkorper (inner body)
PT = pharyngeal thread (central canal)
CC = caudal channel
TN = terminal nuclei
AP = anal pore
R2 = embryonic rectal cells
R3
R4
G1 = embryonic intestinal cell

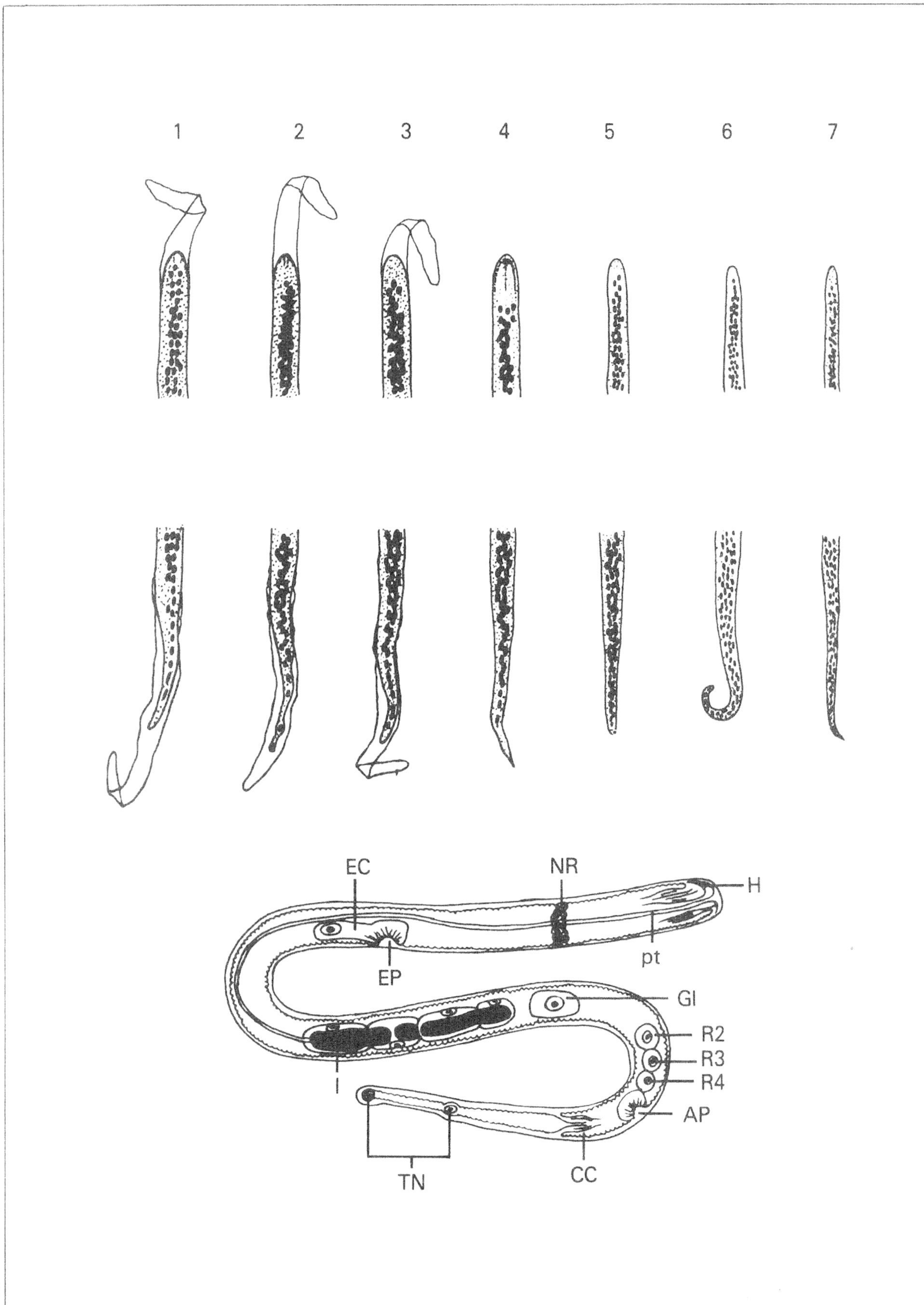


285

## Loa loa

The vector is a day-biting tabanid fly belonging to the genus *Chrysops.* The microfilariae develop in the fat body of the fly and develop into infective larvae in 10 to 12 days.

### *Clinical aspects*

The adult worms develop in connective tissues producing migratory swelling in various parts of the body. The microfilariae are discharged in the blood stream and show a diurnal periodicity. Adult worms may sometimes be seen crossing the conjunctiva. This movement does not generally produce significant pathology but may be alarming to the patient.

### *Diagnosis*

This is based on the clinical picture and the presence of microfilaria in the blood.

## Onchocerca volvulus

The vector is the black fly belonging to the genus *Simulium.* The microfilariae develop in the thoracic musculature of the fly and become infective in 6 to 12 days' time. The adults develop in the subcutaneous tissues and form nodules. Inside the nodule they mate and produce microfilariae which are discharged in the skin and subcutaneous tissue.

### *Clinical aspects*

The main cause of pathology in Onchocerchiasis is due to microfilariae. This stage of the parasite may produce lesions in skin resulting in dermatitis, intradermal oedema and lichenification. The loss of elastic tissue may lead to a condition known as 'hanging groin', in which the skin in the inguinal region hangs in folds.

The eye lesions may start as conjunctivitis, photophobia and lacrimation. Limbitis with brownish pigmentation is often present. In chronic cases punctate keratitis, glaucoma and optic atrophy may occur.

### *Diagnosis*

This is based on examination for microfilariae in skin snip and the detection of adult worms in the nodules.

## Dipetalonema perstans

Is generally regarded as a non-pathogenic filarial worm. The adults are located in body cavities and microfilariae in blood. The vectors are flies belonging to the genus *Culicoides.*

## Dipetalonema streptocerca

Is generally regarded as a non-pathogenic filarial worm. The adults are located in the subcutaneous connective tissues and microfilariae in skin. The vectors are flies belonging to the genus *Culicoides.*

## Mansonella ozzardi

Is generally regarded as a non-pathogenic filarial worm. The adults are located in the peritoneal tissues and microfilariae in blood. The vectors are flies belonging to the genus *Culicoides.*

## Dracunculus mediensis

These are very long worms. The female measures 70 to 120cm and males 12 to 20mm. The females are viviparous and discharge larvae into water through an ulcerated area in the skin. The larvae then swim actively and are ingested by a copepod. In the copepod they become infective third stage larvae in about 12 to 14 days. Man is infected by drinking water containing the copepod. The worms become mature in the body cavities and connective tissue.

### *Clinical aspects*

An ulcerated area appears on the site where the female emerges to discharge larvae. The lesion is frequently seen on the exposed part of the body. Sometimes the worm is unable to reach the skin and dies in the connective tissue giving rise to abscess formation. The dead worm may on occasion become calcified and is then clearly seen on X-rays.

### *Diagnosis*

The lesion produced by *Dracunculus* is obvious after the worm has appeared in the skin. Before the appearance of the worm the diagnosis is not possible.

## Dirofilaria spp

These animal filaria are potentially pathogenic to humans as the adults may develop in various organs of the body. *Dirofilaria immitis,* the heartworm of the dog, has been reported to occur in the lungs, heart and breast tissues of individuals from many parts of the world. In the lungs, they produce circular lesions simulating metastatic carcinoma on X-ray examination. The condition is regarded as a form of occult filariasis as microfilariae are not seen in the blood. Many species of mosquitoes act as a vector and the larval development takes place in the malphigian tubules.

## Angiostrongylus cantonensis [286]

The adult worms live in the pulmonary arteries of the rat. The eggs are deposited in the lungs and the first stage larvae are expelled with the rat faeces. These infect the molluscan intermediate host. The rat is infected by eating molluscs. The larvae then pass through the brain to reach the pulmonary arteries.

Human infection occurs by eating snails, slugs, crustacea or unwashed vegetables. In humans the migration generally ceases at the level of the central nervous system and the worm does not develop to maturity.

### *Clinical aspects*

The disease produced is also known as eosinophilic meningitis as there is high eosinophilia in CSF and blood. The patient has signs and symptoms of meningeal irritation with neck rigidity and headache.

### *Diagnosis*

In endemic areas the clinical picture is suggestive, especially if accompanied by high CSF eosinophilia. The CSF also shows a low sugar and high protein content. Occasionally larval forms of *Angiostrongylus* are recovered from the CSF.

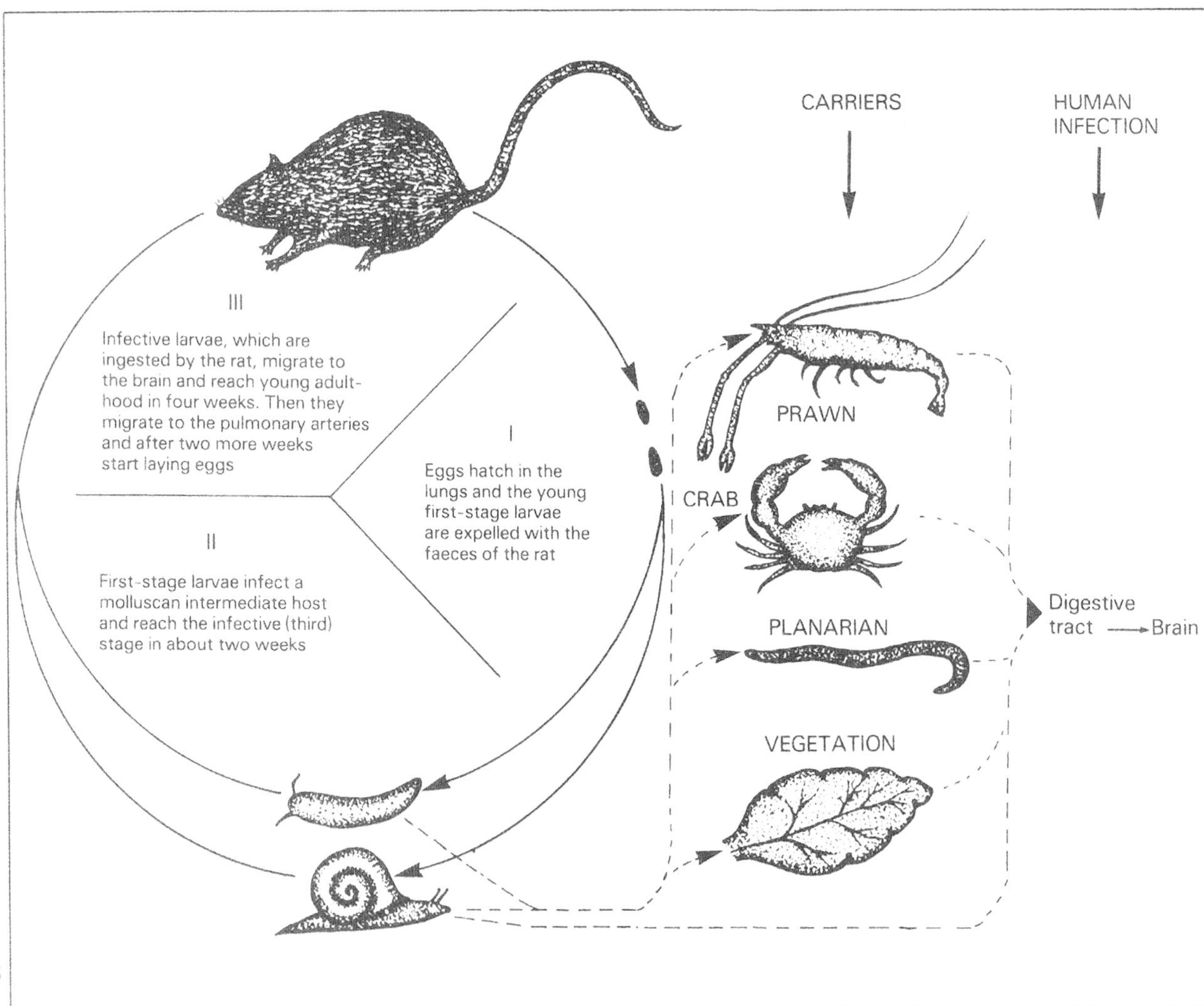


286

Redrawn from Alicata, J.E.: Parasites of Man and Animals in Hawaii (Karger, Basel 1969).

**287 Angiostrongylus cantonensis adults.** These have been dissected out of the lungs of an infected rat. The main characteristic is their 'barber's pole' appearance; i.e. an interweaving of a dark with a white band. This is because the gut is pigmented and the genital structures are whitish.

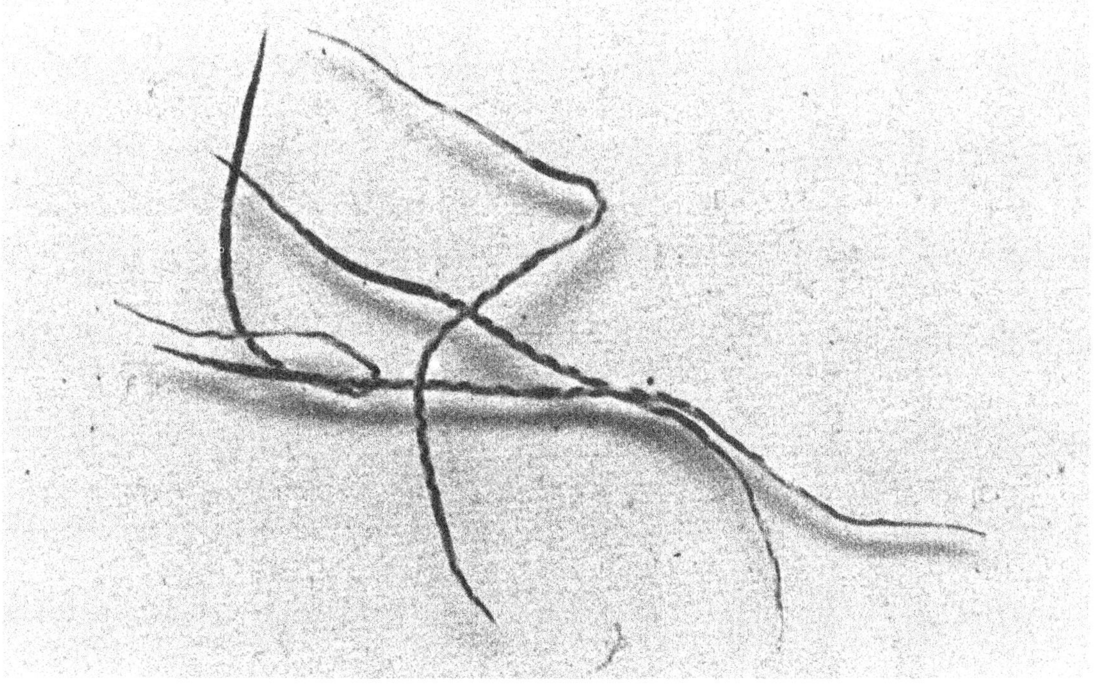

287

## Ascaris lumbricoides [288]

This is a very common parasite of the intestinal tract. It is cylindrical in shape with tapering ends. The female measures 20 to 35cm in length and 3 to 6mm in breadth. The male measures 12 to 31cm in length and 2 to 4mm in breadth. The head has 3 rounded lips which carry minute teeth or denticles along their margins. The fertilised eggs are ovoidal and measure 60 to 75μm by 35 to 50μm. When freshly passed they have a single cell which is surrounded by a thin vitelline membrane. Around the membrane is a thick middle layer which in its turn is surrounded by an irregular albuminous coat. The albuminoid layer is laid down in the uterus and is colourless before coming in contact with faecal matter. In the lumen of the host intestine the eggs acquire a brownish colour due to the bile pigments. The unfertilised eggs are longer and narrower and have disorganised contents.

The eggs become infective in 2 to 3 weeks' time in soil. On being ingested the larva hatches out in the small intestine and enters the blood circulation. In the lungs it undergoes 2 moults and crawls up the bronchial tree to return back to the intestinal tract.

### *Clinical aspects*

The larval migration of *Ascaris* through the liver and lungs produce an inflammatory reaction of a varying degree. As the person becomes hypersensitive even a few migrating larvae could produce asthma like symptoms.

The intestinal symptoms are absent in the majority of cases. Occasionally the worms may migrate to ectopic sites and produce severe disease. The ectopic sites include the appendix, common bile duct and Meckel's diverticulum. In heavy infections intestinal obstruction and intussuception may occur.

### *Diagnosis*

This is based on finding the egg in the faeces. X-rays taken after barium meal sometimes reveal the presence of worms as elongated filling defects and occasionally as linear streaks if the barium has gone into the intestine of *Ascaris.*

## Hepatic ascariasis

This complication of ascariasis occurs when the adult worms migrate into the common bile and hepatic ducts. The worms generally disintegrate in this location releasing a large number of eggs. The eggs become dispersed in the liver tissue and produce multiple abscesses and cholangitis.

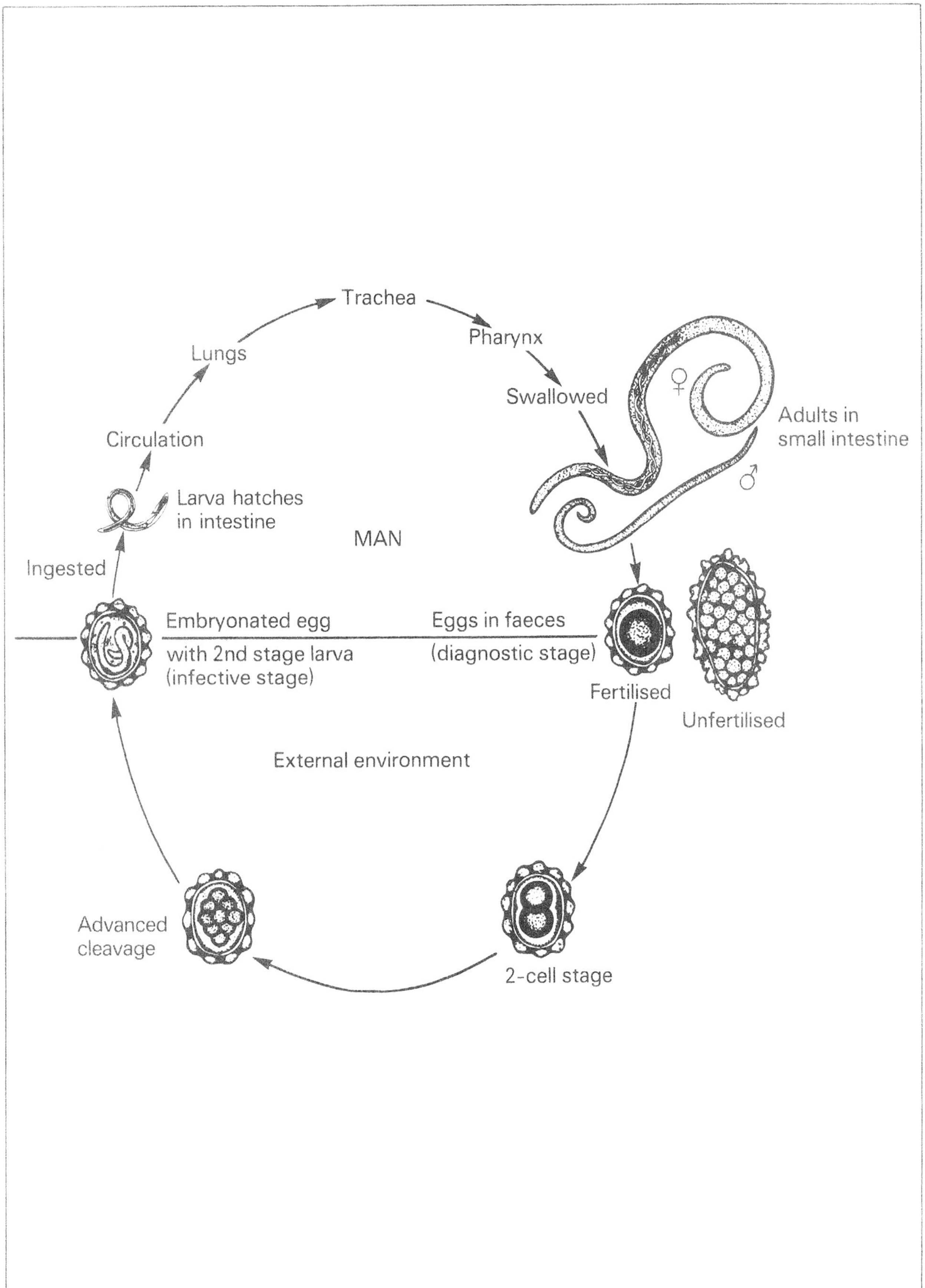
Trachea
Pharynx
Lungs
Swallowed
Circulation
Adults in
small intestine
♀
♂
Larva hatches
in intestine
MAN
Ingested
Embryonated egg
with 2nd stage larva
(infective stage)
Eggs in faeces
(diagnostic stage)
Fertilised
Unfertilised
External environment
Advanced
cleavage
2-cell stage

288

**Trichinella spiralis** [289]

The adults are small nematodes measuring 1.4 to 1.6mm. The males have a pair of conical papillae on sides of the cloacal opening. The female is viviparous and about twice as long as the male. Humans become infected by eating improperly cooked pork containing encysted larvae. The larvae are released in the small intestine and mature into adults. The adults deposit larvae in the mucosa, from where they enter the circulation and encyst in the striated muscles of the body. In nature the cycle is maintained between swines and other carnivores.

*Clinical aspects*

Light infections are generally asymptomatic. Heavy infections may produce symptoms resembling acute food poisoning during the early stage followed at a later date by severe myositis, dyspnoea and oedema of the face.

*Diagnosis*

The clinical picture is usually indicative. Laboratory diagnosis is based on biopsy of muscle and the demonstration of larvae. Various immunological tests also become positive.

**Anisakis sp**

The adults of this parasite normally live in the stomach of fish-eating marine mammals such as dolphins, porpoises and whales. The eggs passed in the faeces of these animals produce second stage larvae which develop in crustaceans into third stage larvae. Further development occurs in fish, which act as a source of infection to humans and marine mammals.

*Clinical aspects*

The symptoms often resemble that of peptic ulcer in cases of gastric involvement.

*Diagnosis*

Is made from a biopsy specimen of stomach or intestine. Gastroscopy is of value in some cases.

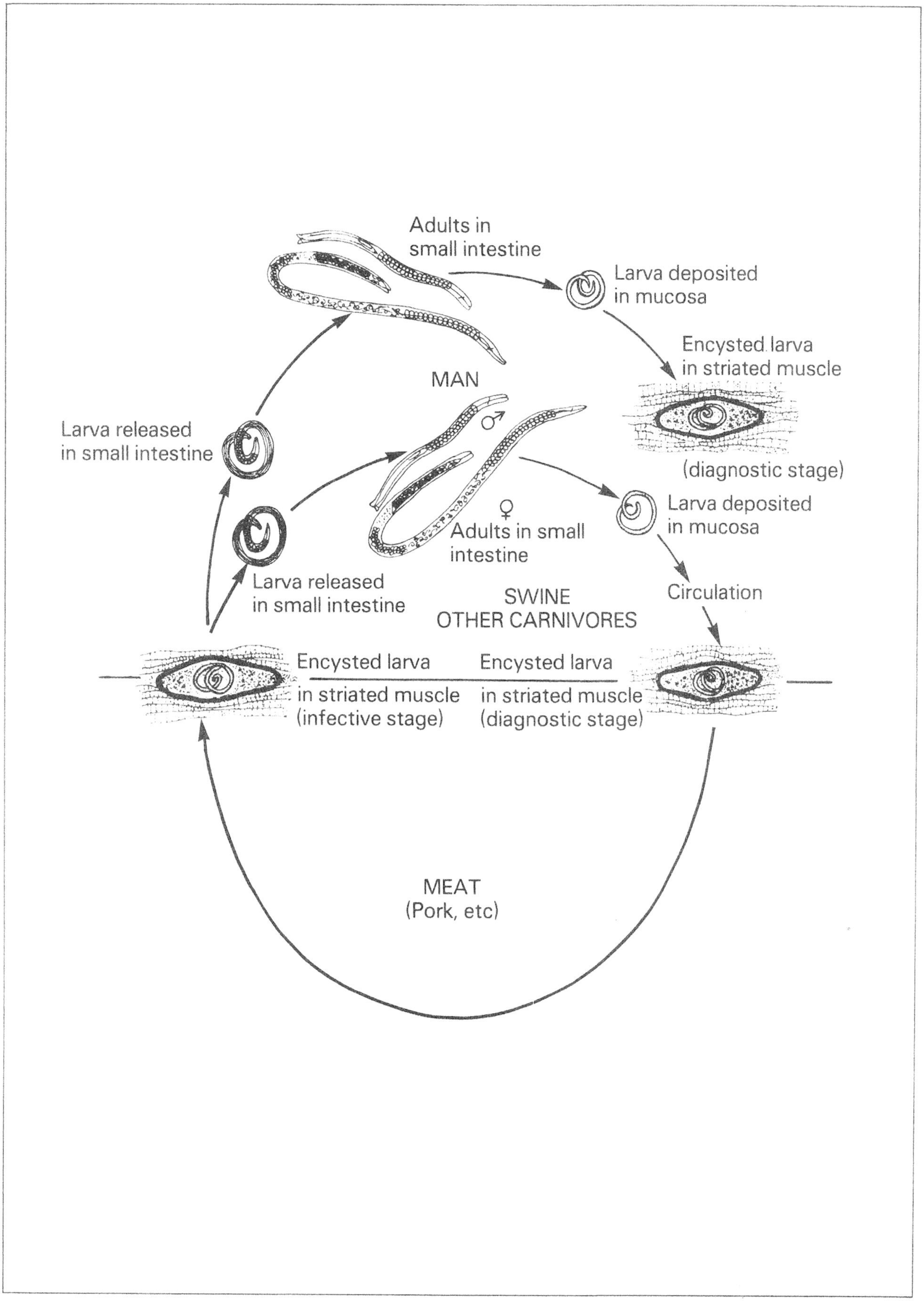
Adults in
small intestine
Larva deposited
in mucosa
Encysted larva
in striated muscle
(diagnostic stage)
MAN
Larva released
in small intestine
♂
♀
Adults in small
intestine
Larva deposited
in mucosa
Larva released
in small intestine
SWINE
OTHER CARNIVORES
Circulation
Encysted larva
in striated muscle
(infective stage)
Encysted larva
in striated muscle
(diagnostic stage)
MEAT
(Pork, etc)

289

## Strongyloides stercoralis [290]

The adult female resides in the mucosa of the small intestine and measures about 2.2mm in length and 30 to 75mm in breadth. It reproduces by parthenogenesis. The rhabditiform larvae are liberated in the intestinal tissues and appear in the faeces. The larvae may develop into filariform stages in the lumen of the bowel and produce autoinfection, or develop into filariform larvae outside the body and initiate the direct cycle, or develop into free living adults outside the body and initiate the indirect or the free living cycle. After entry into the host tissues the filariform larvae pass through the circulation into lungs, migrate up the trachea and are swallowed, to finally mature in the small intestine.

### *Clinical aspects*

Pathological changes may be seen at the site of entry of larvae giving rise to rash and pruritus. There may be pneumonitis during larval migration and intestinal symptoms on maturation. The intestinal symptoms consist of pain in the epigastrium similar to peptic ulcer, diarrhoea, flatulence and vomiting. As autoinfection is common, infection tends to persist for a long time. Occasionally, disseminated strongyloidiasis is seen in immunosuppressed individuals when the larval stages are found in various parts of the body (Purtillo et al., 1974).

### *Diagnosis*

This is based on finding rhabditiform larvae in faeces. Stool culture methods are useful if the larvae are not detected by direct examination.

## Trichostrongylus sp

The adults are 5 to 10mm in length and are located in the small intestine. They are mostly parasites of ruminants and infect humans living in close proximity to these animals. Infection may occur by ingestion of the infective larvae or by penetration of the skin. The eggs are similar to that of hookworms but the ends are more pointed and the ovum is passed in an advanced cleavage (morula) stage.

### *Clinical aspects*

Light infections are asymptomatic. In heavy infections there may be anaemia, abdominal pain and diarrhoea.

### *Diagnosis*

This is based on finding the eggs in the faeces.

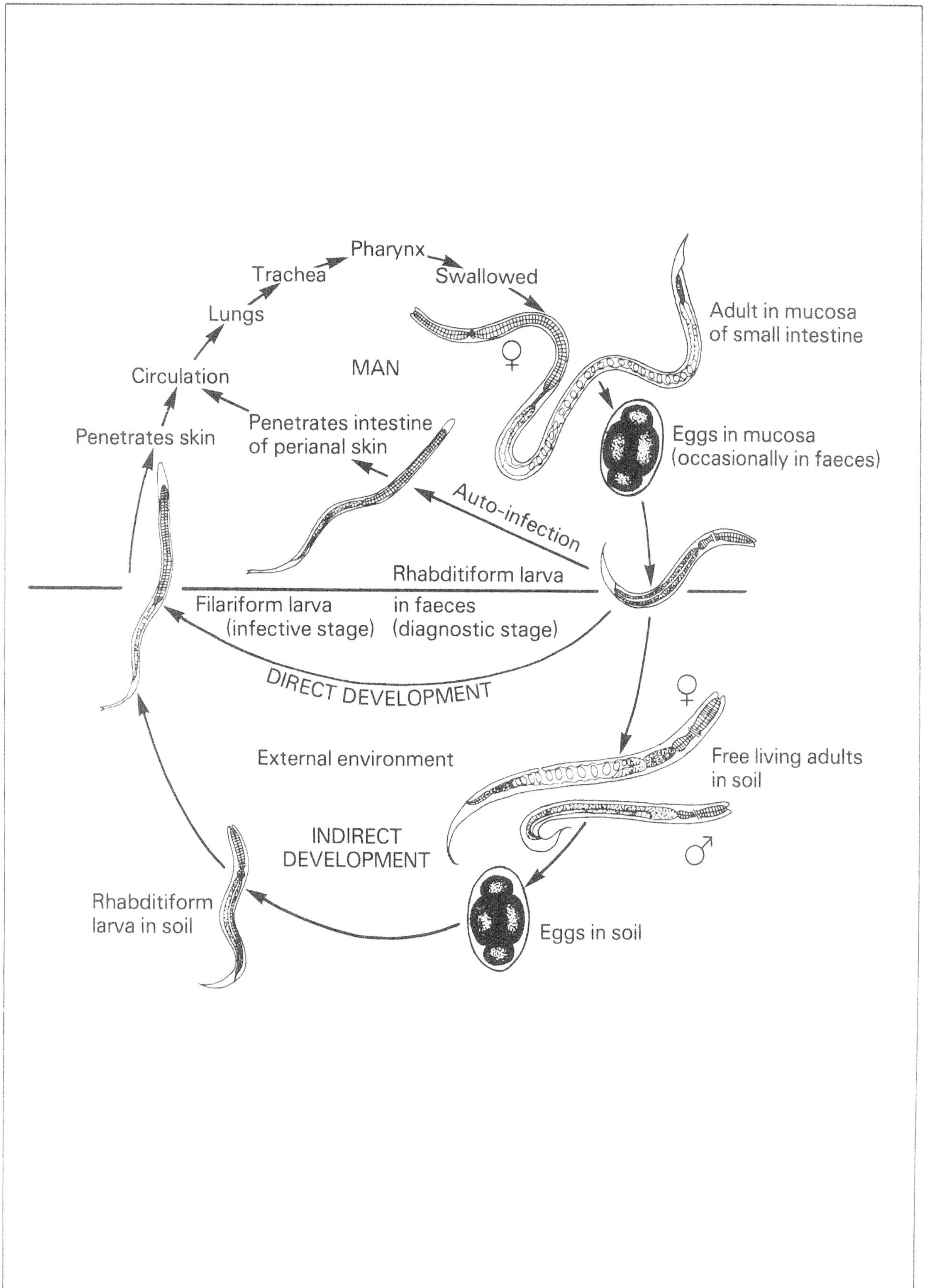
Pharynx
Trachea
Swallowed
Lungs
Adult in mucosa
of small intestine
♀
MAN
Circulation
Penetrates intestine
of perianal skin
Penetrates skin
Eggs in mucosa
(occasionally in faeces)
Auto-infection
Rhabditiform larva
in faeces
(diagnostic stage)
Filariform larva
(infective stage)
DIRECT DEVELOPMENT
♀
Free living adults
in soil
External environment
INDIRECT
DEVELOPMENT
♂
Rhabditiform
larva in soil
Eggs in soil

290

**Trichuris trichiura** [291]

The adult male measures 30 to 45mm and the female 35 to 50mm in length. The anterior three-fifths are thin and elongated and the posterior two-fifths bulbous and fleshy. This gives the parasite a whip like appearance. The adults are located in the caecum and produce barrel shaped eggs which are 22 to 50μm in length. They have a transparent blister like plug at each end and are single celled when freshly passed. In the soil the eggs become infective in about 3 weeks' time. The larvae hatch out on ingestion of the egg, penetrate the mucosa and then develop into adults.

*Clinical aspects*

Light infections are asymptomatic. In heavy infection there is colitis with mucous and blood in faeces. Continuous irritation of the bowel and weakness of the levator ani muscle may result in prolapse of the rectum.

*Diagnosis*

This is based on finding the typical eggs in the faeces. In case of dysentery the faeces also contain Charcot-Leyden crystals and eosinophils.

**Capillaria philippinensis**

The adults are 2 to 4mm long and are located in the mucosa of the small intestine. Human infection probably occurs by the ingestion of fish containing infective larvae. The eggs resemble that of *Trichuris,* but are broader and more ovoid. The lateral plugs do not protrude as in *Trichuris.*

*Clinical aspects*

The symptoms consist of diarrhoea and a sprue-like condition. Autoinfection is common and may result in severe disease with high mortality.

*Diagnosis*

Is based on finding the eggs in the faeces.

**Capillaria hepatica**

Is a common parasite of rats and other rodents. Human cases have been occasionally reported in which liver biopsy showed the presence of eggs. The eggs resemble that of *Trichuris,* but the shell is distinctly striated.

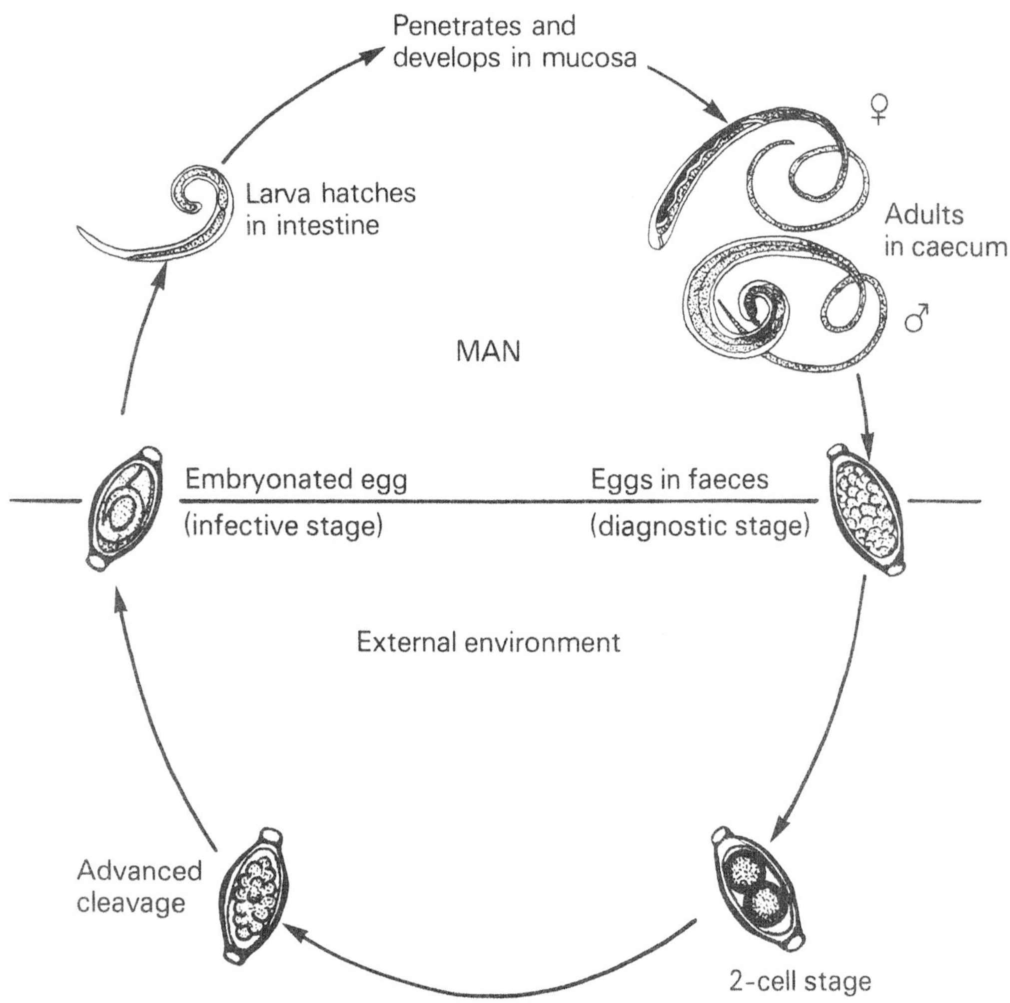
Penetrates and
develops in mucosa
♀
Larva hatches
in intestine
Adults
in caecum
♂
MAN
Embryonated egg
(infective stage)
Eggs in faeces
(diagnostic stage)
External environment
Advanced
cleavage
2-cell stage

291

## Visceral larva migrans [292]

Human infection with *Toxocara* spp, occurs by the ingestion of the embryonated egg of this parasite. This is not uncommon as *Toxocara canis* and *Toxocara cati* are commonly found in dogs and cats all over the world. After the ingestion of the embryonated egg the larva hatches out in the intestine, enters the circulation and migrates to various organs of the body. In humans it does not reach maturity and the condition is known as visceral larva migrans (Beaver, 1969). However, as many nematodes pass through the viscera during their normal phase of migration it is of some importance to be able to differentiate the various species of nematodes at the histological level (Nichols, 1956).

### *Clinical aspects*

Children are most commonly involved because of their close contact with household pets. In the majority of cases there is enlargement of the liver with hypereosinophilia. Occasionally the larvae may enter eye, CNS and other organs of the body.

### *Diagnosis*

This is based on clinical examination and history. Confirmation is possible if the biopsy of the tissues shows the presence of larvae. Serological tests are of some value (Woodruff, 1970).

## Cutaneous larva migrans

The parasites most commonly involved in producing cutaneous larva migrans are animal hookworms, such as *A. braziliense* (dog/cat), *A. caninum* (dog), *Uncinaria stenocephala* (dog) and *Bunostomum phlebotomum* (cattle). In addition *Strongyloides stercoralis* and animal *Strongyloides* spp can also produce this condition.

### *Clinical aspects*

In a typical case there is a linear inflammatory lesion of the skin which moves or migrates. There is often intense pruritus associated with it.

### *Diagnosis*

This is based on clinical examination and history. Biopsy is usually of no value.

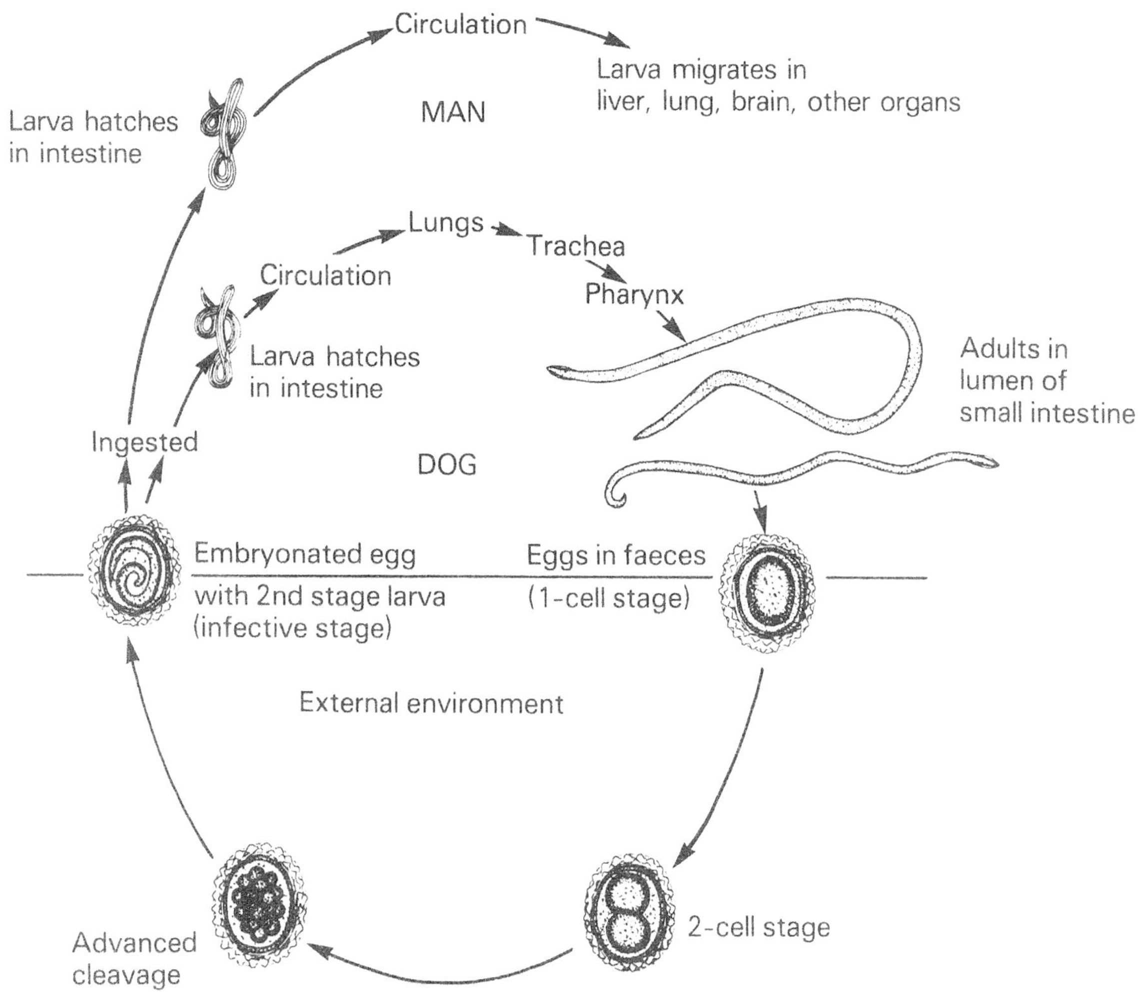
Circulation
Larva migrates in
liver, lung, brain, other organs
MAN
Larva hatches
in intestine
Lungs
Trachea
Circulation
Pharynx
Larva hatches
in intestine
Adults in
lumen of
small intestine
Ingested
DOG
Embryonated egg
with 2nd stage larva
(infective stage)
Eggs in faeces
(1-cell stage)
External environment
Advanced
cleavage
2-cell stage

292

## Enterobius vermicularis [293]

Commonly known as the threadworm. The male is up to 5mm long with a diameter of 0.1 to 0.2mm. The female is up to 13mm long with a diameter of 0.3 to 0.5mm. The cuticle has cervical alae which enables its easy recognition in sections. The gravid female has 2 distended uteri which practically fill the whole body. The male has a single spicule and a curved tail.

The adults are mainly located in the caecal region and the female deposits its eggs on perianal skin by migrating out of the anus. The eggs have a thick shell and are convex on 1 side and flattened on the other. They became infective within a few hours and measure 50 to 60µm. Autoinfection is, therefore, very common. The larvae hatch in the intestine and become adults.

*Clinical aspects*

Pruritus ani is generally the presenting symptom. This causes insomnia and restlessness. The perianal region may become infected by repeated scratching.

*Diagnosis*

This is based on finding the eggs by scotch tape method or by anal swab.

## Gnathostoma spinigerum

These worms are characterised by a distinct head bulb, which is covered by rows of hooks. Adults are located in the tissues of the stomach of carnivores and the eggs passed in the faeces. The eggs are ovoidal and have a single polar thickening or a mucoid plug. It has 2 intermediate hosts in its life cycle. The first intermediate host is *Cyclops* spp and the second intermediate host is the frog or fresh water fish. Humans are infected by eating the second intermediate host.

*Clinical aspects*

A painless subcutaneous swelling is common due to migration of the larva. Occasionally, central nervous system and ocular involvement may occur.

*Diagnosis*

Is based on the clinical aspects, high eosinophilia and positive immunological tests.

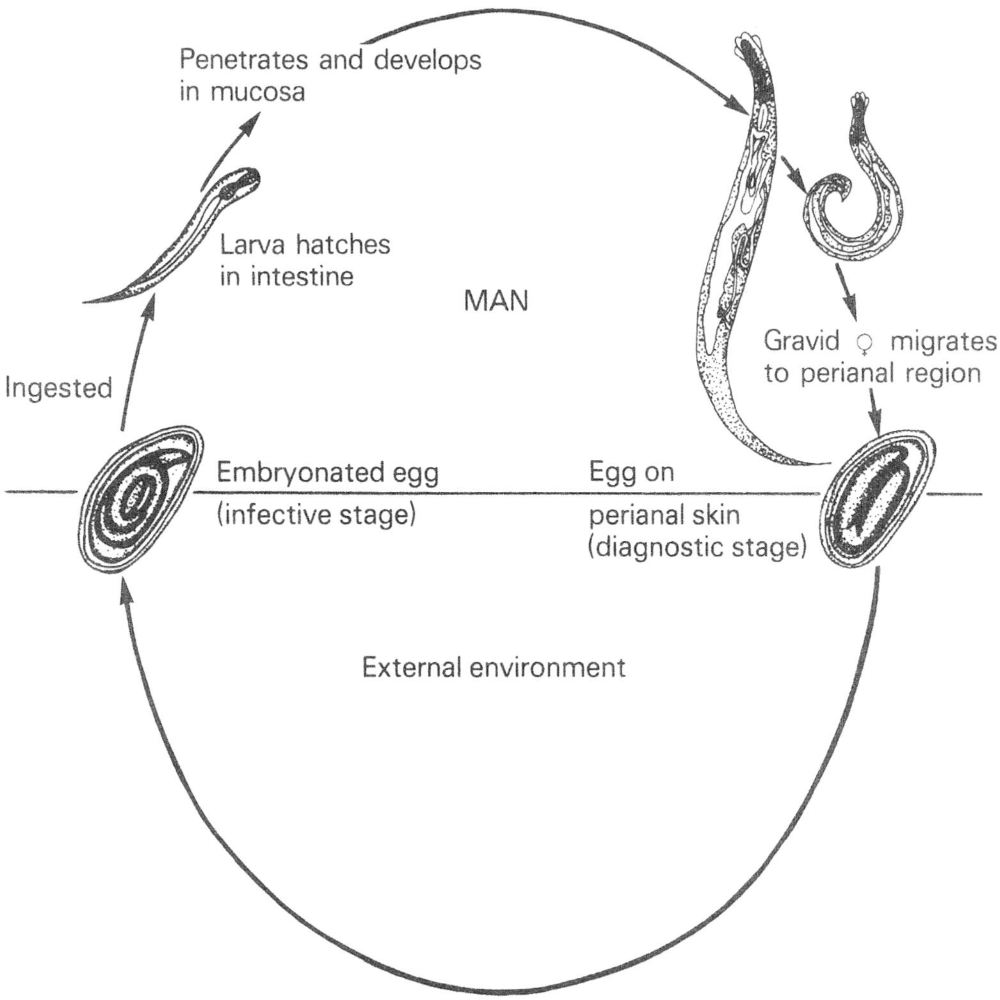
Penetrates and develops
in mucosa
Larva hatches
in intestine
MAN
Ingested
Gravid ♀ migrates
to perianal region
Embryonated egg
(infective stage)
Egg on
perianal skin
(diagnostic stage)
External environment

293

## Hookworms [294]

The common human species are *Necator americanus* and *Ancylostoma duodenale.* In addition *A. ceylanicum* (mainly a parasite of cats) has been reported from man (Chowdhury and Schad, 1972). Two other animal parasites, *A. braziliense* and *A. caninum* infect man producing a form of dermatitis known as 'cutaneous larva migrans' (page 218).

The hookworm species are mainly differentiated by their buccal capsule and the arrangement of rays in the bursa. The eggs of these species are, however, indistinguishable. They are ovoidal in shape, have a thin transparent shell and measure 74 to 76μm by 36 to 40μm. When passed in the faeces they are often in a 4 to 8 celled stage. Within several hours of being passed a multicelled stage and then a rhabditiform larva forms. The larva normally hatches out in the soil and becomes filariform, which is the infective stage.

The filariform larva penetrates the skin and via the circulation enters the lungs, crawls up the trachea and is swallowed back to mature in the small intestine. In the small intestine it attaches to the mucosa by its buccal capsule.

### *Clinical aspects*

Light infections are generally asymptomatic. In heavy infections macrocytic hypochromic anaemia is produced. The blood loss is due to sucking by the worm and also due to continuing haemorrhage at the site of attachment (Kalkofen, 1970). It is estimated that a *Necator* produces a loss of 0.03 to 0.05ml of blood/day and an *Ancylostoma* 0.15ml of blood/day.

### *Diagnosis*

This is based on finding the eggs in the faeces. An estimation of worm load is often necessary to make any correlation of anaemia with hookworm infection.

### References


Beaver, P.C.: The nature of visceral larva migrans. Journal of Parasitology 55: 3 (1969).

Bird, A.F.: The structure of nematodes, pp.1-317 (Academy Press, New York and London 1971).

Chowdhury, A.B. and Schad, G.A.: *Ancylostoma ceylanicum:* A parasite of man in Calcutta and environs. American Journal of Tropical Medicine and Hygiene 21: 300 (1972).

Denham, D.A.: The diagnosis of filariasis: in Diagnosis of Parasitic Diseases, p.95 (Proceedings of an International Colloquium, Prince leopold Institute, Antwerp, 6-8 December 1974).

Kalkofen, U.P.: Attachment and feeding behaviour of *Ancylostoma caninum.* Zeitschrift fur Parasitenkunde 33: 339 (1970).

Nichols, R.L.: The etiology of Visceral Larva Migrans. I. Diagnostic morphology of infective second-stage *Toxocara* larvae. II. Comparative larval morphology of *Ascaris lumbricoides, Necator americanus, Strongyloides stercoralis* and *Ancylostoma caninum.* Journal of Parasitology 42: 349, 363 (1956).

Purtillo, D.T.; Meyers, W.M. and Connor, D.H.: Fatal strongyloidiasis in immunosuppressed patients. American Journal of Medicine 56: 488 (1974).

Simpson, M.G. and Laurence, B.R.: Histochemical studies on microfilariae. Parasitology 64: 61 (1972).

Tongu, Y.: Ultrastructural studies on the microfilaria of *Brugia malayi.* Acta Medica Okayama, 28: 219 (1974).

Woodruff, A.W.: Toxocariasis. British Medical Journal 3: 663 (1970).

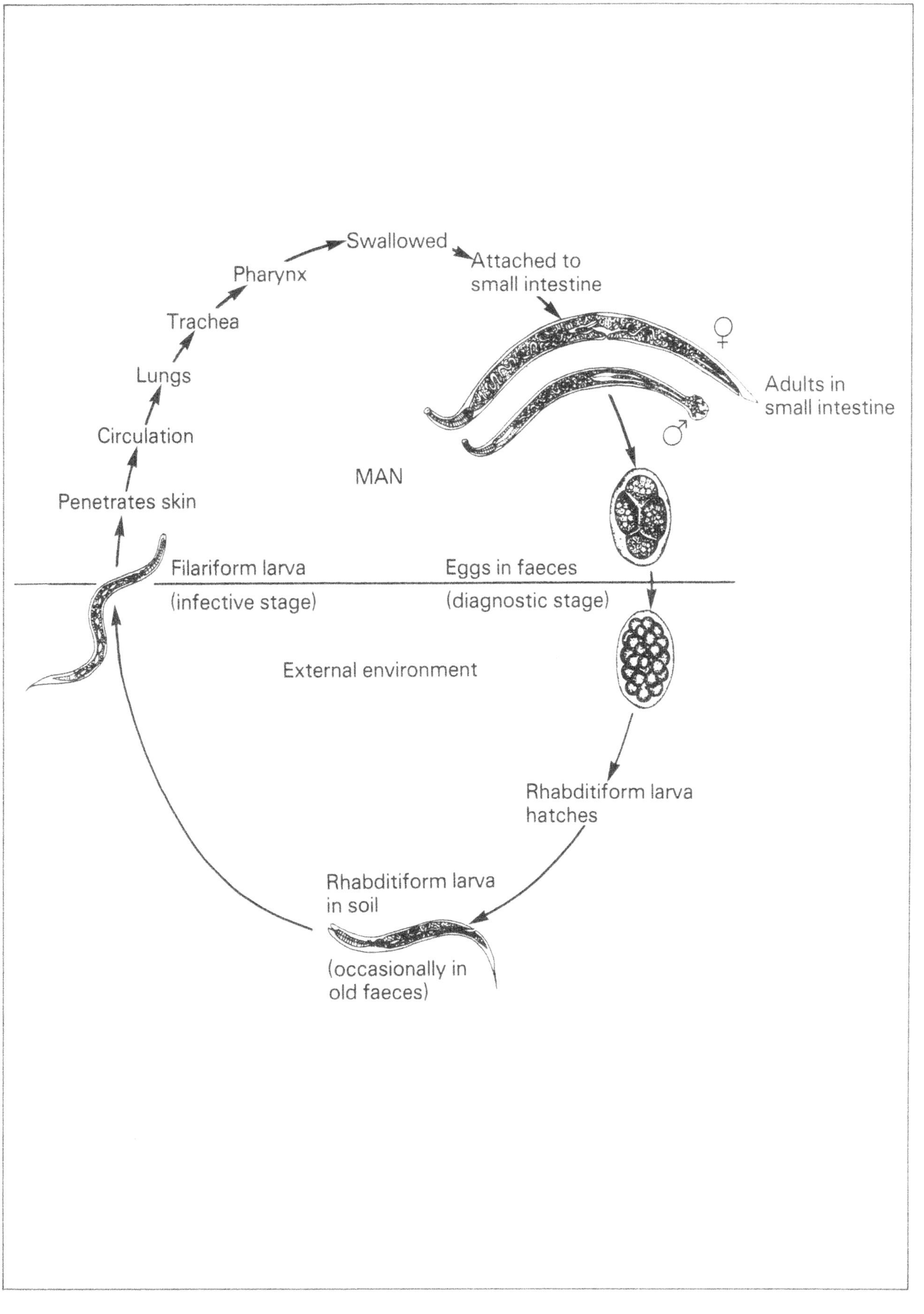
Swallowed
Attached to small intestine
Pharynx
Trachea
Lungs
Circulation
Penetrates skin
Adults in small intestine
MAN
Filariform larva
(infective stage)
Eggs in faeces
(diagnostic stage)
External environment
Rhabditiform larva hatches
Rhabditiform larva in soil
(occasionally in old faeces)

294

**295a 3rd stage larva of Ancylostoma exsheathing.** Note the sharply pointed tail and the unstriated sheath. In the case of *Strongyloides* the tail is notched and in the case of *Necator* the sheath is striated. Phase contrast. X100.

**295b Differentiation of human hookworm with Strongyloides larvae.**

| *Strongyloides* (S) | *Hookworm* (H) |
|---|---|
| **2nd stage larva** | |
| a) Rhabditiform oesophagus with a small buccal capsule (BC) | a) Rhabditiform oesophagus with a large buccal capsule (BC) |
| b) Genital primordium large (GP) | b) Genital primordium small (GP) |
| **3rd stage larva** | |
| a) Oesophagus filariform extending to approximately 40% of the total length | a) Oesophagus filariform extending to approximately 25% of the total length |
| b) No sheath | b) Sheath |
| c) Tail forked | c) Tail pointed |

**Differentiation of infective (3rd stage) of larvae of** *Ancylostoma duodenale* **with** *Necator americanus* at high magnification.

| | *A. duodenale* | *N. americanus* |
|---|---|---|
| Oesophageal spears (OS) | Not prominent and tend to close at the anterior end | Prominent and tend to open at the anterior end |
| Sheath | Smooth | Marked striations especially at posterior end |

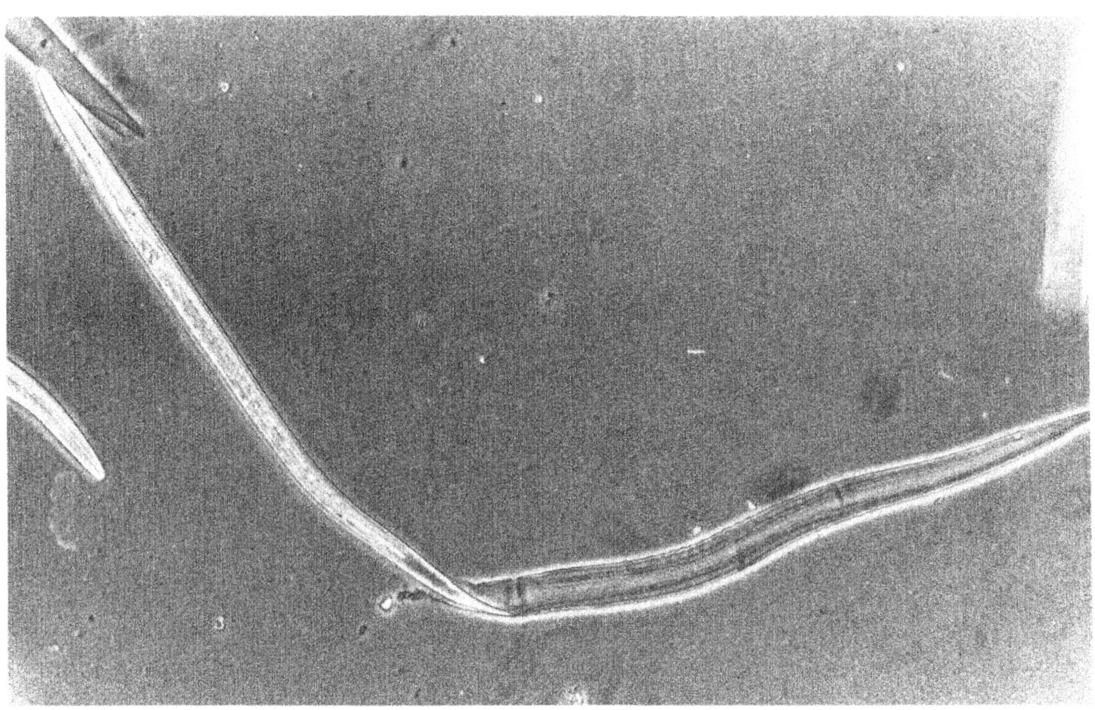

295a

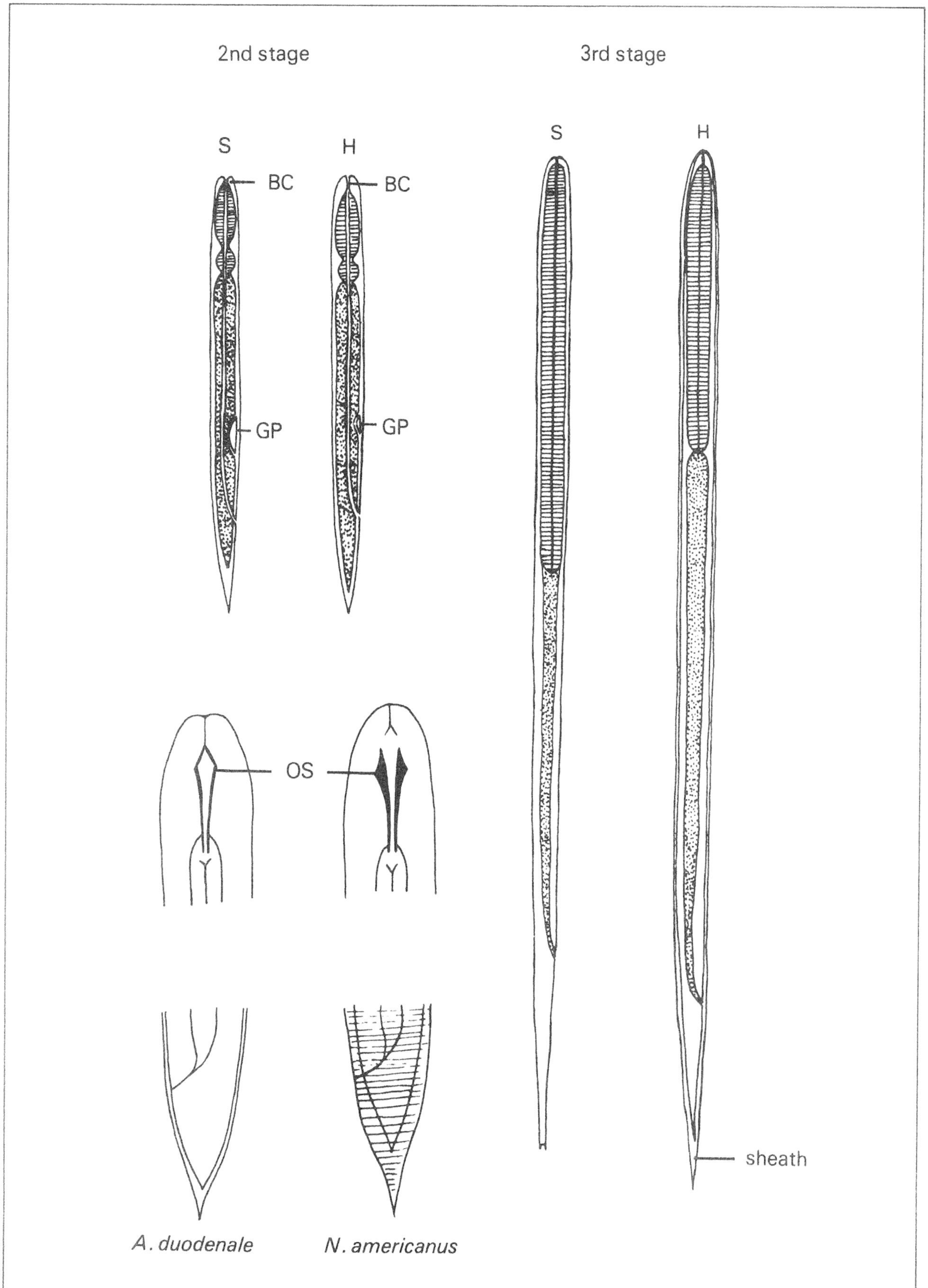
2nd stage
3rd stage
S
H
S
H
BC
BC
GP
GP
OS
sheath
A. duodenale
N. americanus

295b

**296, 297 Morphological difference between adult N. americanus and A. duodenale** (Courtesy of Dr Yukio Yoshida).

| N. americanus (296) | *A. duodenale* (297) | |
|---|---|---|
| **Male** | **Male** | |
| a) A pair of ventral cutting plates instead of teeth<br>b) In lateral view a pair of subventral, sublateral and subdorsal lancets | Buccal capsule: | a) Two pairs of ventral teeth<br>b) Accessory toothlet at the inner margin of the inner ventral teeth<br>c) Two triangular subventral lancets |
| a) Dorsal ray deeply clefted<br>b) Tip of each branch bidigitate | Bursa: | a) Dorsal ray not deeply clefted<br>b) Tip of each branch tridigitate |
| a) Tips united to form a hook | Spicule: | a) Tips not united |
| **Female** | | **Female** |
| a) Mucron absent | | a) An elongated process (mucron) at the caudal end |

*Necator americanus*
VCP = ventral cutting plate
SVL = subventral lancet
SLL = sublateral lancet
SDL = subdorsal lancet
D = dorsal ray
ED = externo dorsal ray
PL = postero lateral ray
ML = medio lateral ray
EL = externo lateral ray
V = ventral ray

*Ancylostoma duodenale*
AT = accessory toothlet
VT = ventral tooth
SVL = subventral lancet
D = dorsal ray
ED = externo dorsal ray
PL = postero lateral ray
ML = medio lateral ray
EL = externo lateral ray
V = ventral ray

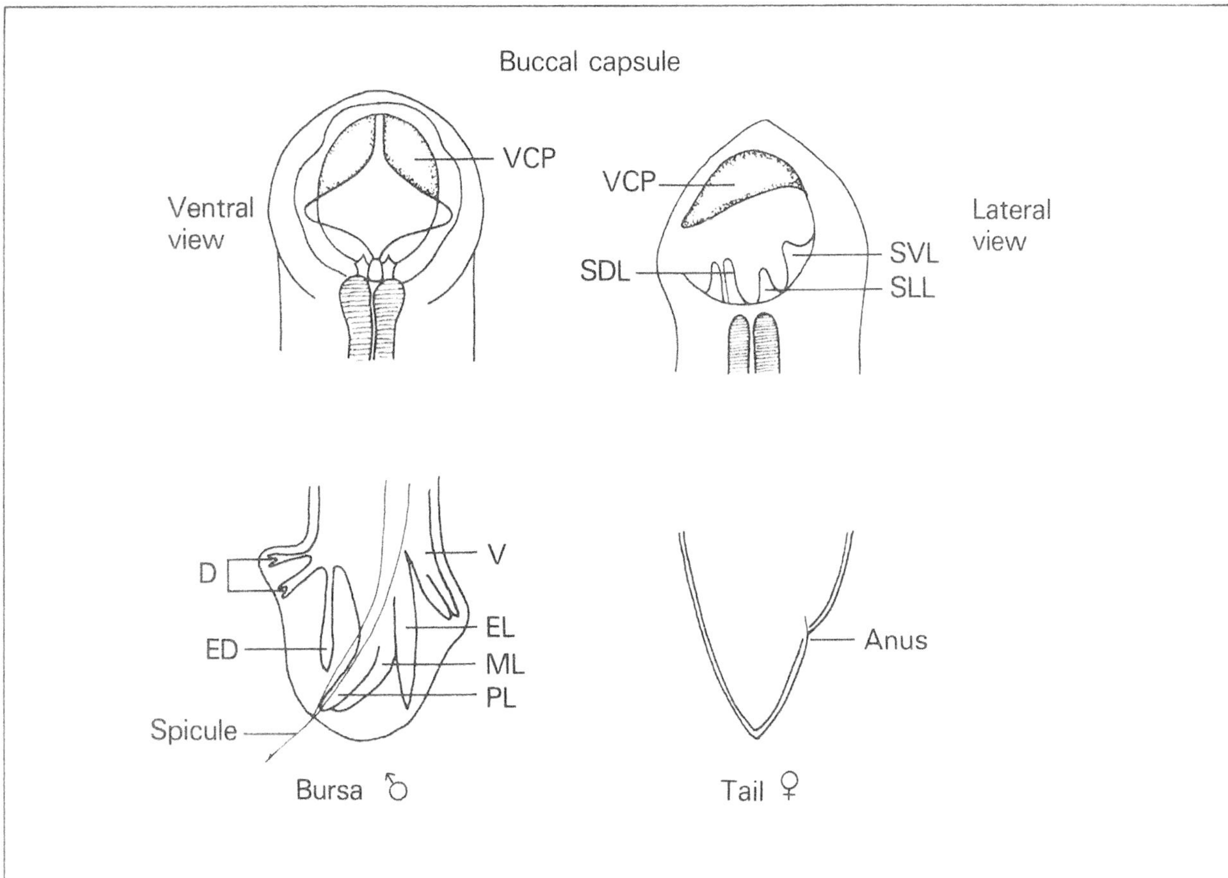
Buccal capsule
VCP
Ventral view
VCP
Lateral view
SDL
SVL
SLL
D
V
ED
EL
ML
PL
Spicule
Anus
Bursa ♂
Tail ♀

296

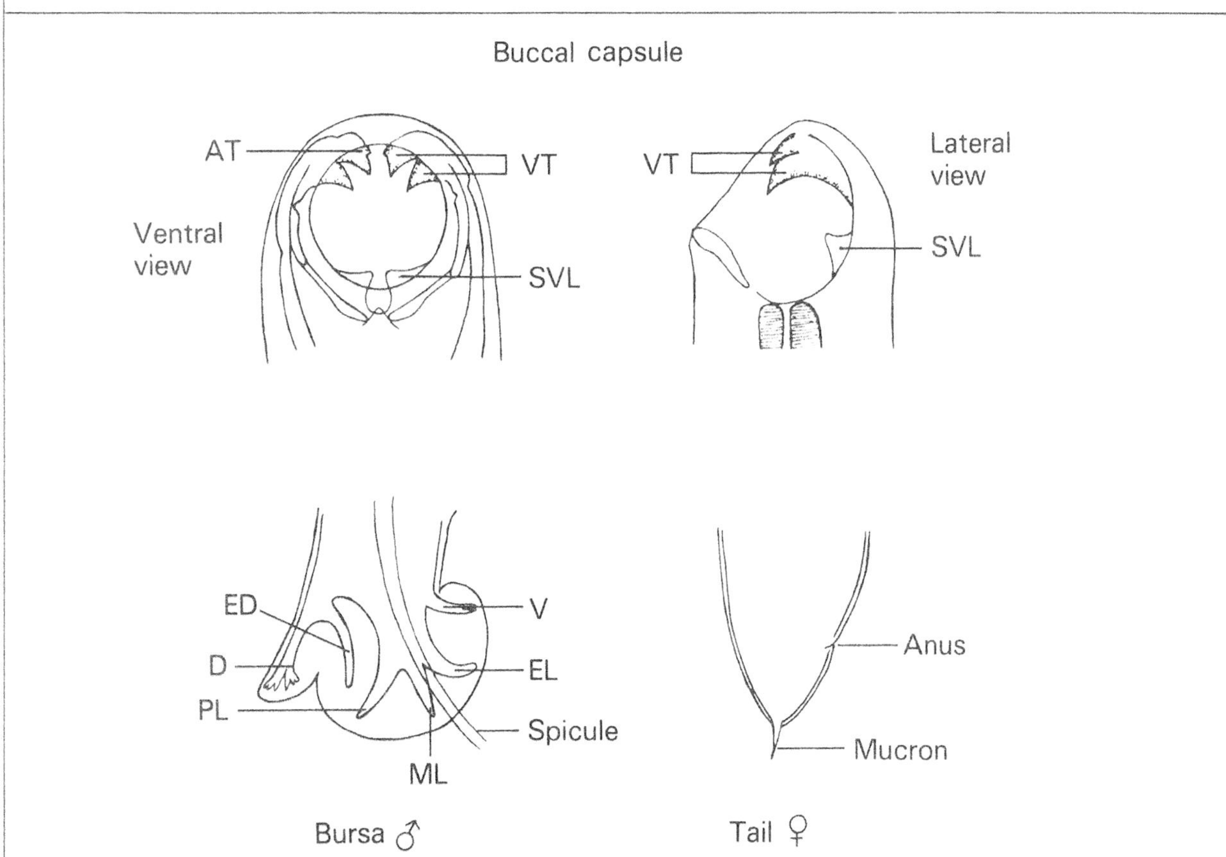
Buccal capsule
AT
VT
VT
Lateral view
Ventral view
SVL
SVL
ED
V
D
EL
PL
Spicule
ML
Anus
Mucron
Bursa ♂
Tail ♀

297

**298 Cross-section of Ascaris to show the cuticle hypodermis and the somatic musculature.** Note the muscle cells are polymyarian-coelomyarian type. X100. Enlarged by 23.4.

**299 Cross-section of Ascaris to show the uterus.** Columnar epithelial cells line the uterine wall. The eggs almost fill the lumen of the uterus. X100. Enlarged by 5.4.

**300 Cross-section of Ascaris to show the ovaries.** The ovaries have an epithelial wall and radially arranged oogenia. These remain attached until they mature. After maturation they pass down the oviduct to the uterus. Lying above the ovaries is the intestinal tract. X100. Enlarged by 5.4.

**301 Cross-section of Ascaris to show** the tri-radiate lumen of the intestinal tract. The tract is lined by long columnar epithelium, which has an inner border of microvilli (bacillary layer). X100. Enlarged by 23.4.

**302 Fertilised eggs of Ascaris dissected from the uterus.** As these eggs have not been exposed to intestinal contents no faecal debris is seen and the outer mammillated layer is completely intact. Interference contrast. X400. Enlarged by 5.4.

**303 Unfertilised eggs of Ascaris dissected from the uterus.** Note the difference in shape and the disorganised contents. Interference contrast. X400. Enlarged by 5.4.

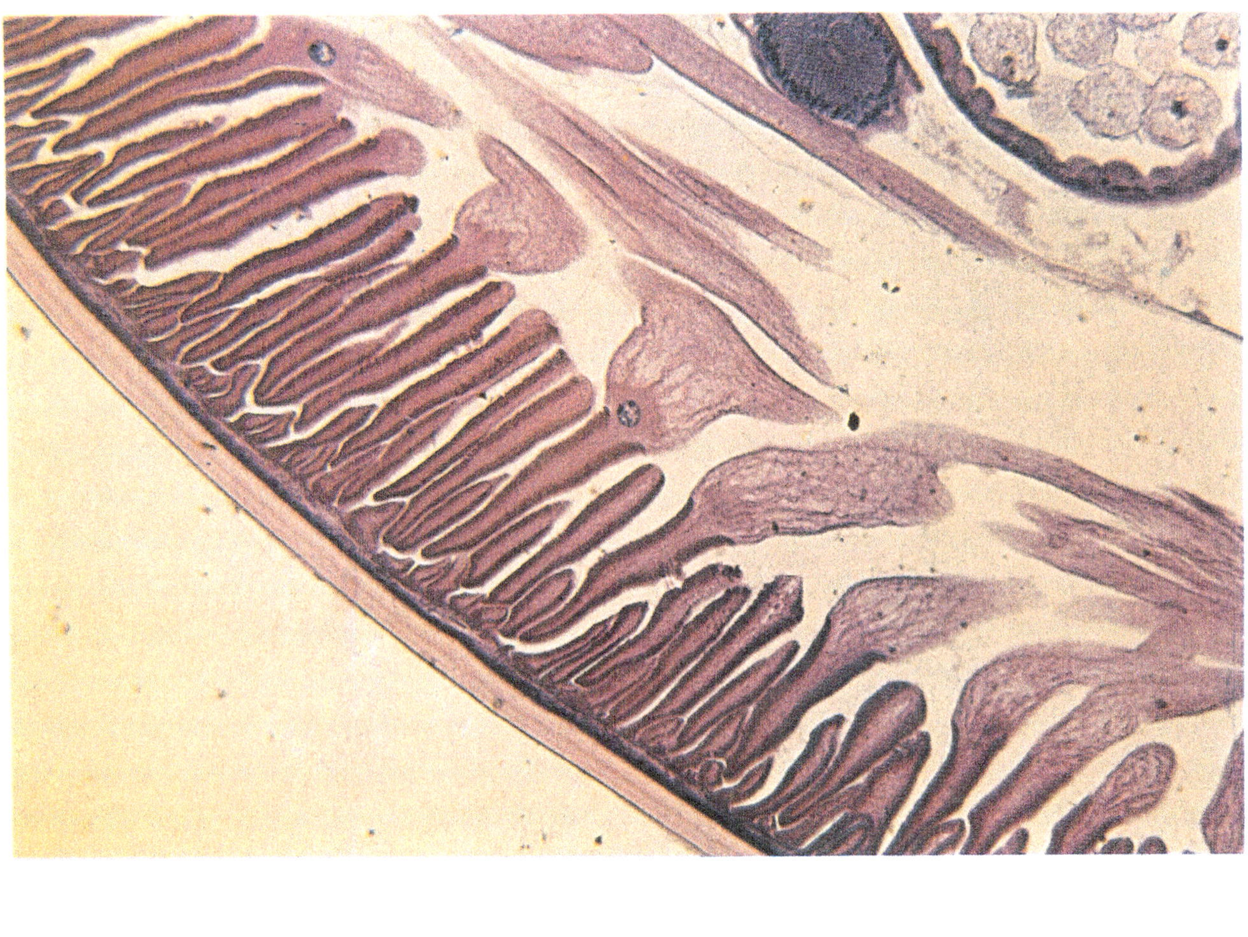

298

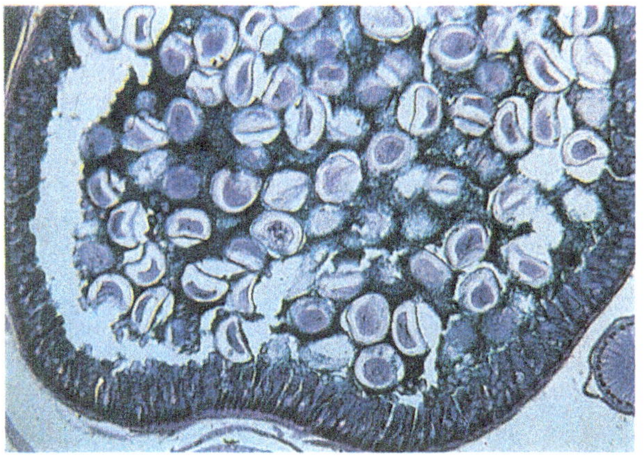
299

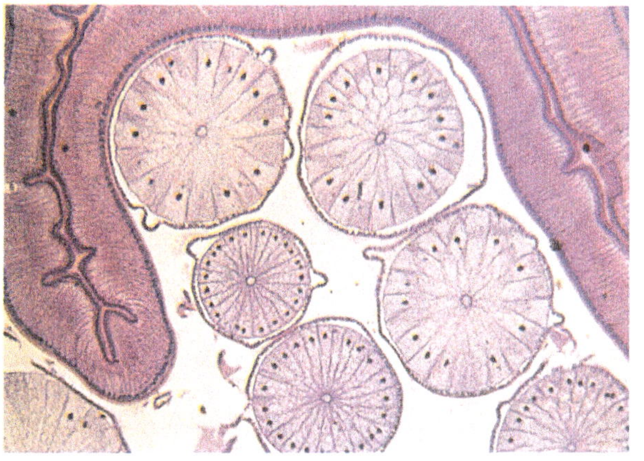
300

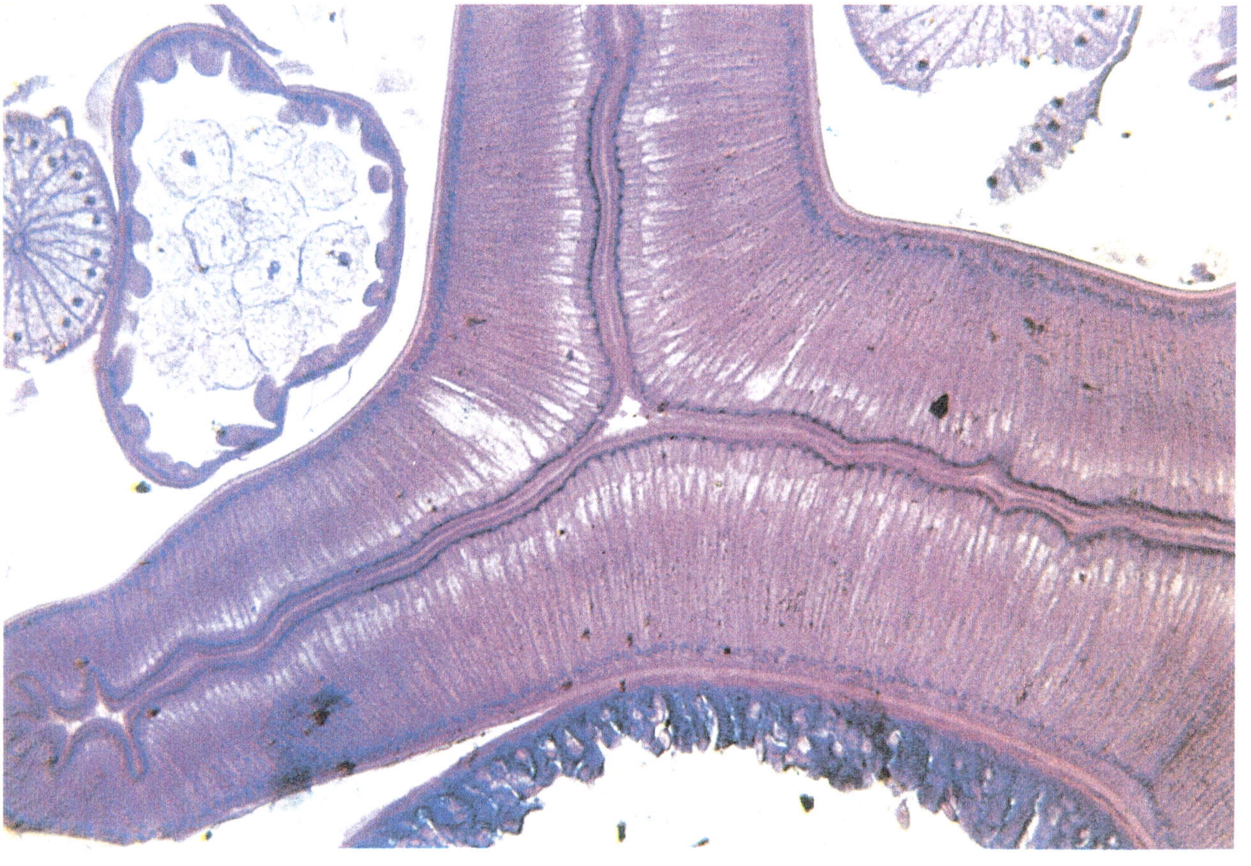
301

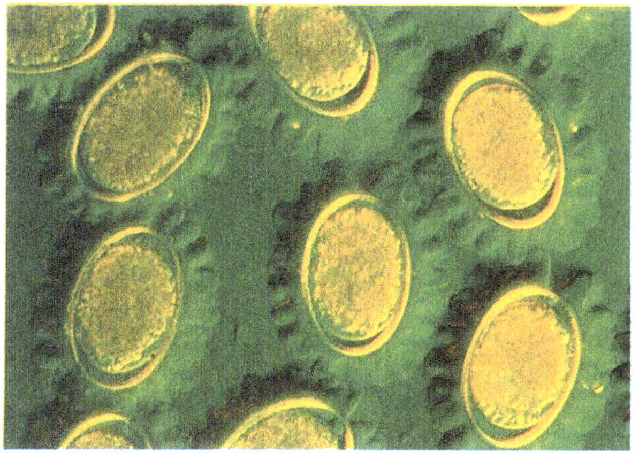
302

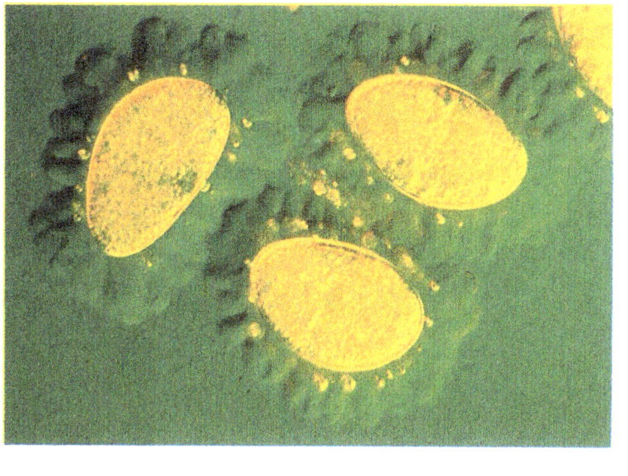
303

**304 Freshly passed Ascaris eggs from faeces.** The eggs may appear from light to dark brown in colour. Interference contrast. X400. Enlarged by 5.4.

**305 Ascaris eggs showing cleavage.** This process normally occurs in the soil and leads to the formation of the larva. Interference contrast. X400. Enlarged by 5.4.

**306 Embryonated eggs of Ascaris.** The mammillated layer has been removed by chemical treatment to allow better visualisation of the contents. A larva is escaping from the egg in the centre. This larva has moulted as its sheath is left behind in the egg. Interference contrast. X400. Enlarged by 5.4.

**307 3rd stage larvae of Ascaris in the lungs.** In the centre of each larva is the patent gut. On the sides of the gut are two circular structures which are the excretory columns. The cuticle shows lateral alae and there is a marked cellular reaction around the larvae. H and E. X400. Enlarged by 9.6.

**308 Ascaris eggs in liver.** This is postmortem tissue from a case of hepatic ascariasis. The eggs are in a single celled stage and the mammillated coat around them has disappared. H and E. X100. Enlarged by 9.6.

**309 Tissues from the same case.** PAS. X400. Enlarged by 9.6.

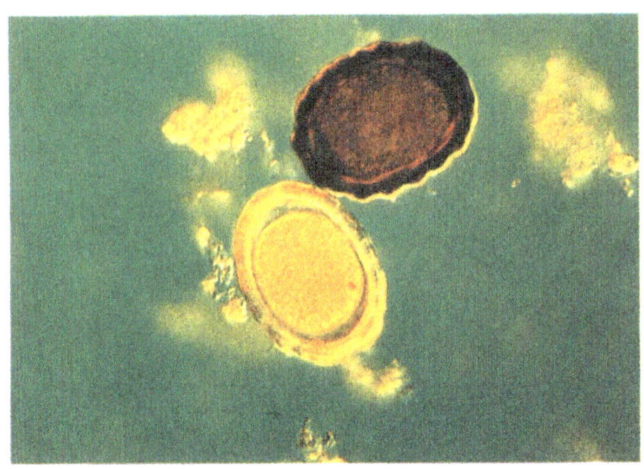

304

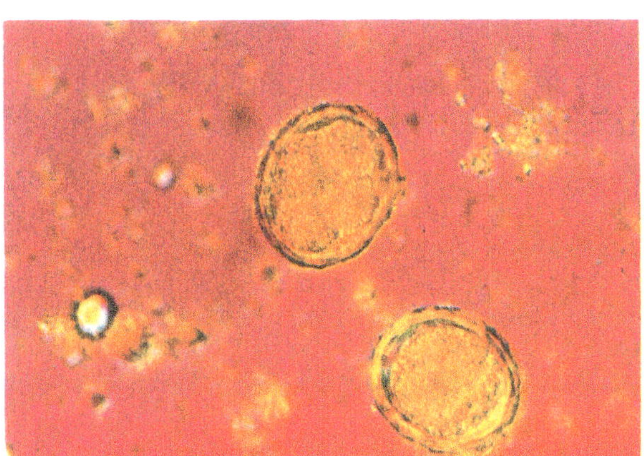

305

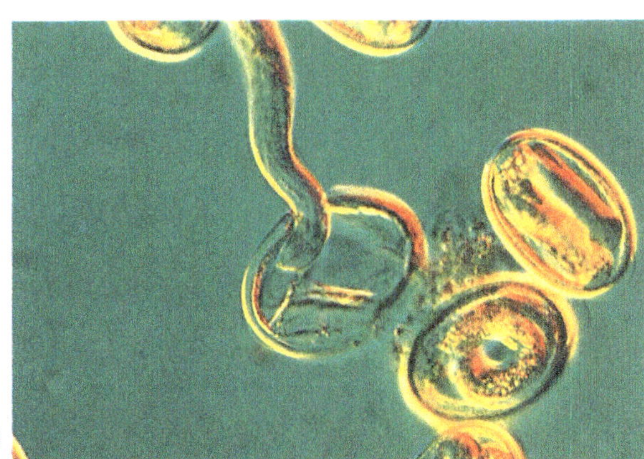

306

307
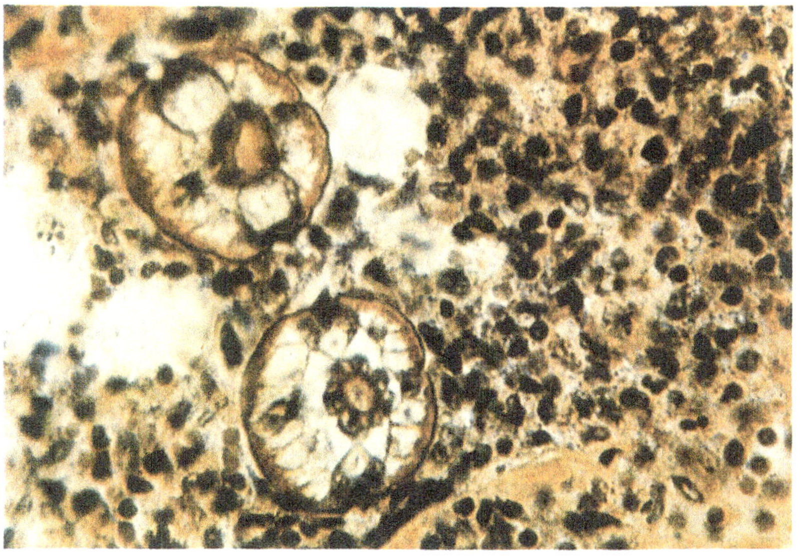

308
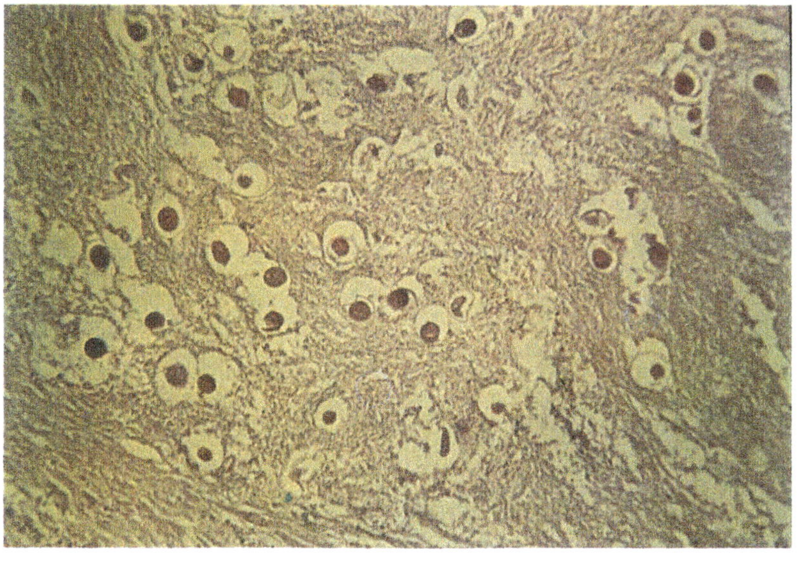

309
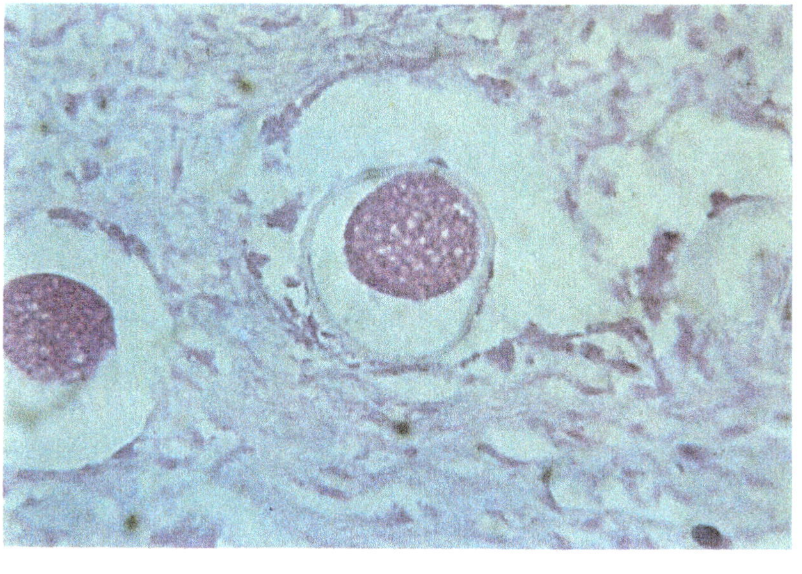

**310 Toxocara canis egg with a newly hatched larva.** This larva has moulted (2nd stage), as its sheath is lying next to it. Note the egg's surface is pitted, which is a characteristic of the genus *Toxocara.* Interference contrast. X400. Enlarged by 23.4.

**311 Toxocara canis larva in liver** from a human case of visceral larva migrans. Note the smaller diameter as compared to *Ascaris,* large excretory columns and non patent inconspicuous gut. There is marked cellular reaction around the larvae. H and E. X400. Enlarged by 5.4. (Courtesy of Dr Paul C. Beaver).

**312 Toxocara canis granuloma in monkey liver.** In chronic infection the larva becomes surrounded by a layer of fibrous tissue and forms a non-inflammatory stable lesion. H and E. X400. Enlarged by 5.4. (Courtesy of Dr Paul C. Beaver).

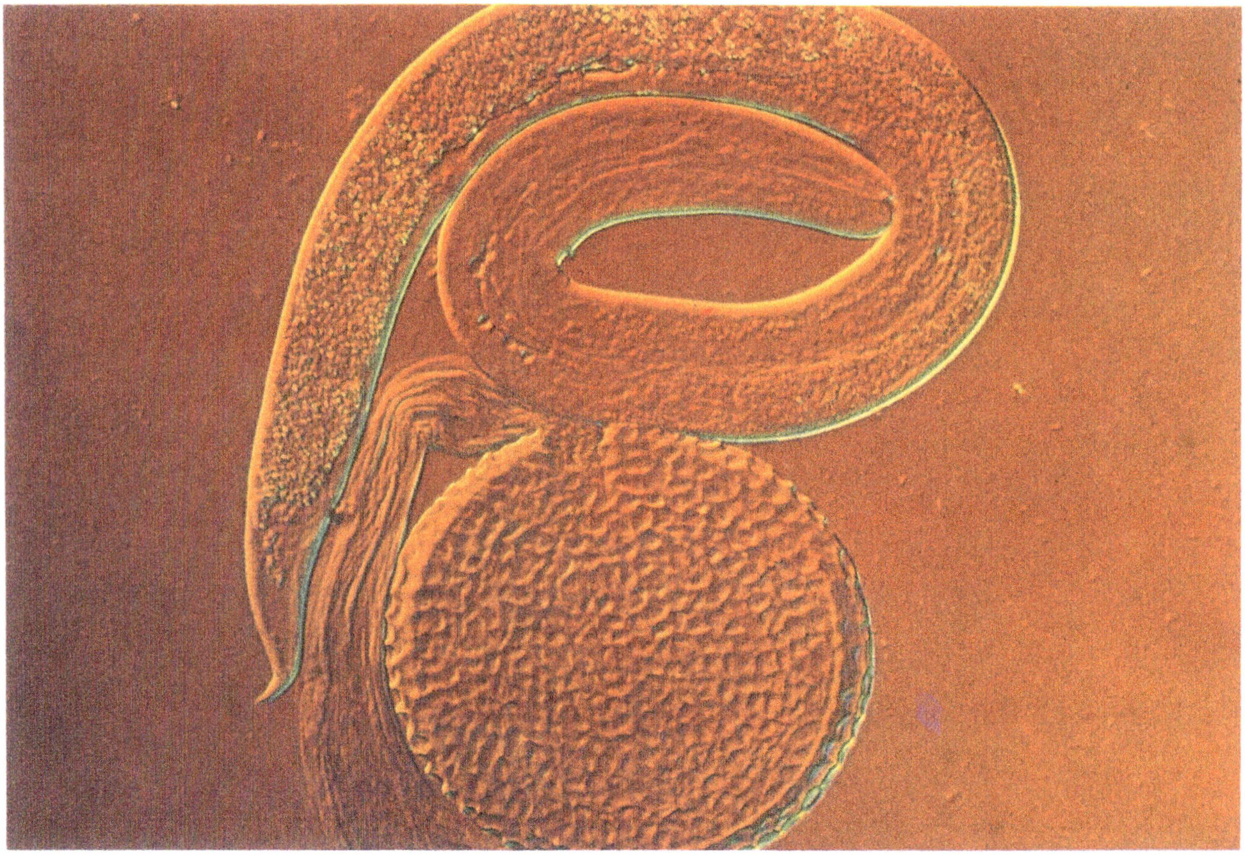

310

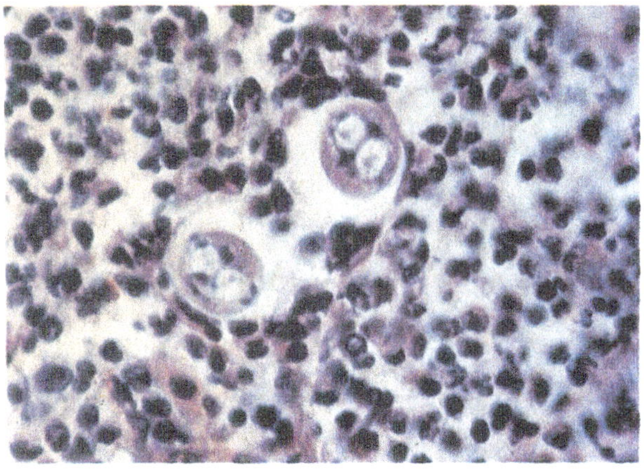

311

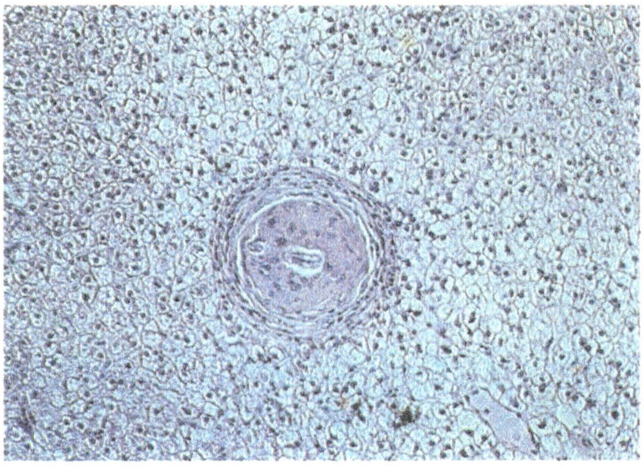

312

**313 Enterobius vermicularis adult.** The cephalic alae are clearly seen at the anterior end. The cuticle and the alae are transversely striated. Corpus, isthmus and the oesophageal bulb are also visible. Interference contrast. X50. Enlarged by 23.4.

**314 Enterobius in the lumen of appendix.** Section has cut the worm at various levels. In some only the intestine is seen in the centre, and in others both the intestine and gonads can be seen. The parasite shows characteristic lateral alae and platymyarian type muscles. H and E. X100. Enlarged by 5.4.

**315 Anal smear showing large numbers of Enterobius eggs.** In the background are also two *Ascaris* eggs. X100. Enlarged by 5.4.

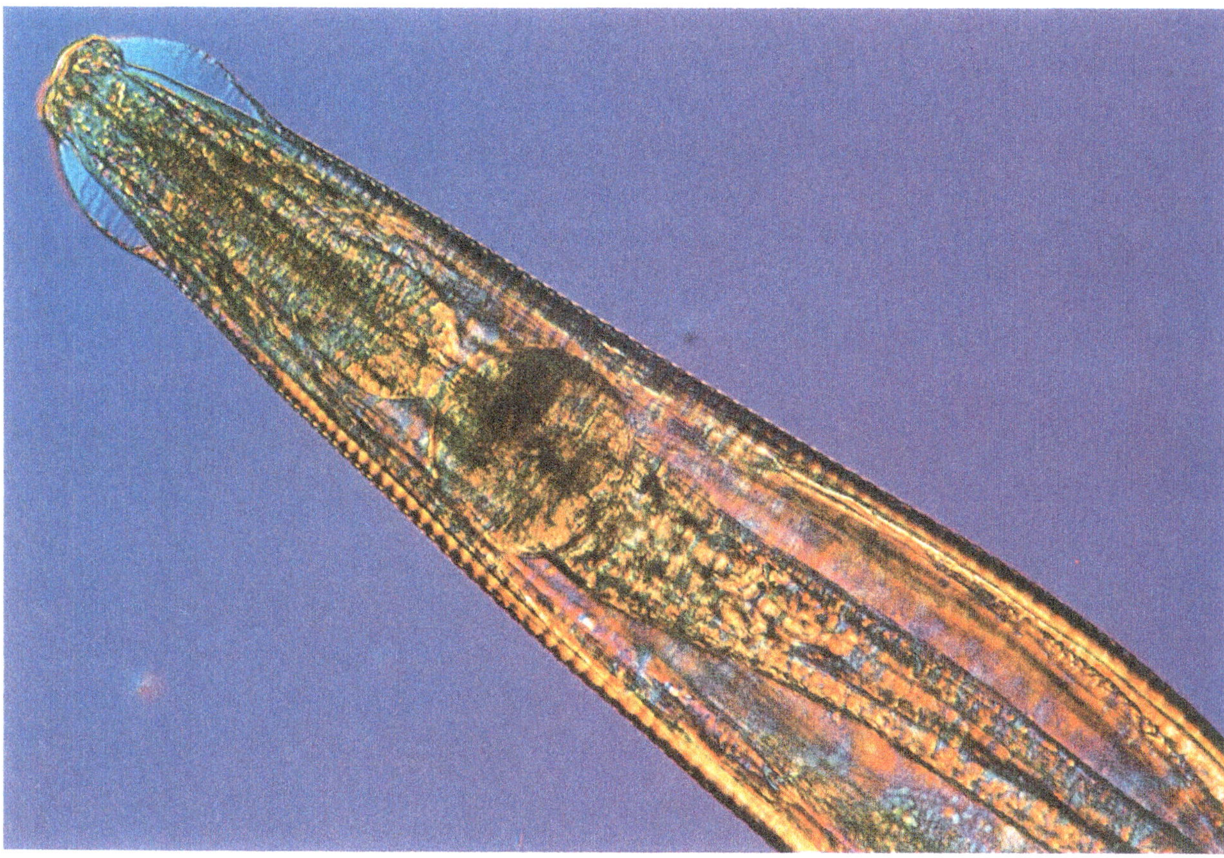

313

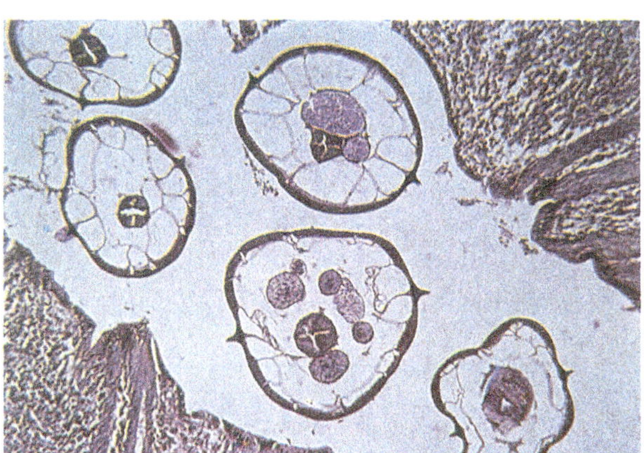

314

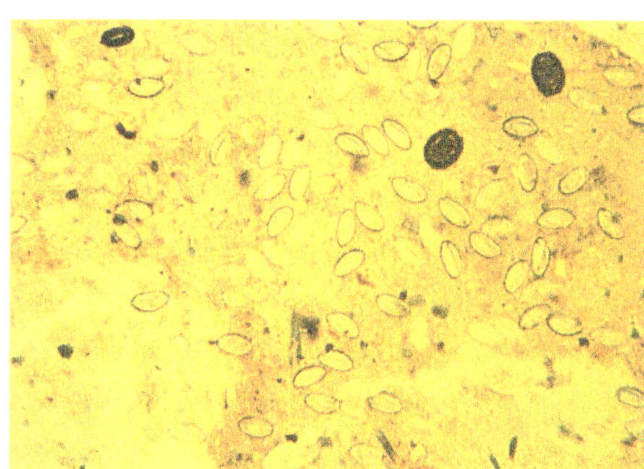

315

**316 Trichuris adult showing the stichosome** in the antérior (thinner) portion of the parasite. The stichosome is made up of a series of large cells (stichocytes) arranged transversely. Interference contrast. X50. Enlarged by 5.4.

**317 Trichuris adult male showing the curved tail** with a long spicule, lying close to the ventral surface. Interference contrast. X50. Enlarged by 5.4.

**318 Trichuris adult male showing the spicule protruded.** In this genus the spicule is covered by a protrusible sheath. Interference contrast. X50. Enlarged by 5.4.

**319 Trichuris lying in the superficial layers of the intestinal epithelium.** The parasite forms a small tunnel in which the anterior end is embedded. H and E. X100. Enlarged by 5.4.

**320 Trichuris sectioned longitudinally in the intestinal epithelium.** The stichosome is clearly seen. The darkly staining rounded body is the nucleus of a stichocyte. H and E. X400. Enlarged by 5.4.

**321 Freshly passed Trichuris trichiura eggs in faeces.** As in the case of *Ascaris* the eggs may show a varying degree of pigmentation. Interference contrast. X400. Enlarged by 5.4.

**322 An embryonated egg of Trichuris trichiura.** Interference contrast. X400. Enlarged by 5.4.

**323 Trichuris attached to the mucous membrane of the large intestine.** A few haemorrhagic areas can be seen near the worm. Photographed through a proctoscope.

**324 Prolapsed rectum due to heavy trichuriasis.** Worms can be seen attached to the mucous membrane. (Courtesy of Dr S.L. Lee and Amer. J. Trop. Med.).

316

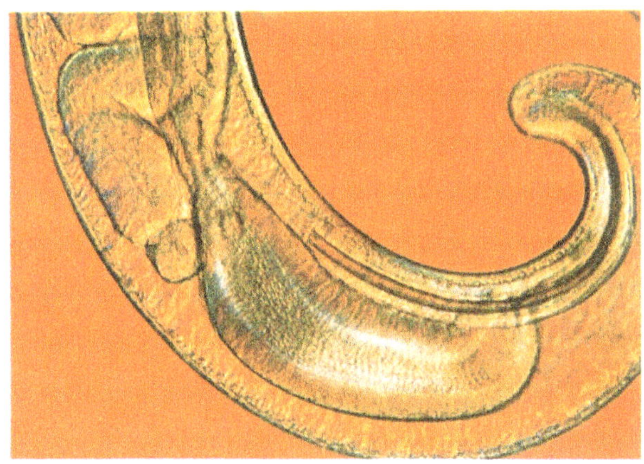
317

318

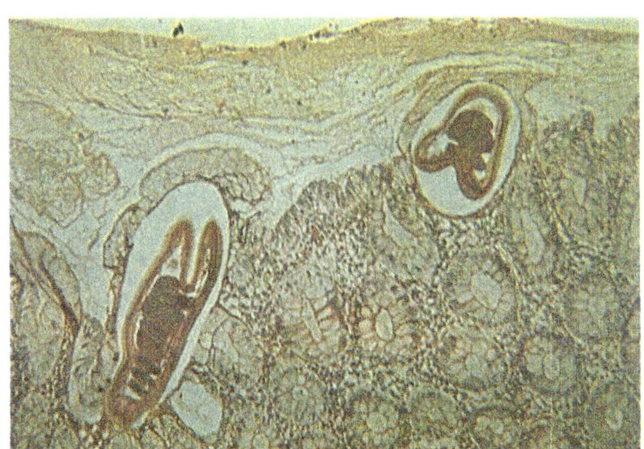
319

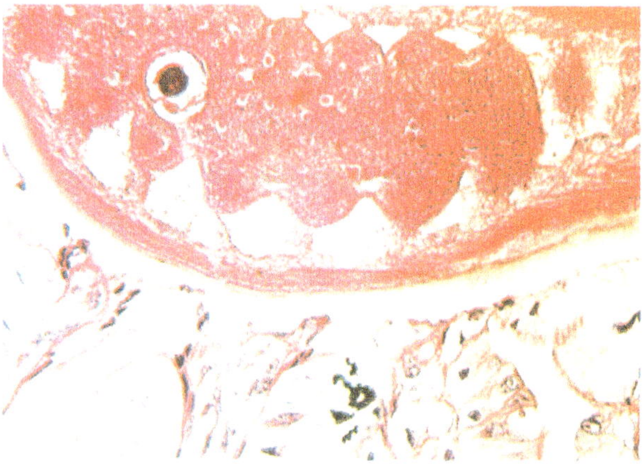

320

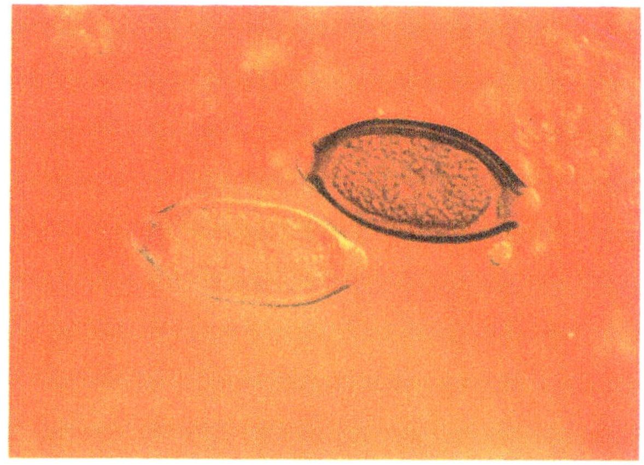

321

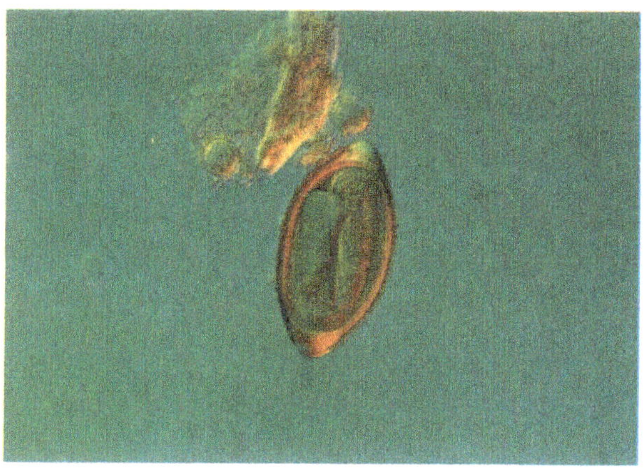

322

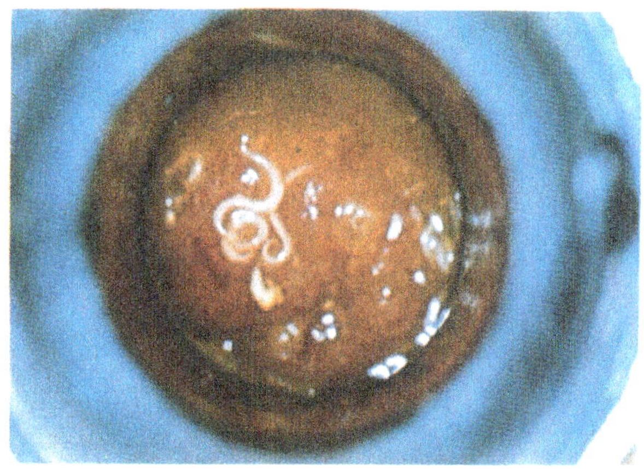

323

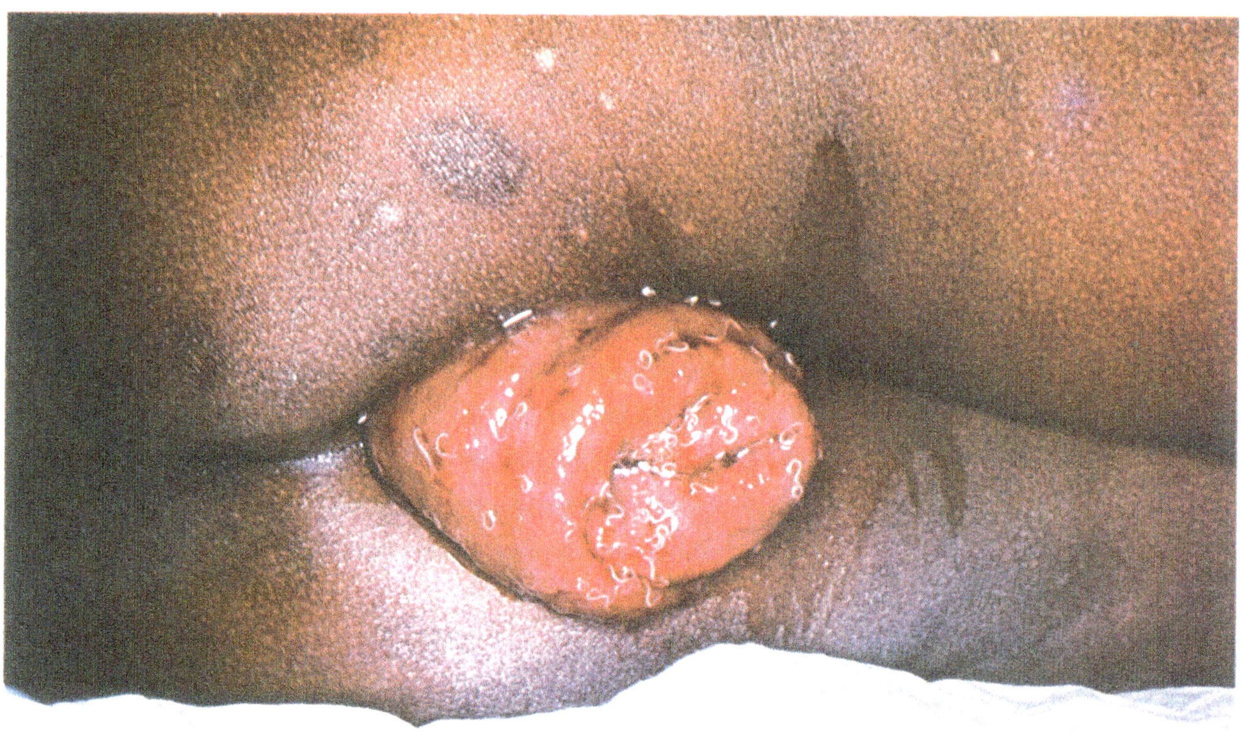

324

**325 Egg of Capillaria philippinensis.** The egg is broader than *Trichuris* and the lateral plugs are not as distinct. Interference contrast. X400. Enlarged by 5.4.

**326 Egg of Capillaria hepatica.** The egg shell is distinctly striated. Interference contrast. X400. Enlarged by 5.4.

**327 Eggs of Capillaria hepatica in the liver.** In this location they are generally grouped in clumps and are surrounded by fibrous tissue. H and E. X400. Enlarged by 5.4.

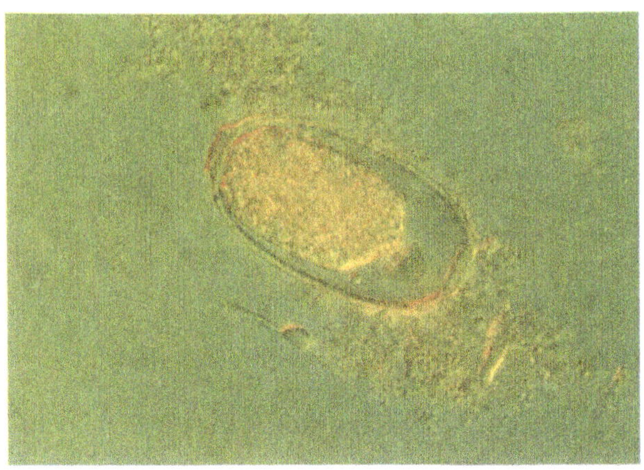
325

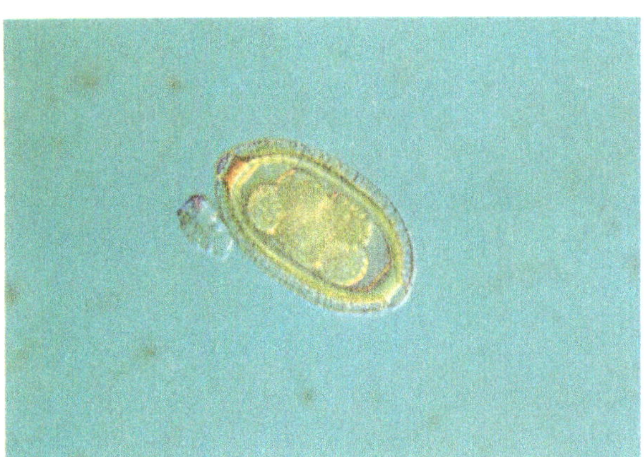
326

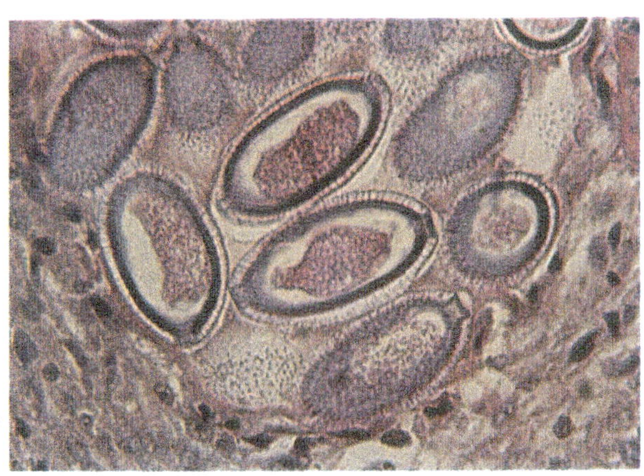
327

**328 Trichinella spiralis larvae** in muscle of infected rat. H and E. X400. Enlarged by 5.4.

**329 Trichinella spiralis larva in a squash preparation** of the infected muscle. The globular structures at the terminal end are remnants of degenerated muscle. Interference contrast. X400. Enlarged by 23.4.

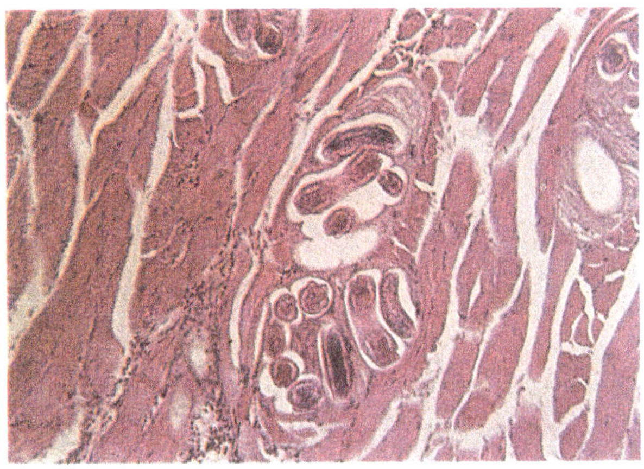

328

329

**330 Hookworm buccal capsule.** In this case *Ancylostoma caninum* showing 3 pairs of teeth. Interference contrast. X100. Enlarged by 23.4.

**331 A hookworm attached to the mucous membrane.** In this case *A. caninum.* Note the large haemorrhagic area around the site of attachment.

**332 Section of a hookworm attached to the mucous membrane.** The section has gone through the buccal capsule and shows the host tissues which are sucked in by the muscular oesophagus. H and E. X100. Enlarged by 5.4.

**333 Hookworm egg.** Morphologically it is not possible to differentiate between *A. duodenale* and *N. americanus.* Interference contrast. X400. Enlarged by 5.4.

**334 Hookworm egg containing an infective larva.** Interference contrast.

**335 Infective larva of** *Necator americanus* showing distinct striations on the sheath. Anoptral (negative) phase. X100. Enlarged by 5.4.

330

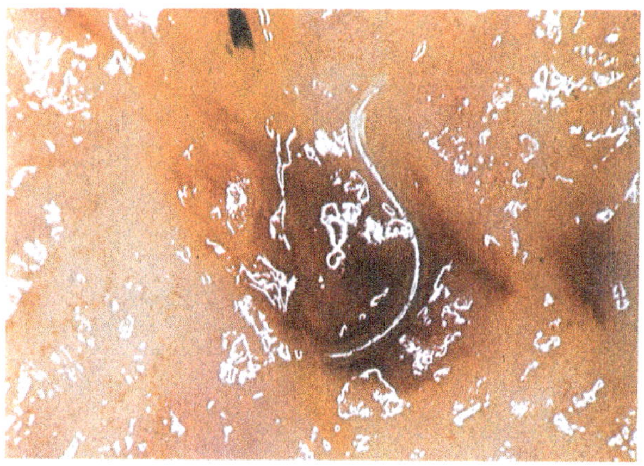
331

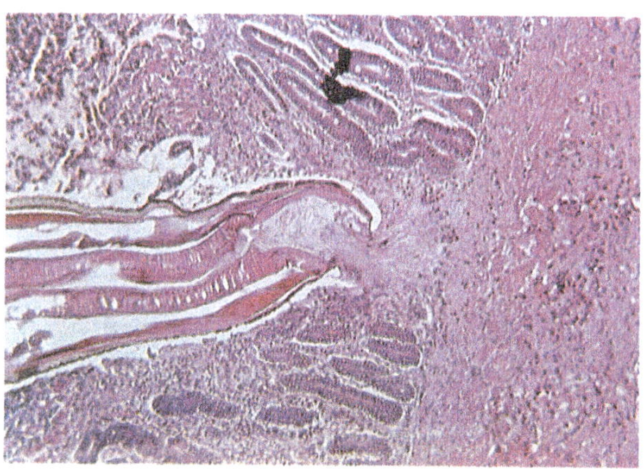
332

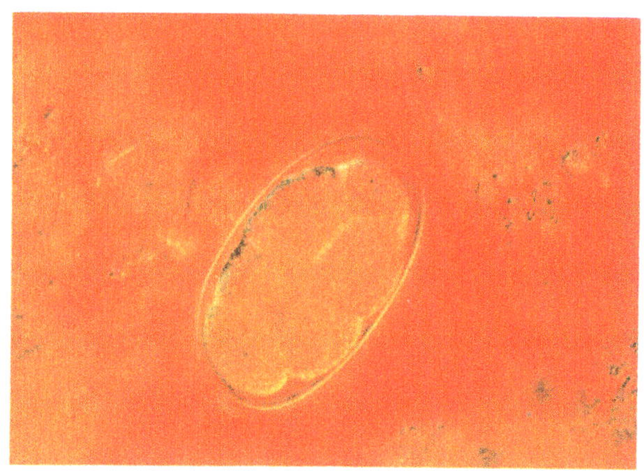
333

334

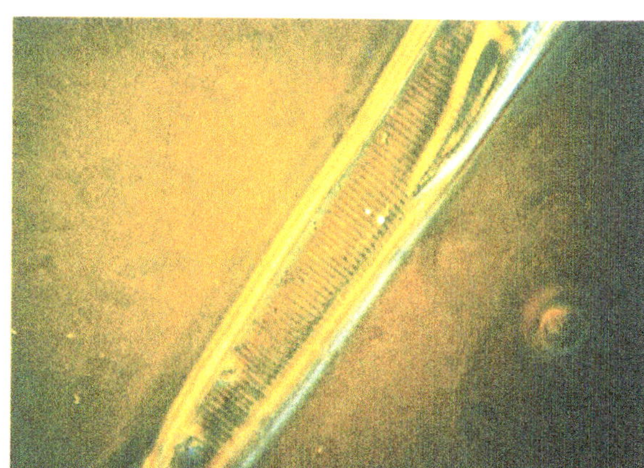
335

**336 Freshly passed larva of Strongyloides stercoralis** showing the rhabditiform oesophagus. Interference contrast. X100. Enlarged by 9.6.

**337 Rhabditiform larva of Strongyloides** showing the large genital primordium which appears as a disc like structure in between the cuticle and the internal organs. Interference contrast. X400. Enlarged by 9.6.

**338 Tail of the filariform larva of Strongyloides** showing the notched end. Interference contrast. X400. Enlarged by 9.6.

**339 Egg of Trichostrongylus sp.** The ends are more pointed than that of a hookworm and the ovum is in an advanced cleavage stage. Interference contrast. X400. Enlarged by 9.6.

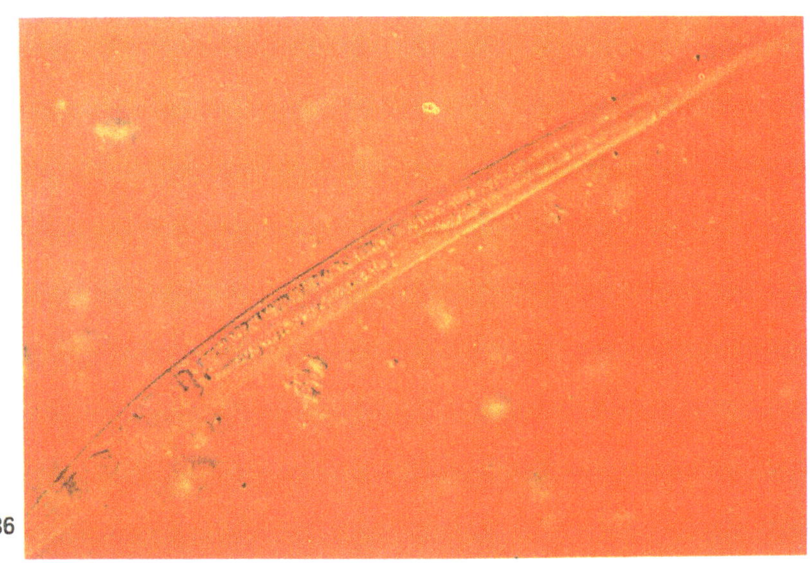

336

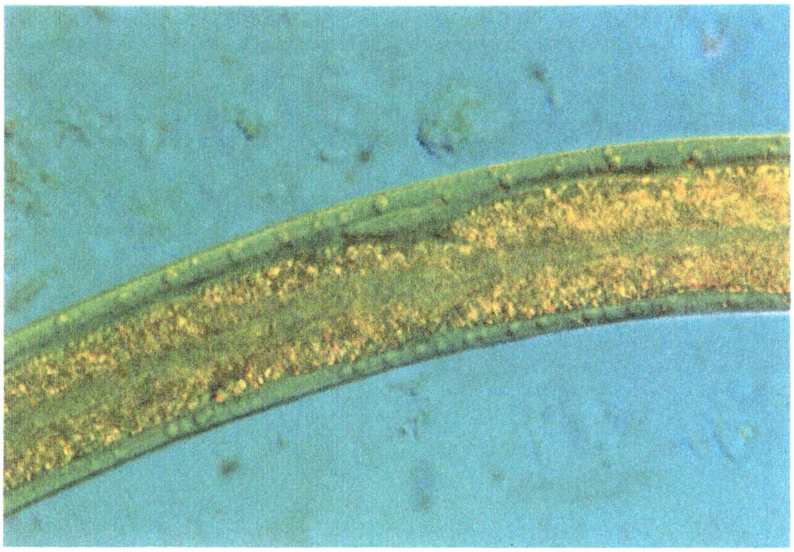

337

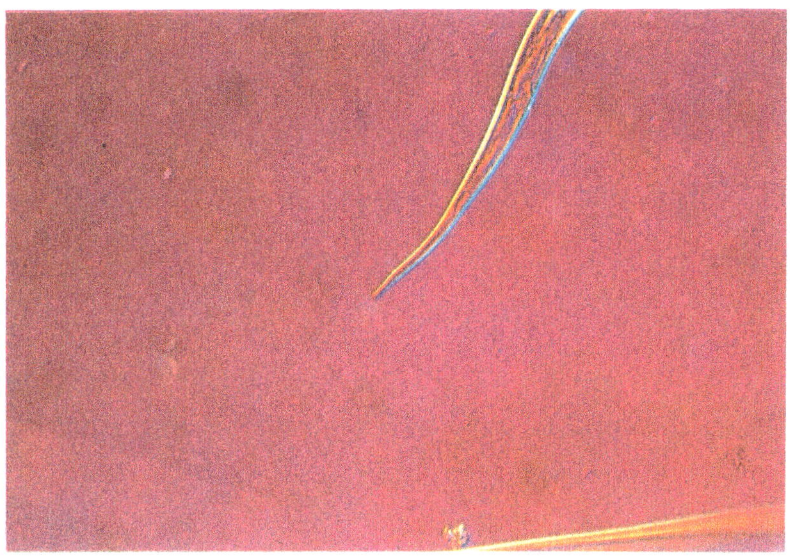

338

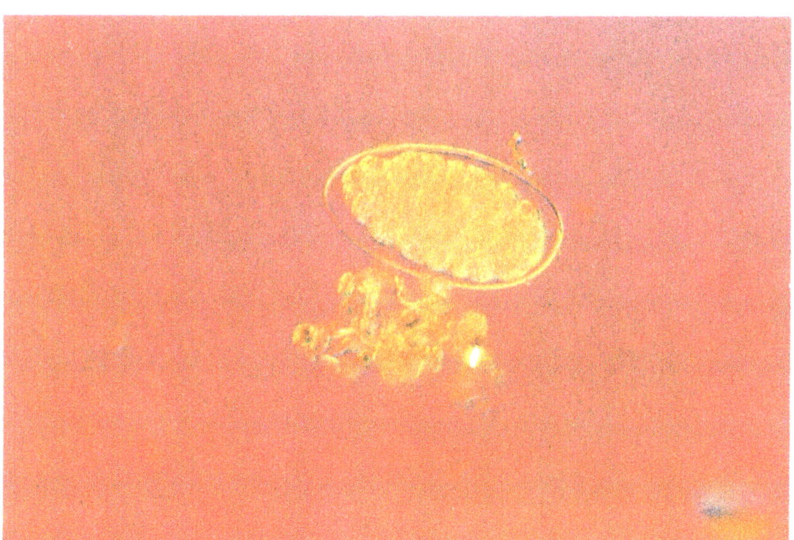

339

**340 Cutaneous larva migrans**. This patient had a large number of thin serpiginous tracts on his legs. Species diagnosis of causative agent is not possible in this condition by clinical picture alone. (Courtesy of Dr V.S. Rajan).

**341 Cutaneous larva migrans on hand**. A single long tract which was partly crusted due to scratching of the skin. (Courtesy of Dr V.S. Rajan).

**342 Cutaneous larva migrans on the buttock of a child**. The lesion was thick and the skin was partly broken in places. Secondary bacterial infection is common in this location. (Courtesy of Dr V.S. Rajan).

**343 Cutaneous larva migrans on the back of the leg**. In this case 2 tracts could be seen. This patient was a dog breeder and circumstantial evidence indicates that the infection was due to dog hookworms.

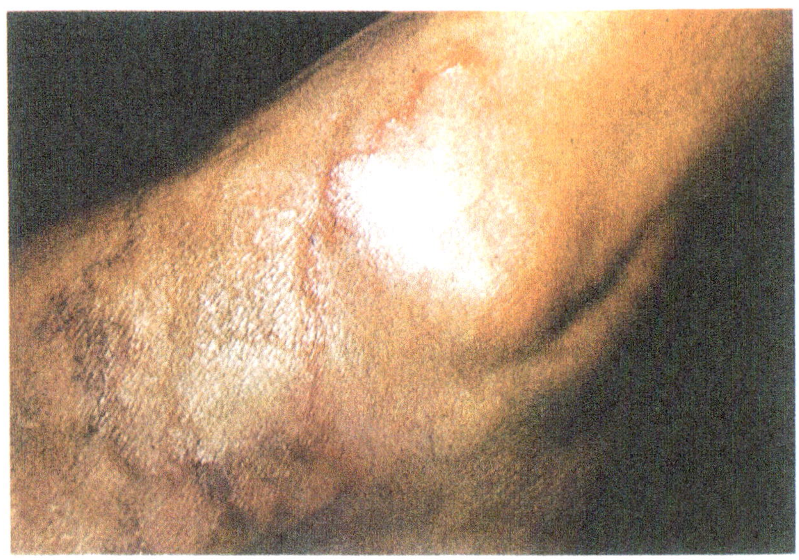

340

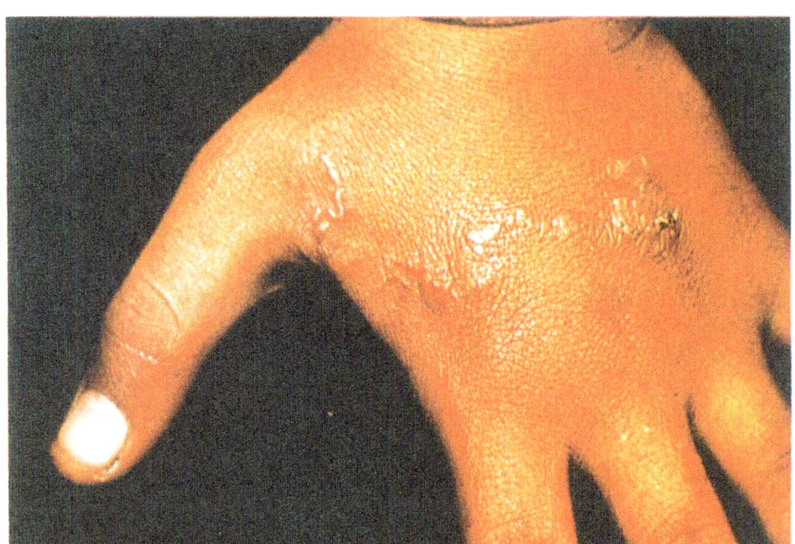

341

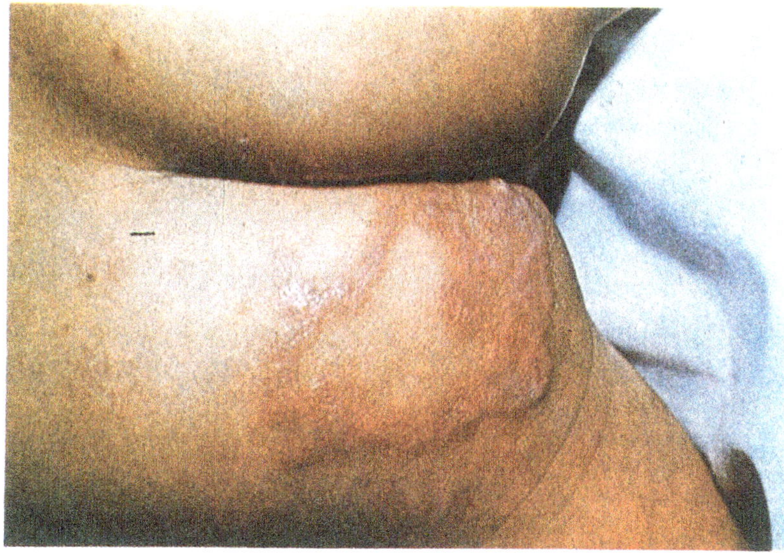
342

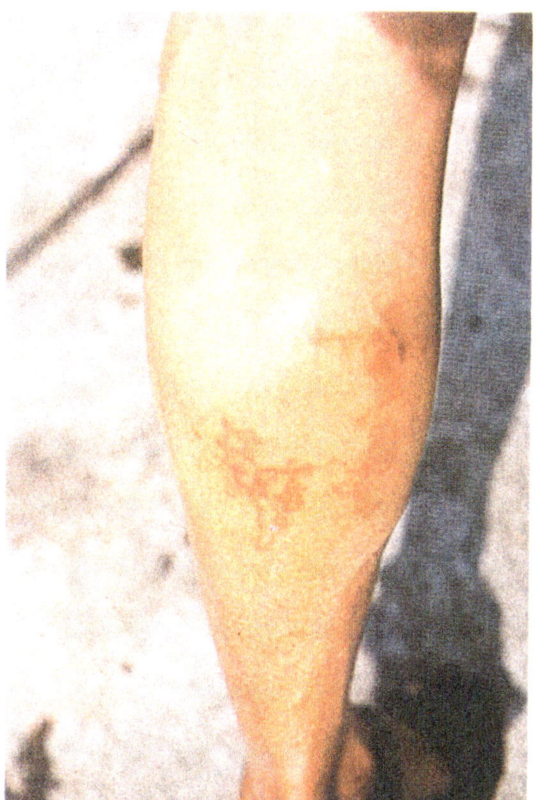
343

**344 Gnathostoma spinigerum.** Adult worms are robust with a length of 11 to 25mm. One or both ends may be curved ventrally.

**345 Gnathostoma 3rd stage larva,** lateral view showing the head bulb which is covered with hooks. X50. Enlarged by 9.6.

**346 Gnathostoma larvae in a Cyclops (freshwater copepod),** which acts as intermediate host of this parasite.

**347 Gnathostoma eggs are ovoidal** and have a mucoid plug at one end. Interference contrast. X100. Enlarged by 9.6.

344

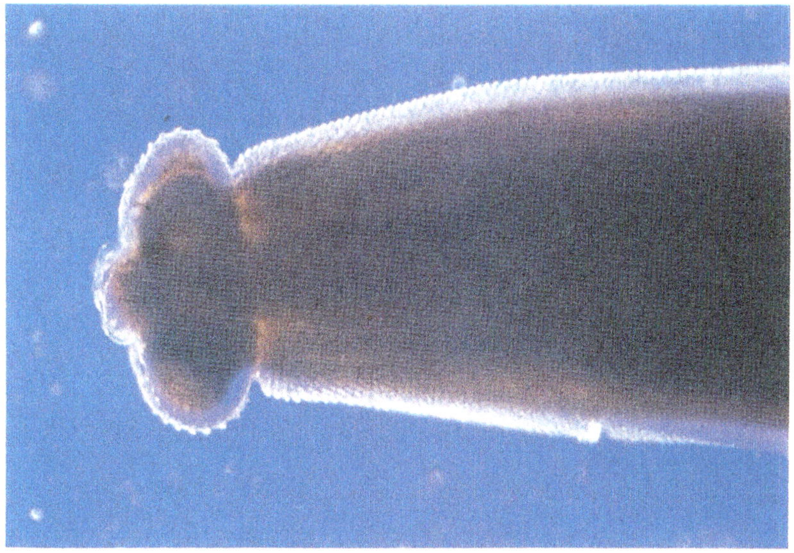

345

346

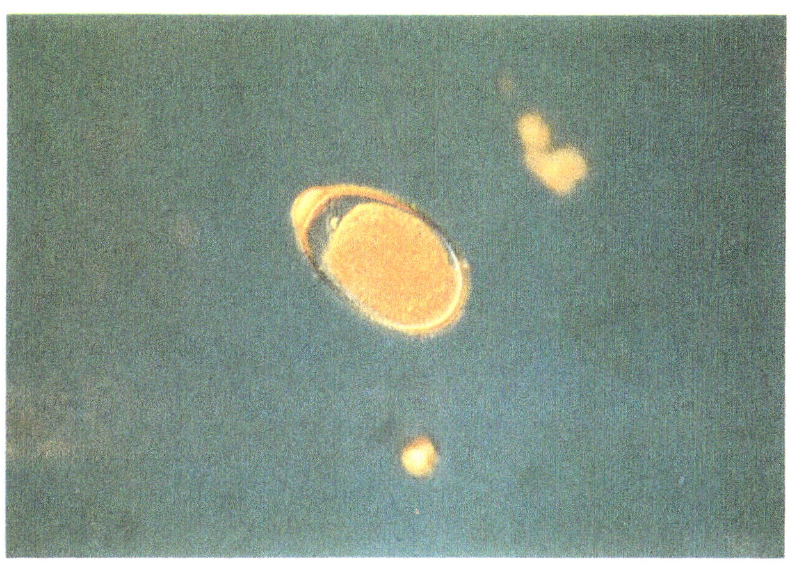

347

**348 Angiostrongylus cantonensis young adults** in the subarachnoid space of the cerebellum in an experimentally infected rat. Note the presence of haemorrhages on the brain surface because of heavy infection.

**349 Angiostrongylus cantonensis** in the subarachnoid space adjacent to cerebellum. H and E. X100. Enlarged by 9.6.

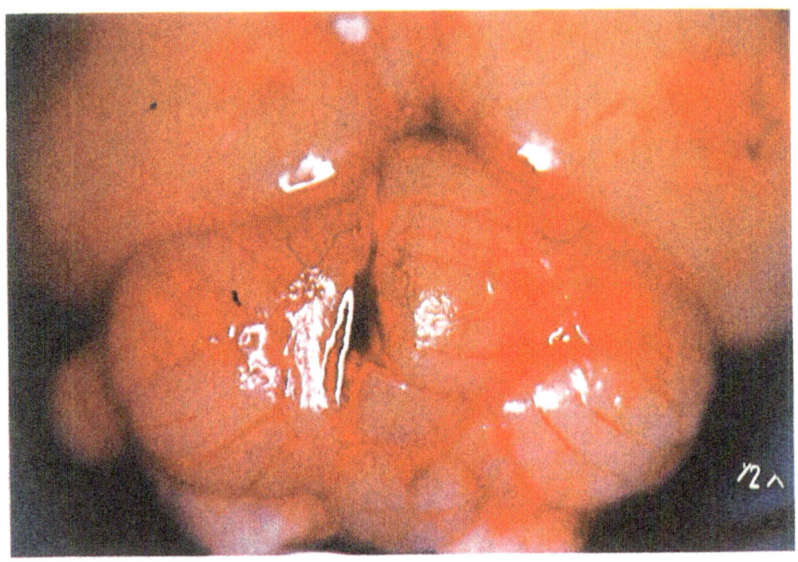

348

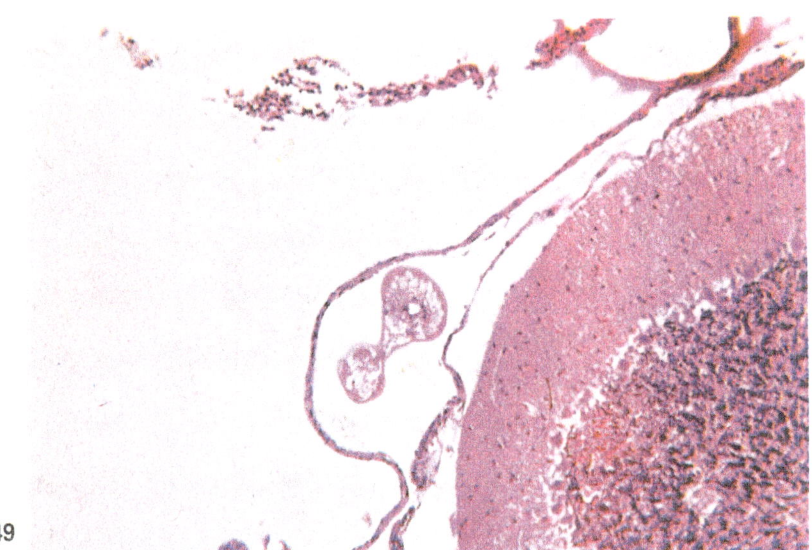

349

**350 Anisakis larva from the ileum of a man from Japan,** acquired by eating raw fish. The larva is characterised by the prominent lateral chords which split at their anterior end. (Courtesy of Dr Paul C. Beaver and Dr I. Miyazaki).

**351 Dracunculus medinensis adult** extracted from a patient by the conventional method of rolling it on a stick. A part of the worm broke during this procedure and is lying in front.

**352 Dracunculus larvae from the fluid discharged** by the adult female. These are generally curved and have a long pointed tail.

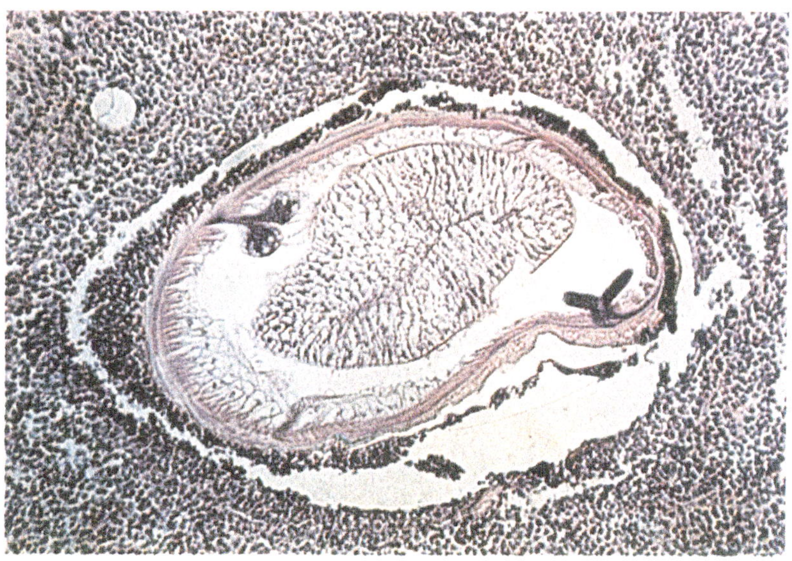
350

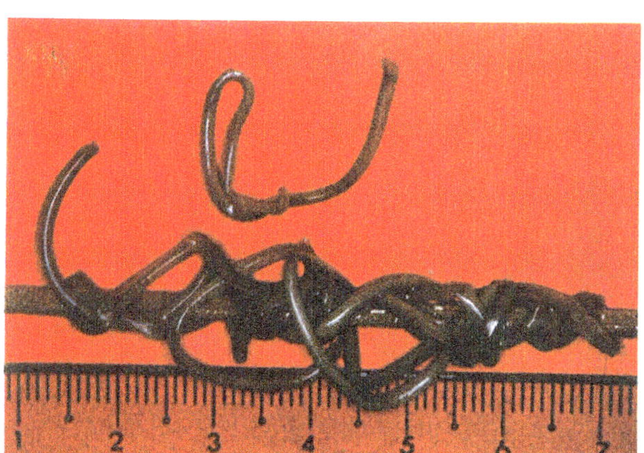

351

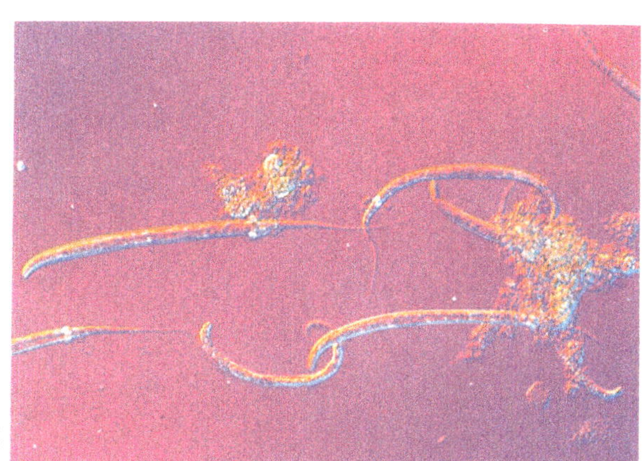
352

**353 Microfilaria of Wuchereria bancrofti** stained with haematoxylin. Sheath is visible at the posterior end. X400. Enlarged by 9.6.

**354 Microfilaria of Wuchereria bancrofti** stained with Giemsa. Sheath is not stained, but appears as an empty space. X400. Enlarged by 9.6.

353
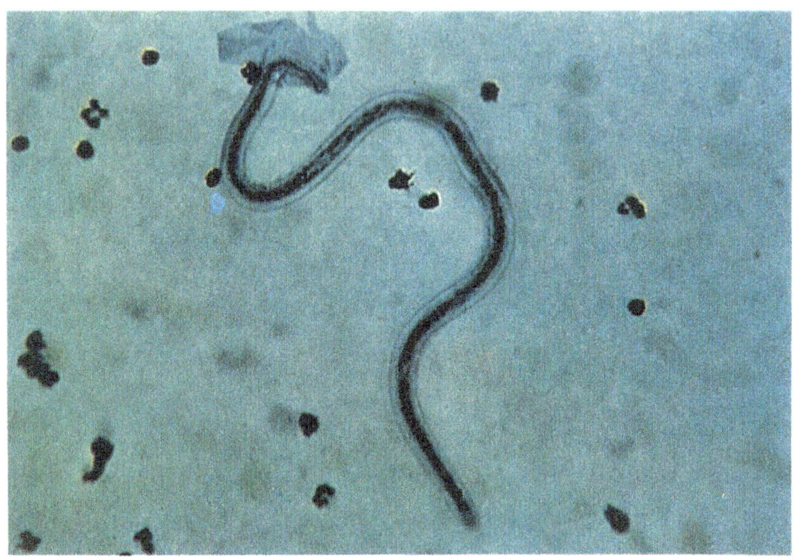

354
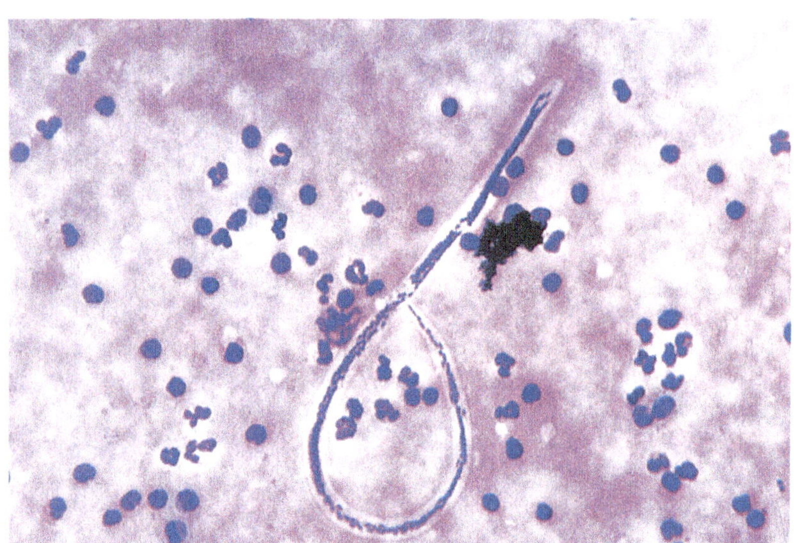

**355 Microfilaria of Brugia malayi** stained with Giemsa. The sheath takes a dark pink stain. X400. Enlarged by 5.4.

**356 Microfilaria of Brugia malayi,** showing the terminal nuclei and the darkly staining sheath. Giemsa. X800. Enlarged by 5.4.

**357 Microfilaria of Brugia malayi,** showing the terminal nuclei. Acridine orange. X800. Enlarged by 5.4.

**358 Microfilaria of Onchocerca volvulus.** Note the distinct cephalic space which is slightly bulbous. Haematoxylin. X400. Enlarged by 5.4.

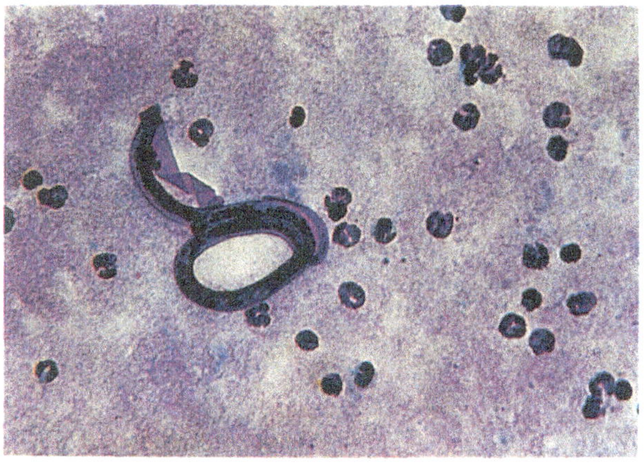

355

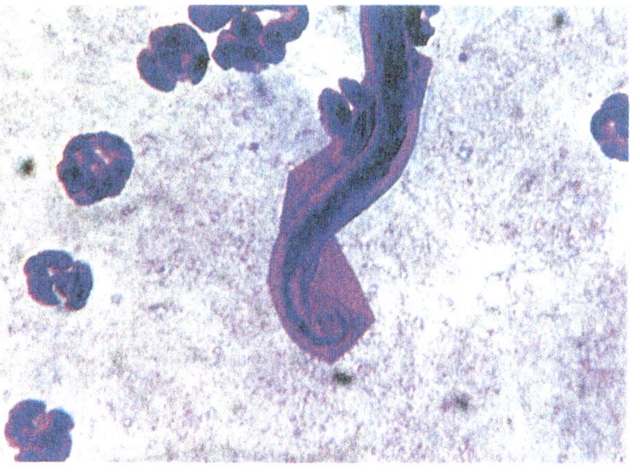

356

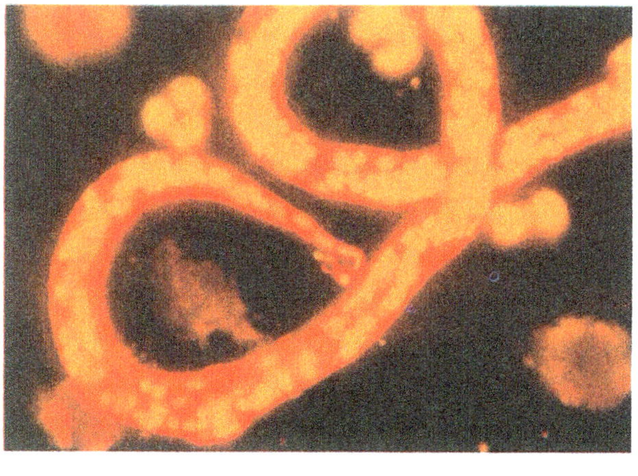

357

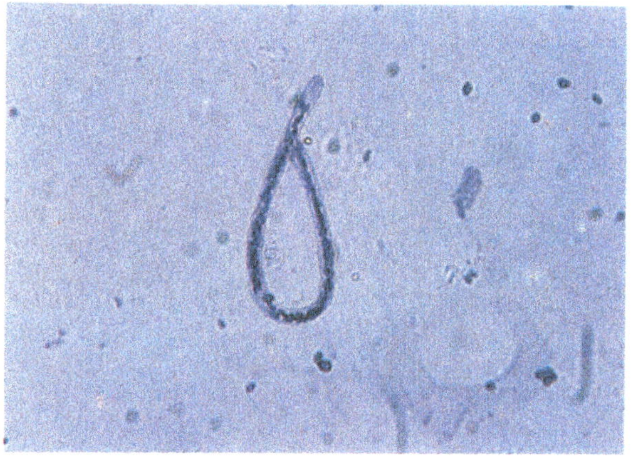

358

**359 Microfilariae of Loa loa and Dipetalonema perstans** lying in the same field. *Loa loa* is thicker and partially visible. Giemsa. X400. Enlarged by 9.6. (Courtesy of Dr A.A. Buck).

**360 Brugia malayi larva developing** in the thoracic musculature of a mosquito. H and E. X100. Enlarged by 9.6.

359
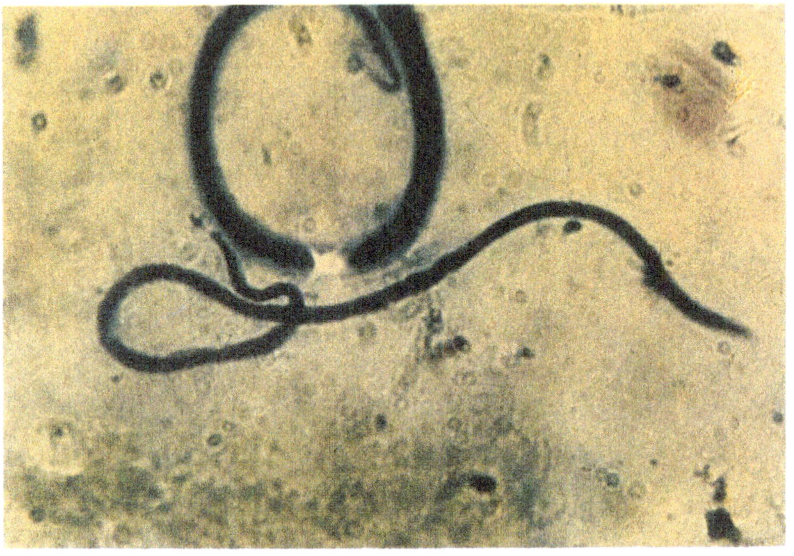

360
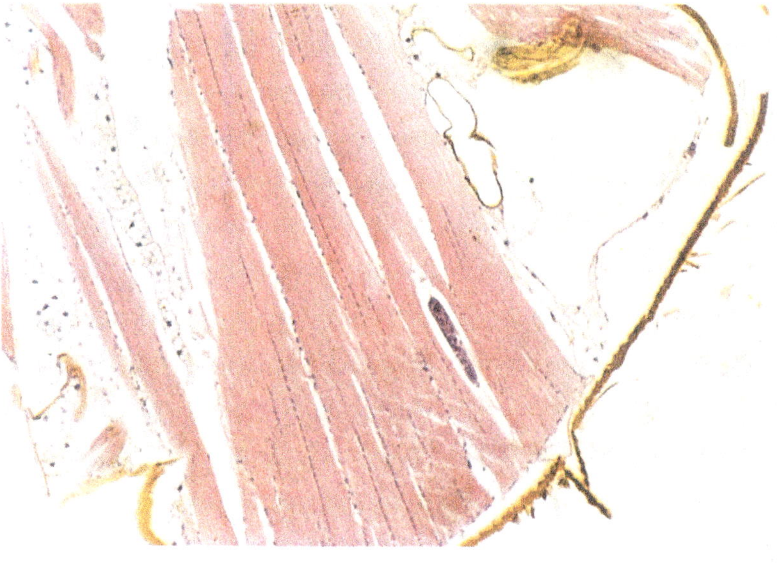

**361 A large number of 3rd stage larvae** of a filaria sp in the head region of a mosquito. H and E. X100. Enlarged by 5.4.

**362 A 3rd stage larva of a filaria sp** lying longitudinally in the proboscis of a mosquito. H and E. X100. Enlarged by 5.4.

**363 A large number of 3rd stage larvae** of a filaria sp emerging from the proboscis of a mosquito. Interference contrast. X100. Enlarged by 5.4.

**364 A 3rd stage larva of Wuchereria bancrofti** showing the 3 caudal papillae. In *B. malayi* the caudal papillae are indistinct. Interference contrast. X100. Enlarged by 5.4.

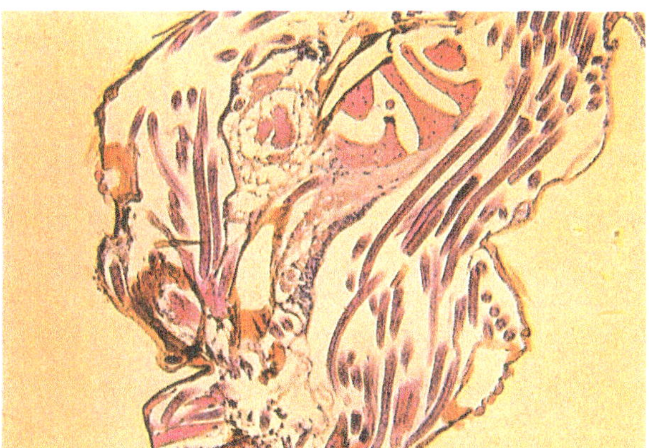

361

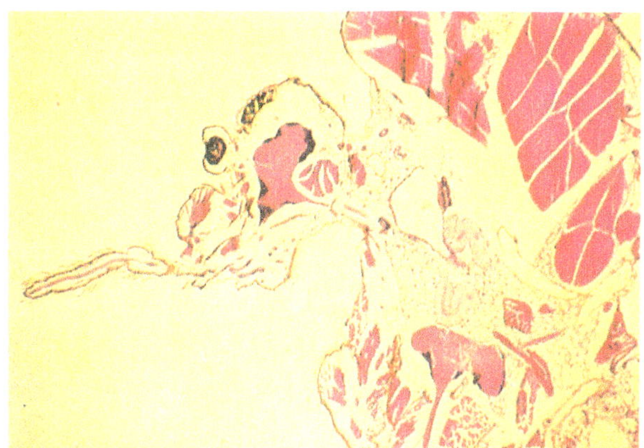

362

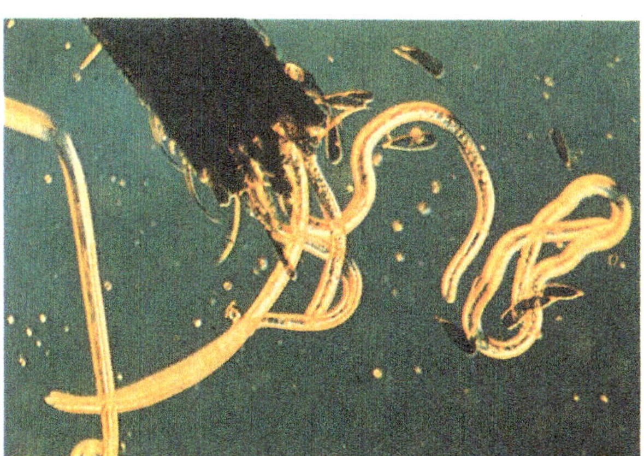

363

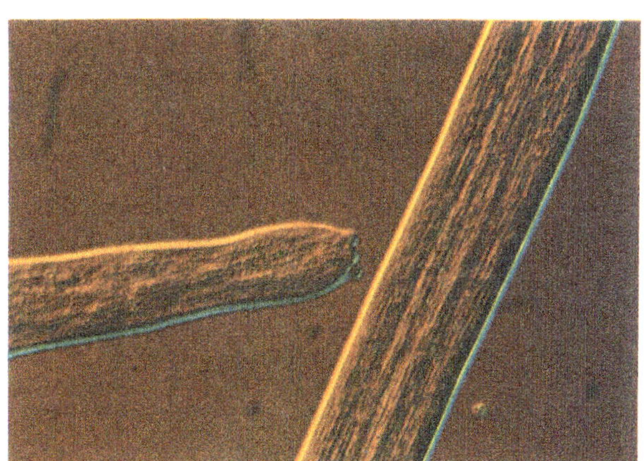

364

**365 Adult Wuchereria bancrofti in a lymph node.** The paired uteri with microfilariae at different stages of development can be seen. H and E. X100. Enlarged by 5.4.

**366 Adult Onchocerca volvulus in a nodule.** Worm has been cut at different levels. Note the presence of fibrous tissue in the background. H and E. X100. Enlarged by 5.4.

**367 Dirofilaria immitis developing** in the malphigian tubules of a mosquito. The 2 infected tubules are dilated and contain a large number of larvae. Interference contrast. X100. Enlarged by 5.4.

**368 Adult Dirofilaria immitis** in the lungs of an experimentally infected monkey. This is an occasional parasite of man. Note the large size of the worm, its thick cuticle and well developed somatic muscles. H and E. X100. Enlarged by 5.4. (Courtesy of Dr M.M. Wong).

**369 Microfilaria in the lung of a monkey** surrounded by a large number of eosinophils. In humans similar pathology is observed in the case of 'eosinophilic lung'. H and E. X400. Enlarged by 5.4.

**370 Simulium sp sitting on skin.** Note the small size of the fly and its black thorax.

**371 Chrysops sp sitting on skin.** Note the large size of the fly and its ornate body and wings.

**372 Skin snip from a case of Onchocerciasis** showing the emergence of a microfilaria in the surrounding fluid. X100. Enlarged by 5.4. (Courtesy of Dr A.A. Buck).

**373 Adult Loa loa lying in the conjunctival sac.** This parasite is motile but generally does not cause much damage. (Courtesy of Department of Ophthalmology, University of Tubingen).

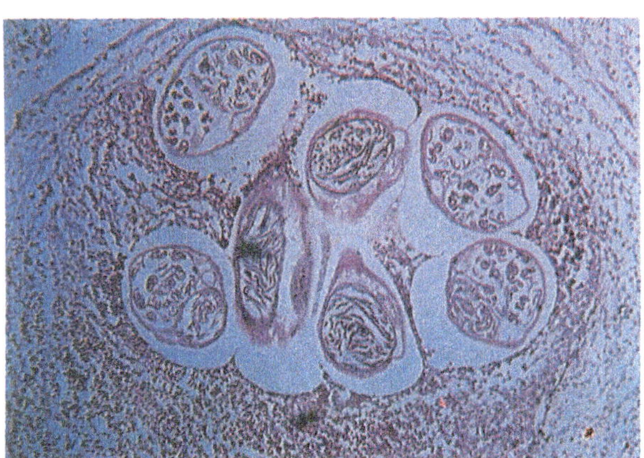

365

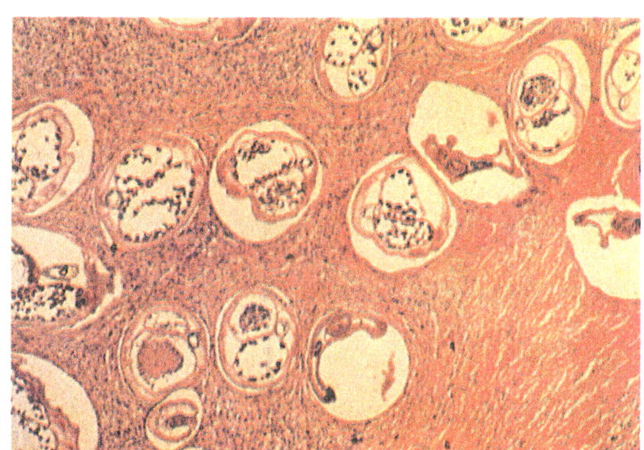

366

367
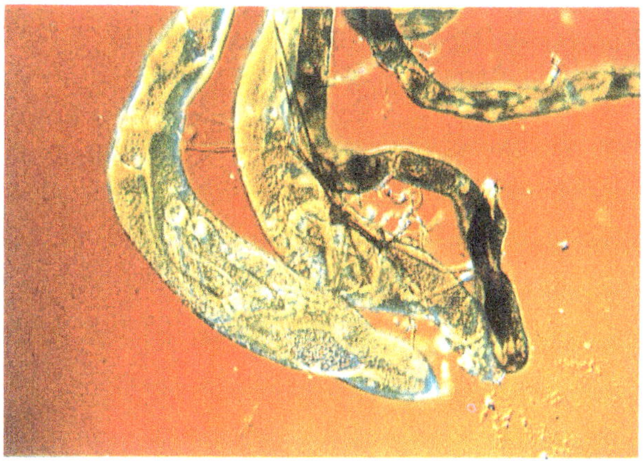

368
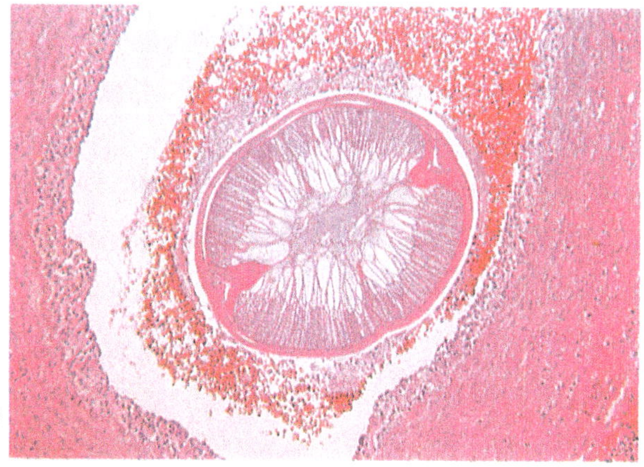

369
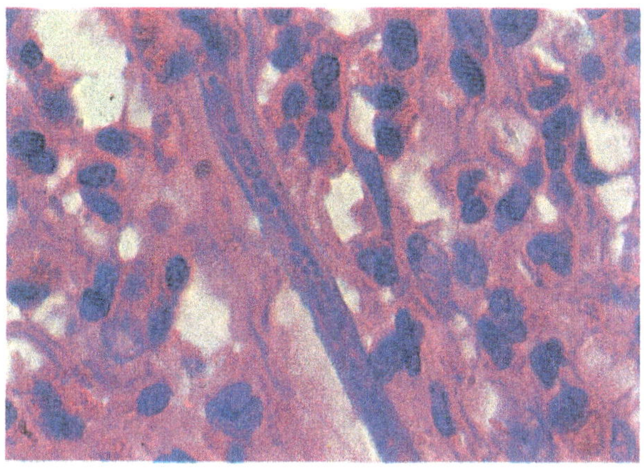

370
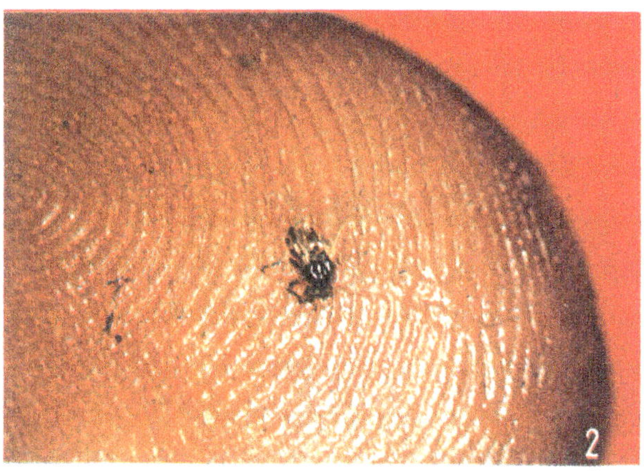

371

372
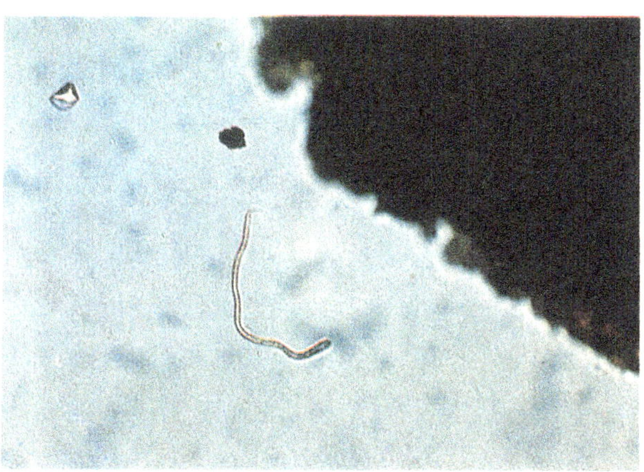

373
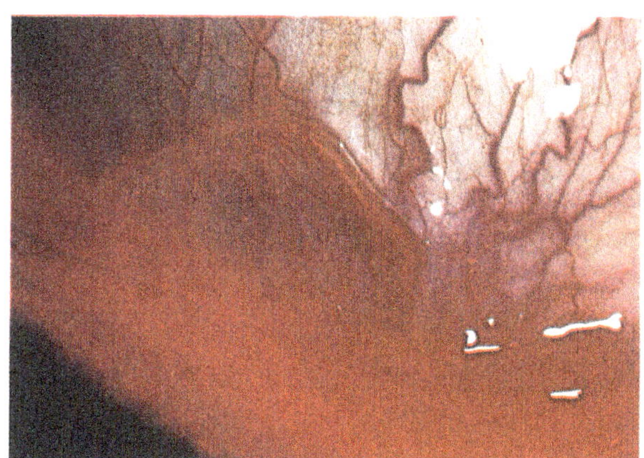

**374 Onchocercal dermatitis.** Note the presence of wrinkled appearance at the lower back. (Courtesy of Dr P. Wenk).

**375 Onchocercal dermatitis.** Late stage with severe wrinkling and hanging of the skin into loose folds. Involvement of the inguinocrural glands produce the 'hanging groin' appearance. (Courtesy of Dr A.A.Buck).

**376 Early stage of onchocercal keratitis** with limbal involvement. (Courtesy of Dr P. Wenk).

**377 Advanced stage of onchocercal keratitis** with diffuse corneal opacity. (Courtesy of Dr P. Wenk).

**378 Late stage of onchocercal keratitis** with total corneal opacity and blindness. (Courtesy of Dr P. Wenk).

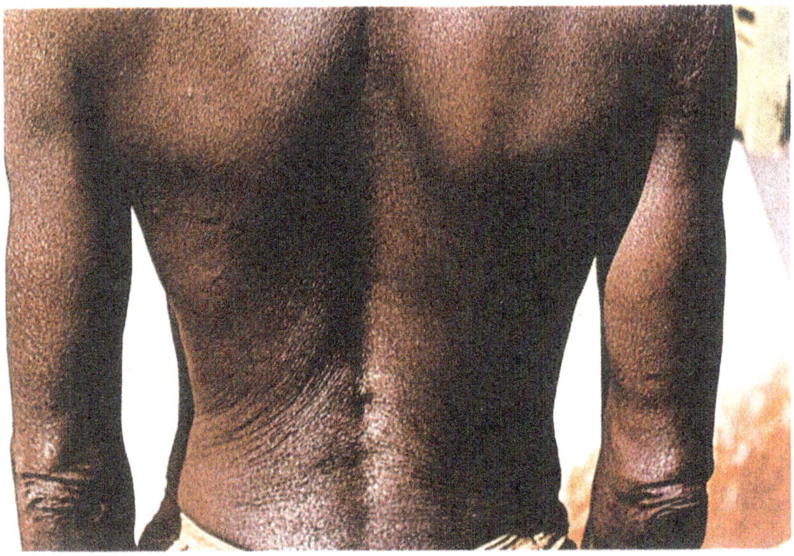

374

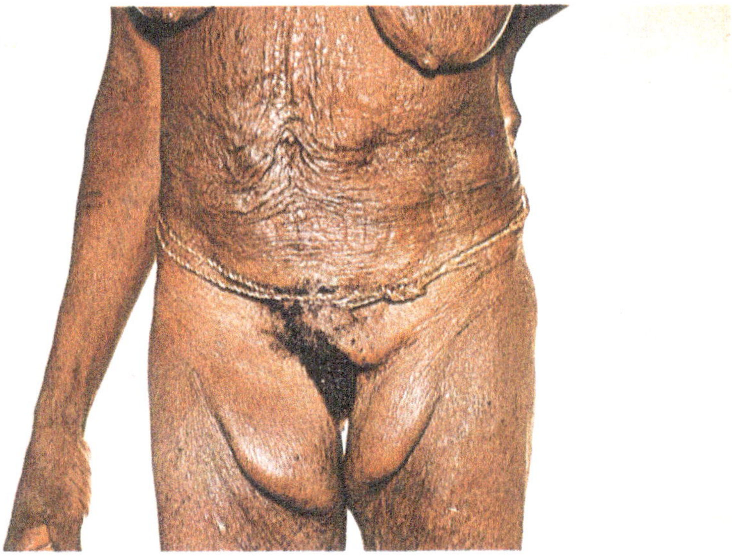

375

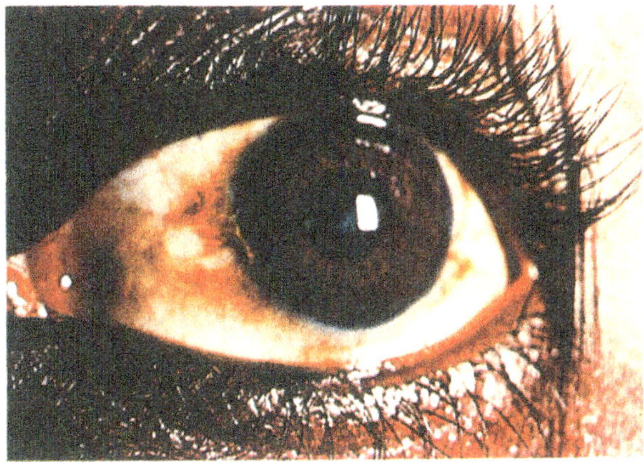
376

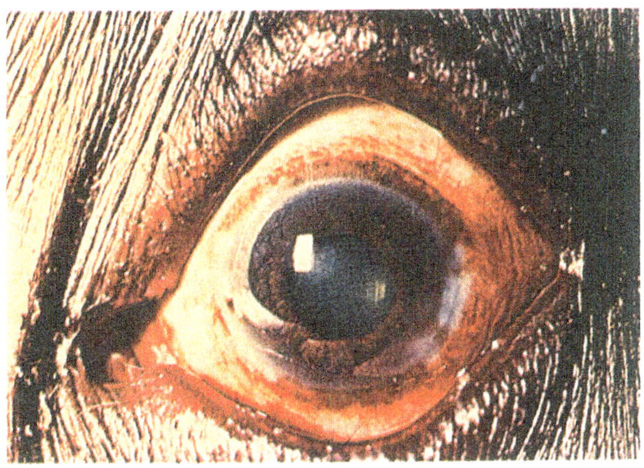
377

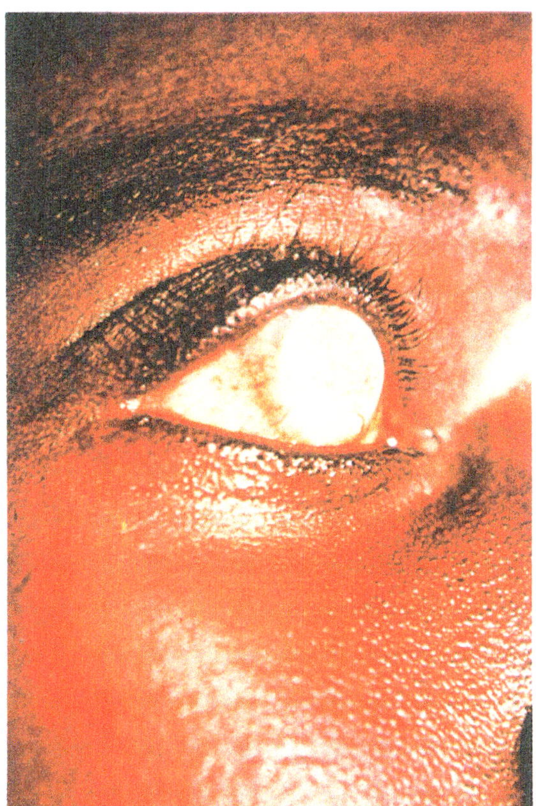
378

**379 Wuchereria bancrofti infection.** The genital involvement is more common than in *B. malayi.* In this case there is swelling of the penis and early hydrocele. (Courtesy of Dr T. Dondero).

**380 Wuchereria bancrofti infection.** Hydrocele and lymph scrotum. (Courtesy of Dr T. Dondero).

**381 Brugia malayi infection.** Involvement of lower limb is more common than in *W. bancrofti.* In early stages there may be slight oedema, which may not be noticed by the patient. However, pitting is usually demonstrable.

**382** Same case showing pitting.

**383 Elephantiasis due to Brugia malayi.** Pitting does not occur in this stage (solid oedema).

**384 Same case showing papillae formation** on the skin of the toes, which is often seen in chronic infection.

**385 Elephantiasis due to Brugia malayi,** complicated by severe dermatitis and secondary bacterial infection. (Courtesy of Dr T. Dondero).

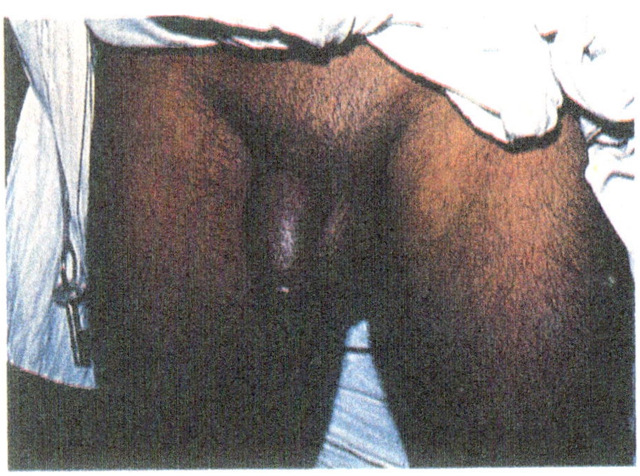

379

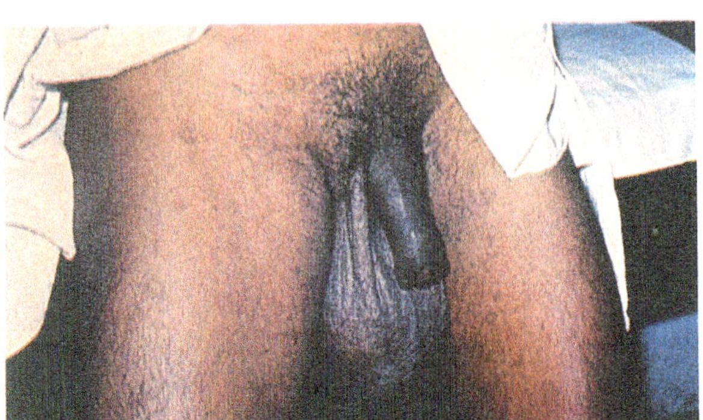

380

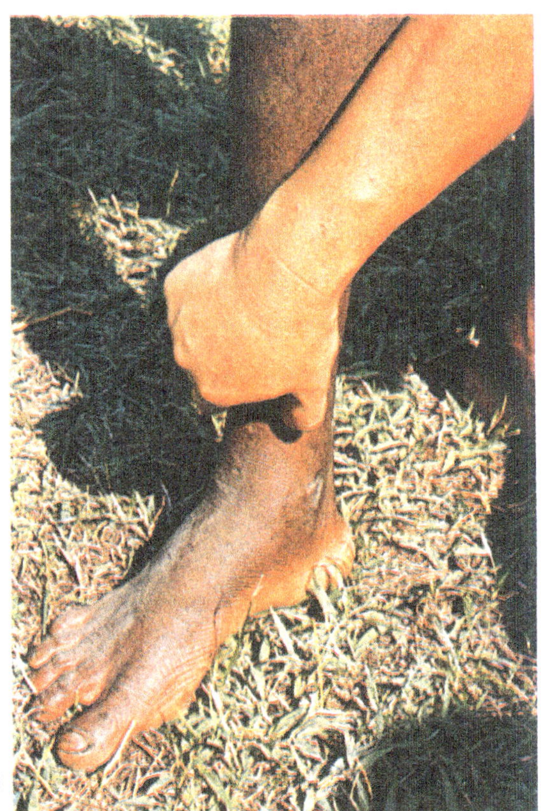

381

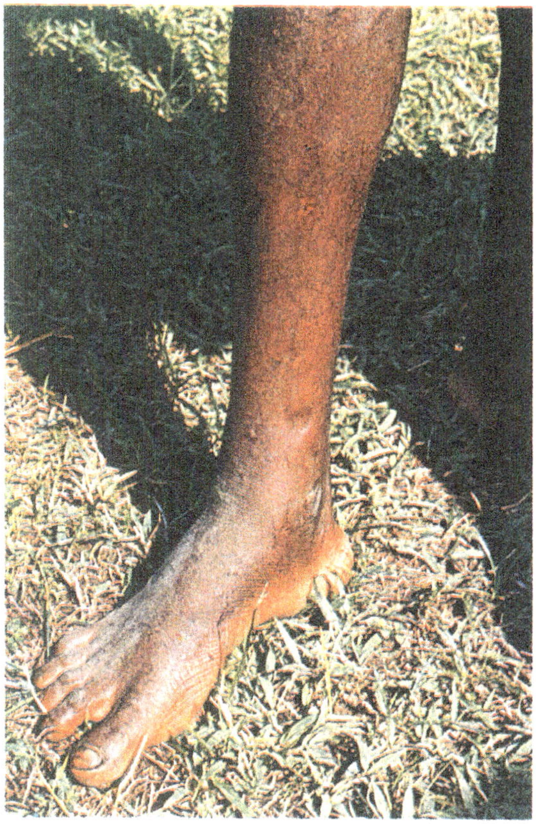

382

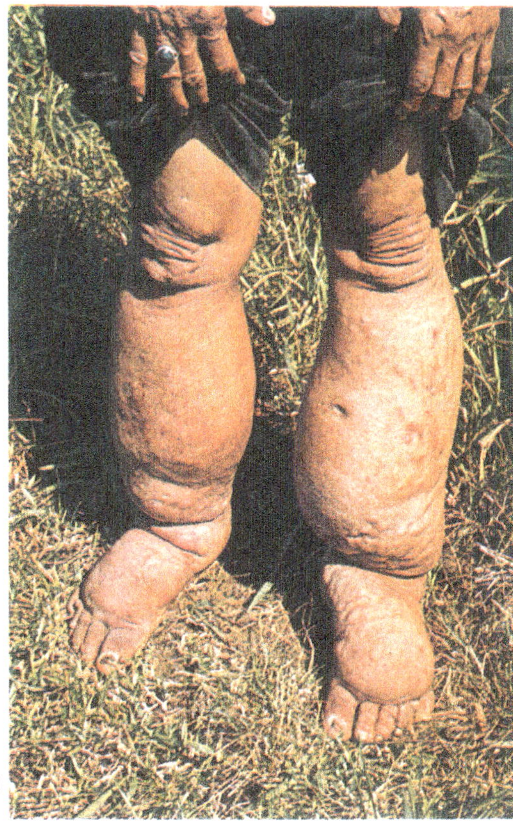

383

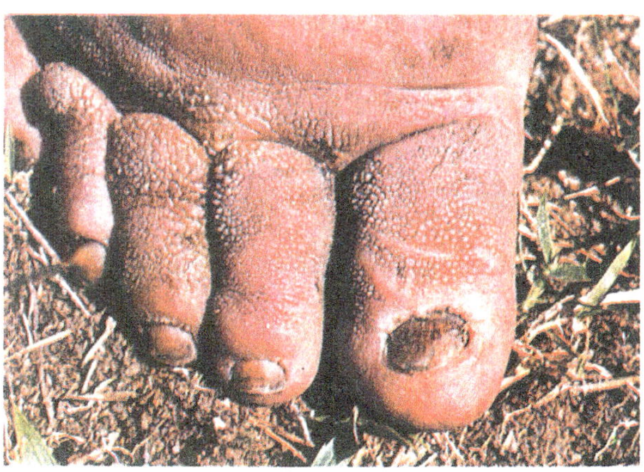

384

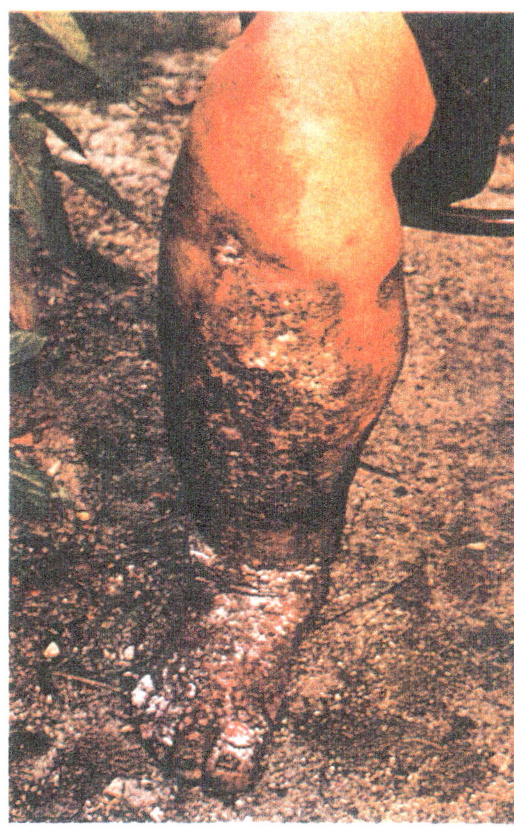

385

# Part III: Arthropods

# Section 13 Arthropods

## Classification of Medically Important Arthropods

| Class | Order | Family | Common Name |
|---|---|---|---|
| Insecta | *Diptera* | Simuliidae<br>Psychodidae<br>Culicidae<br>Ceratopogonidae<br>Tabanidae | Black flies<br>Sand flies<br>Mosquitoes<br>Biting midges<br>Horse flies<br>Deer flies |
| | | Gasterophilidae<br>Oestridae<br>Muscidae | Bot flies<br>Warble flies<br>House flies<br>Tsetse flies etc |
| | | Calliphoridae<br><br>Hippoboscidae | Flesh flies<br>Blow flies<br>Keds<br>Louse flies |
| | *Hemiptera* | Reduviidae<br>Cimicidae | Assassin bug<br>Bed bugs |
| | *Siphonaptera* | Pulicidae | Fleas |
| | *Anoplura* | Pediculidae | Sucking lice |
| | *Hymenoptera* | Apidae<br>Vespidae<br>(many other families) | Honey bee<br>Wasps |
| | *Orthoptera*<br>*Coleoptera* | Blattidae<br>Staphylinidae<br>Cantharidae<br>Scarabaeidae | Cockroaches<br>Beetles<br>Beetles<br>Beetles |
| Arachnida | *Acarina* | Ixodidae<br>Argasidae<br>Sarcoptidae<br>Trombiculidae<br>Demodicidae | Hard ticks<br>Soft ticks<br>Itch mites<br>Chiggers<br>Follicle mites |
| | *Araneida* | Many families | Spiders |
| | *Scorpionida* | Buthidae<br>Scorpionidae<br>Centuridae | Scorpions |
| | *Pentastomida* | Linguatulidae<br>Porocephalidae | Tongue worms |
| | **Subclass** | | |
| Myriapoda | Diplopoda<br>Chilopoda | | Millipedes<br>Centipedes |
| Crustacea | Includes lobsters, crabs and water fleas which may act as intermediate host of helminths | | |

## Terms used in Relation to Arthropods

**Complete metamorphosis** (holometabolous) — the cycle of development involves egg, larva, pupa and adult. Each stage is morphologically distinct from the other.

**Incomplete metamorphosis** (hemimetabolous) — the morphological difference between adults and immature stages are mainly of size and sexual maturity. Each stage, therefore, resembles the other morphologically.

**Nymph** — an immature stage in insects undergoing incomplete metamorphosis.

**Ecdysis** — shedding of larval skin during transformation from one stage to another.

**Emergence** — the appearance of winged insect from the pupal stage.

**Maggot** — common name given to larva of some Diptera.

**Exoskeleton** — external skeleton.

**Sternum** — the ventral side (underside) of thorax.

**Metathorax** — last or 3rd thoracic segment.

**Tergum** — a dorsal plate on the thoracic segment.

**Scutellum** — a small posterior section of the tergum.

**Prothorax** — the anterior segment of thorax.

**Pleuron** — a lateral part of a segment.

**Labrum** — upper lip which forms roof of the mouth.

**Mandibles** — a pair of upper jaws.

**Maxillae** — a pair of lower jaws.

**Labium** — lower lip which forms floor of the mouth.

**Hypopharynx** — tongue shaped structure which lies between the labrum and the labium. It is connected with the salivary duct.

**Epipharynx** — an organ attached to the inner part of labrum.

**Antennae** — the segmented appendages located in the lateral side of the head.

**Labial palpi** — a pair of appendages arising from each side of the labium.

**Cerci** — two slender appendages at the terminal abdominal segment.

**Coxa** — the proximal segment of the leg.

**Claw** — a sharp curved appendage at the distal end of leg.

**Chelicerae** — grasping or cutting appendages in place of jaws, found in Arachnida.

**Hypostome** — the lower lip which takes the form of an elongated structure armed with teeth or hooks, found in ticks.

**Pedipalp** — a second pair of appendages which arise from the cephalothorax, found in Arachnida.

**Capitulum** — the 'false-head', found in Arachnida. The basal portion of this structure is 'basis capituli'.

**Palmate hairs** — fan shaped hairs on the dorsal surface of Anopheline larva.

**Sensilla** — hairs which have sensory function.

**Seta** — hair like structures which are hollow internally.

**Haltere** — club shaped structure in place of the second pair of wings, seen in Diptera.

**Compound eye** — a large number of visual elements grouped together on each side of the head.

**Ocellus** — the simple eye.

**Tracheae** — ringed breathing tubes.

**Spiracles** — respiratory openings.

**Gill** — a specialised respiratory organ seen in aquatic insects.

**Pseudotracheae** — small tubes on labella found in houseflies and Tabanidae through which fluid is sucked.

**Ovipositor** — a tubular structure through which eggs are laid.

**Costa** — thickening of the anterior edge of the wing.

**Subcosta** — a vein lying posterior to the costa.

**Longitudinal veins** — a series of six veins following the subcosta. These are numbered in sequence.

**Cell** — an area of wing enclosed by veins.

**Fat body** — group of cells which are widely distributed in the body and act as a food store. They contain protein, fats and glycogen.

**Spermatheca** — a sperm storing organ found in the lower genital tract of the female.

**Pygidium** — a pincushion like structure seen on the ninth segment of fleas. It is believed to have a sensory function.

**Myiasis** — a condition in which infestation of human and animal tissues occurs with fly maggots.

**Vector** — transmitter of infection. Generally refers to an arthropod.

**Notum** — the dorsal part of a segment.

**Pronotum** — the dorsal surface of the first thoracic segment.

**Mesonotum** — the dorsal surface of the second thoracic segment.

**Pronotal comb** — conspicuous spines seen on the pronotum of some fleas.

**Scutum** — a chitinous shield seen on the dorsum of Acarina.

**Festoons** — notches seen at the posterior part of the body of hard ticks.

## Human Vector-borne Infections

| Infection (Disease) | Causative Agent | Vector (common name) |
|---|---|---|
| **Viral Infections** | | |
| | *(Togaviridae)* | |
| Yellow fever | Yellow fever virus | Mosquitoes |
| Dengue haemorrhagic fever | Dengue virus | Mosquitoes |
| Japanese encephalitis | JE virus | Mosquitoes |
| Murray valley encephalitis | MVE virus | Mosquitoes |
| Western equine encephalomyelitis | WEE virus | Mosquitoes |
| Eastern equine encephalomyelitis | EEE virus | Mosquitoes |
| Chikungunya | Chikungunya virus | Mosquitoes |
| O'Nyong-Nyong | O'Nyong-Nyong virus | Mosquitoes |
| Tick-borne encephalitis (Central European) | CET virus | Ticks |
| Tick-borne encephalitis (Russian Far East) | FER virus | Ticks |
| Kyasanur forest disease | KFD virus | Ticks |
| Omsk haemorrhagic fever | OHF virus | Ticks |
| | *(Reoviridae)* | |
| Colorado tick fever | CTF virus | Ticks |
| | *(Unclassified Arboviruses)* | |
| Rift valley fever | Rift valley virus | Mosquitoes |
| Epidemic haemorrhagic fever | EHF virus | Mite (?) |
| Sandfly fever | Sandfly fever virus | Sandflies |
| **Bacterial and Rickettsial Infections** | | |
| Tick typhus | *Rickettsia rickettsi* | Ticks |
| Queensland tick typhus | *R. australis* | Ticks |
| Q fever | *Coxiella burnetti* | Ticks |
| Relapsing fever | *Borrelia duttoni* | Ticks |
| Tularaemia | *Pasteurella tularensis* | Ticks |
| Scrub typhus | *R. tsutsugamushi* | Mites |
| Epidemic typhus | *R. prowazeki* | Lice |
| Relapsing fever | *B. recurrentis* | Lice |
| Epidemic (murine) typhus | *R. mooseri* | Fleas |
| Plague | *Yersinia pestis* | Fleas |
| Oraya fever | *Bartonella bacilliformis* | Sandflies |
| **Protozoal Infections** | | |
| Malaria | *Plasmodium* spp | Mosquitoes |
| Kala-azar | *Leishmania donovani* | Sandflies |
| Oriental sore | *L. tropica* | Sandflies |
| Chaga's disease | *Trypanosoma cruzi* | Triatomid bugs |
| E. African trypanosomiasis | *T. rhodesiense* | Tsetse flies |
| W. African trypanosomiasis | *T. gambiense* | Tsetse flies |
| Babesiosis | *Babesia* spp | Ticks |
| **Helminthic Infections** | | |
| Filariasis | *Wuchereria bancrofti* | Mosquitoes |
| Filariasis | *Brugia malayi* | Mosquitoes |
| Filariasis | *Dirofilaria* spp | Mosquitoes |
| Filariasis | *Dipetalonema perstans* | Biting midges |
| Filariasis | *D. streptocerca* | Biting midges |
| Filariasis | *Mansonella ozzardi* | Biting midges |
| Onchocerciasis | *Onchocerca volvulus* | Black flies |
| Loiasis | *Loa loa* | Deer flies |

## General Characters of Arthropoda

The Arthropods are characterised by the possession of an exoskeleton and chitinised appendages. The internal structure consists of a haemocoele which is filled with haemolymph. The organs float in the haemolymph. In most arthropods the body is divisible into a head, thoracic and abdominal regions. In the class Insecta the thorax is divisible into 3 segments. These segments bear appendages used for locomotion.

The head is generally spherical and joined to the thorax by a thin neck. It bears a pair of eyes which are often compound with many facets. The 2 main appendages coming out of the head are a pair of antennae.

The mouthparts consist of labrum, a labium, a pair of maxillae and a pair of mandibles, in the class Insecta and a pair of chelicerae and a pair of pedipalpi in the Arachnida.

The nervous system consists of a brain and a pair of nerve cords which lie ventrally. The circulatory system consists of a tubular heart which lies dorsally and pumps blood into the haemocoele. The excretory system consists of malphigian tubules which collect the excretory products from the haemocoele and discharge them into the hindgut.

Arthropods may transmit disease by the following methods:-

1. Mechanical transmission — In this case the disease producing agent does not multiply e.g. *Salmonella* transmission by house fly. The arthropod is not an integral part of the life cycle of the microorganism.
2. Biological transmission — In this case the arthropod is an integral part of the life cycle. The disease producing agent may develop in the following manner:
   a) Only multiplication e.g. *Yersinia pestis* in fleas.
   b) Only development e.g. *Wuchereria bancrofti* in mosquitoes.
   c) Multiplication and development e.g. *Plasmodia* in mosquitoes.

## Ticks

These are of two kinds, the Argasidae or soft ticks, and the Ixodidae or hard ticks. The main morphological difference between the two groups is:-

| *Argasidae* | *Ixodidae* |
|---|---|
| a) Without dorsal shield (scutum) | a) With dorsal shield (scutum) |
| b) Capitulum located on the ventral surface and does not project anteriorly | b) Capitulum on anterior margin |

The capitulum in the ticks is a characteristic organ in that it bears a conspicuous hypostome which has large recurved teeth. This allows penetration and firm attachment to the host tissues. Ticks usually hide in cracks and crevices during the daytime. The female produces eggs in very large numbers as the chances of survival are low. The larva which emerges has 3 legs. This becomes a nymph with 4 legs after moulting. The nymph moults to become an adult. In some species there are many nymphal stages before the adult stage is reached. All stages are blood sucking.

In addition to being a vector of certain infections, ticks also produce local and systemic damage. The local inflammatory lesion is produced as a result of trauma caused by the hypostome and sharply toothed chelicerae. Systemic damage is produced by some fraction in tick's saliva which blocks the release of acetylcholine. This leads to an ascending flaccid motor paralysis, commonly known as the 'tick paralysis'.

**386 Morphological features** of hard ticks (Redrawn from Medical Veterinary Protozoology: Adam, Paul and Zaman; Churchill Livingstone, Edinburgh 1971. Original J.A. Campbell).

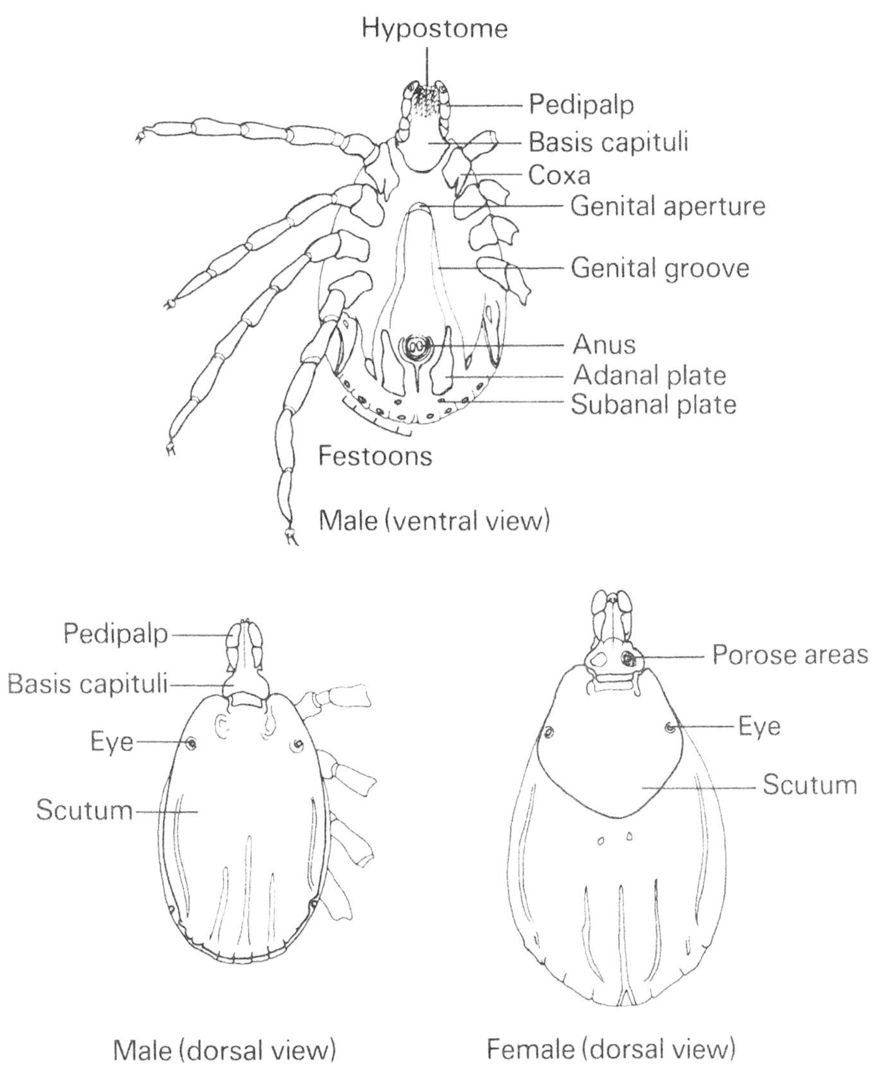


386

## Mites

These are usually small to microscopic. The cephalothorax and the abdomen are fused without any line of demarcation. The mouthparts have chelicerae. Three major groups are of medical importance. These are: (a) trombiculid mites; (b) sarcoptic mites, and (c) house dust mites.

### *Trombiculid Mites*

They live in soil and vegetation. Only the larval stage attacks animals or humans. They feed on tissue fluids and lymph by embedding their mouthparts into the skin. After feeding they drop off to moult into nymphs and then into adults. The rickettsial organisms are carried by these mites and pass through the egg from one generation to another. This process being known as the transovarial transmission.

### *Sarcoptic Mites*

These belong to the species *Sarcoptes scabiei* and produce a contagious disease, scabies. Clinically the condition is characterised by a papular rash which appears over the body, particularly involving the interdigital areas, the axilla, the genitalia, the buttocks, the ulnar surface of the forearms and the wrist. Itching, which is often worse at night, occurs after the patient has become sensitised to the mite. This may take 4 to 6 weeks after the infection. The papular lesions may become vesicular and secondary bacterial infection may result in ulceration. Diagnosis is usually made on clinical grounds. Skin scrapings may show the mites, although this procedure is not always successful.

### *Dust Mites*

A number of species of mites are found in dust. These are microscopic and release antigenic material in the respiratory passages on being inhaled. These mites are now regarded as an important cause of asthma in man.

**387a Trombicula akamushi larva** (dorsal view).
**387b Sarcoptes scabiei** (male, ventral view).

a

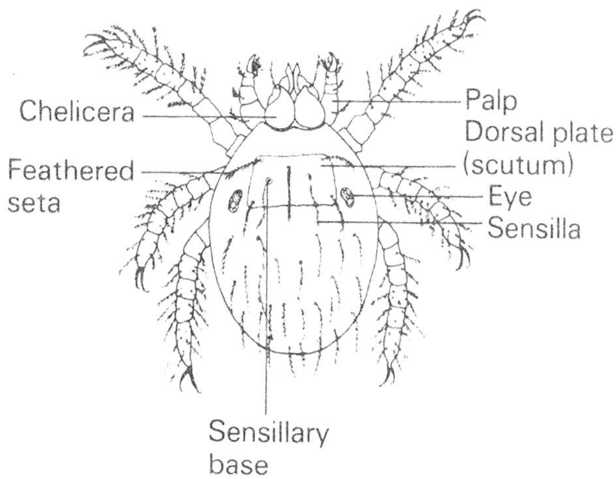


b

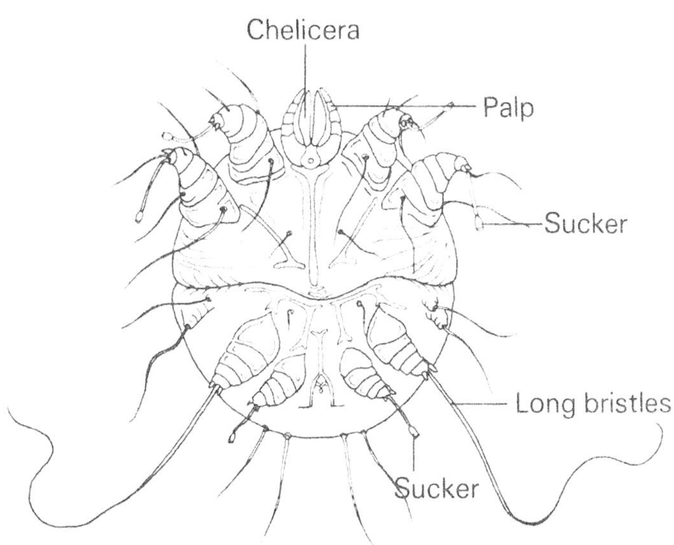


*Sarcoptes scabiei* (male, ventral view)

387

## Bugs

The common bed bugs are *Cimex lectularius* (temperate regions) and *Cimex hemipterus* (tropical regions). The body of these animals are flattened dorso ventrally and covered by bristles. A 4 jointed antenna comes out from the head and the pronotum is expanded laterally to partly surround the head. The legs have distinct terminal claws. In the female, the posterior end of the abdomen is rounded while in the male, it is pointed.

The proboscis is short and lies folded on the ventral surface of the head. The fertilised female lays eggs after a blood meal. The eggs take 7 to 9 days to hatch. They pass through 5 ecdyses. All the nymphal stages thus produced are haematophagous. The bite of bed bugs can be a source of severe skin irritation and discomfort. In the laboratory they are able to carry many pathogens but in nature are not known to be a vector of any disease.

**388 Cimex reticularis.**

388

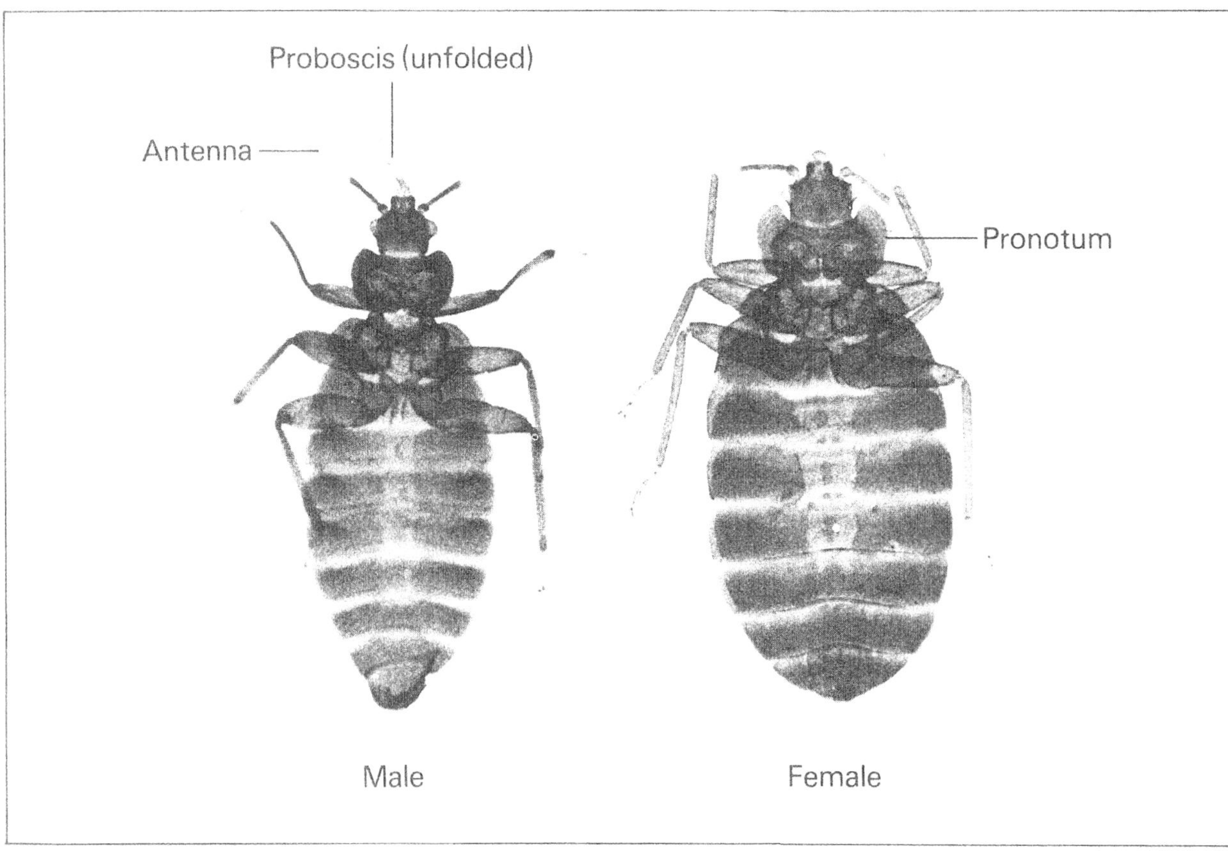

## Lice

Three species parasitise man. These are *Pediculus humanus* var corporis (body louse), *P. humanus* var capitis (head louse) and *Phthirus pubis* (pubic louse). The head louse is found on hairs. It lays eggs on the shaft of the hair and firmly cements it. The nymph hatches in about 5 days and the adults develop in about 16 days. The body louse oviposits on clothing. The lice migrate to the new host when there is hyperpyrexia or if the person dies. This causes easy dissemination of infections.

**389 Pediculus humanus and Phthirus pubis**. Lice commonly found on man.

389 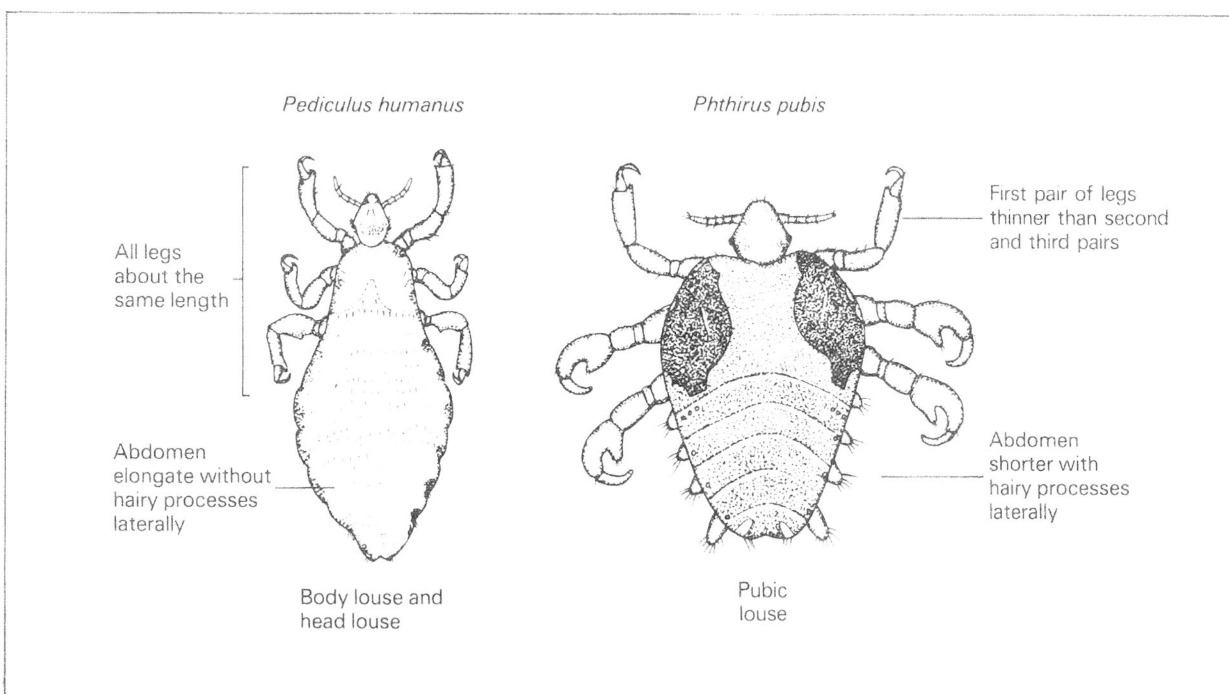

## Fleas

These are laterally flattened and without wings. The antennae lie in a groove in the head region. The legs are long and muscular which allows them to jump. The eggs are laid in burrows and the larvae are maggot-like. The adults feed at periodic intervals and do not stay on their host all the time. The fleas of medical importance are *Pulex irritans* (common human flea), *Ctenocephalides canis* and *C. felis* (dog and cat flea) and *Xenopsylla cheopis* (oriental rat flea), as these are vectors of plague. On ingestion of *Yersinia,* the bacteria multiply in such large numbers in the stomach of the flea that the oesophagus is blocked. Such a flea is known as a 'blocked' flea. Fleas when 'blocked' move from person to person in an effort to obtain a satisfying blood meal. This causes rapid transmission of infection.

## House Fly and Related Species

These are of medical importance for 2 reasons:
a) as mechanical carriers of microorganisms, and
b) as causative agents of myiasis.

The maggots of these flies are distinguished by the morphology of their posterior spiracles (stigmata). Myiasis is generally classified into External myiasis when there is invasion of dermal tissues and Internal myiasis when there is invasion of intestinal or genito urinary systems. In the tropics the important myiasis producing genera are *Dermatobia, Callitroga, Chrysomyia, Lucilia, Sarcophaga* and *Cordylobia.*

**390 Generalised adult flea.**

390

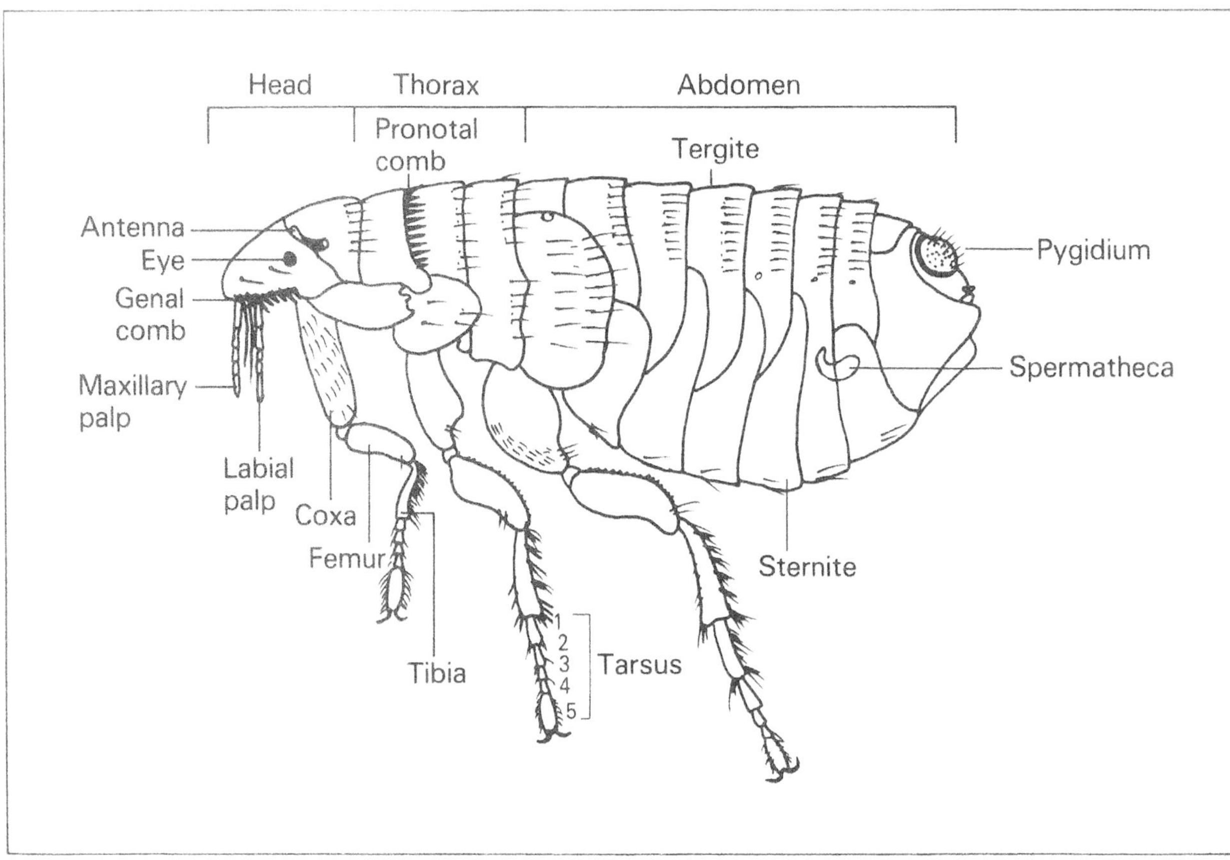

## Sand Flies

These are small flies measuring about 5mm in length. The wings and the body are heavily covered with fine hairs. Only one sub-family, *Phlebotominae* is of medical importance.

The female lays her eggs in cracks and holes in the ground. The larva hatches out in 1 to 2 weeks. It has 3 to 4 long caudal bristles. Pupation occurs for about 1 to 2 weeks. The adults are poor fliers and have a restricted range of flight. Only the female takes a blood meal.

## Black Flies

These are small flies measuring 1 to 5mm and generally have a strongly humped appearance. Only the female sucks blood and it lays the eggs below the surface of streams and rivers. The larvae which hatch out are always found in water. They attach themselves to submerged objects and trap food by a pair of feeding brushes at their anterior end. Pupae also remain submerged in water. The fly, therefore, emerges under water and then comes to the surface. The most important species is *Simulium damnosum.*

## Tsetse Flies

These are robust, yellowish to brownish black coloured flies measuring 6 to 15mm in length. Both sexes suck blood. They are easily recognised by a conspicuous proboscis which projects anteriorly. In the case of tsetse the larva is full grown in the uterus before it leaves the mother. Larviposition occurs under vegetation, caves or in tree holes. The larva then burrows in the soil or debris and turns into a barrel shaped pupa. The adult fly emerges in about a month after pupation. The medically important species are *Glossina palpalis, G. tachinoides, G. morsitans, G. swynnertoni* and *G. pallidipes.*

## Biting Midges

These are the smallest of biting flies measuring between 1 to 4mm. The female is a blood feeder and deposits her eggs on plants or on objects close to or covered with water. The larva is worm-like with a distinct pigmented head. Medically the most important genus is *Culicoides.*

## Venomous Insects

In addition to the commonly known venomous insects such as wasps, bees and spiders, the Chilopoda (centipede) and scorpions are of medical importance in the tropics. In Chilopoda the body is made up of a large number of segments. Each segment bears a pair of jointed legs. The paired antennae are long and taper to a fine end. The poison glands are embedded in the fat body and are connected to the first pair of legs which are modified to form poison claws (maxillipeds). A centipede may vary from a few inches to almost a foot in length.

The scorpions are easily recognised by their large pedipalps which end in pincers for holding the prey. The tail has 6 segments, the last of these bears the poison sac and the sting. They are viviparous and are more commonly found in dusty and dry places. The venom can produce severe local reaction and occasionally, circulatory failure and death.

## Mosquitoes

These belong to the family Culicidae. This is a very large family containing approximately 31 genera and many hundreds of species. The genera of greatest medical importance are *Anopheles, Culex, Aedes* and *Mansonia.* Mosquitoes have elongated mouthparts with only the female capable of taking a blood meal. The wings are covered with scales and have veins running longitudinally. The proboscis is made up of a pair of mandibles, maxillae and hypopharynx which are enclosed in the labium which forms a protective sheath. Eggs are laid on water or on a moist substrate. In the case of *Culex* they are laid close to each other to form rafts. In the case of *Mansonia* the female attaches her eggs to the underside of the leaves of *Pistia* which floats on the surface of water. The larvae of *Mansonia* have special chitinous siphons which are inserted into the roots of another aquatic plant, *Eichhornia.* The adult mosquitoes emerge from the pupae and mate when they are 1 to 2 days' old. The female generally takes a blood meal every 4 or 5 days after which she lays eggs.

**391 Generalised adult mosquito.**

**392 Comparison of heads of male and female Anopheles and Culex mosquitoes.**

**393 Mosquito mouth parts,** side view and cross-section.

**394 Comparison of morphology of Aedes and Anopheles larva.**

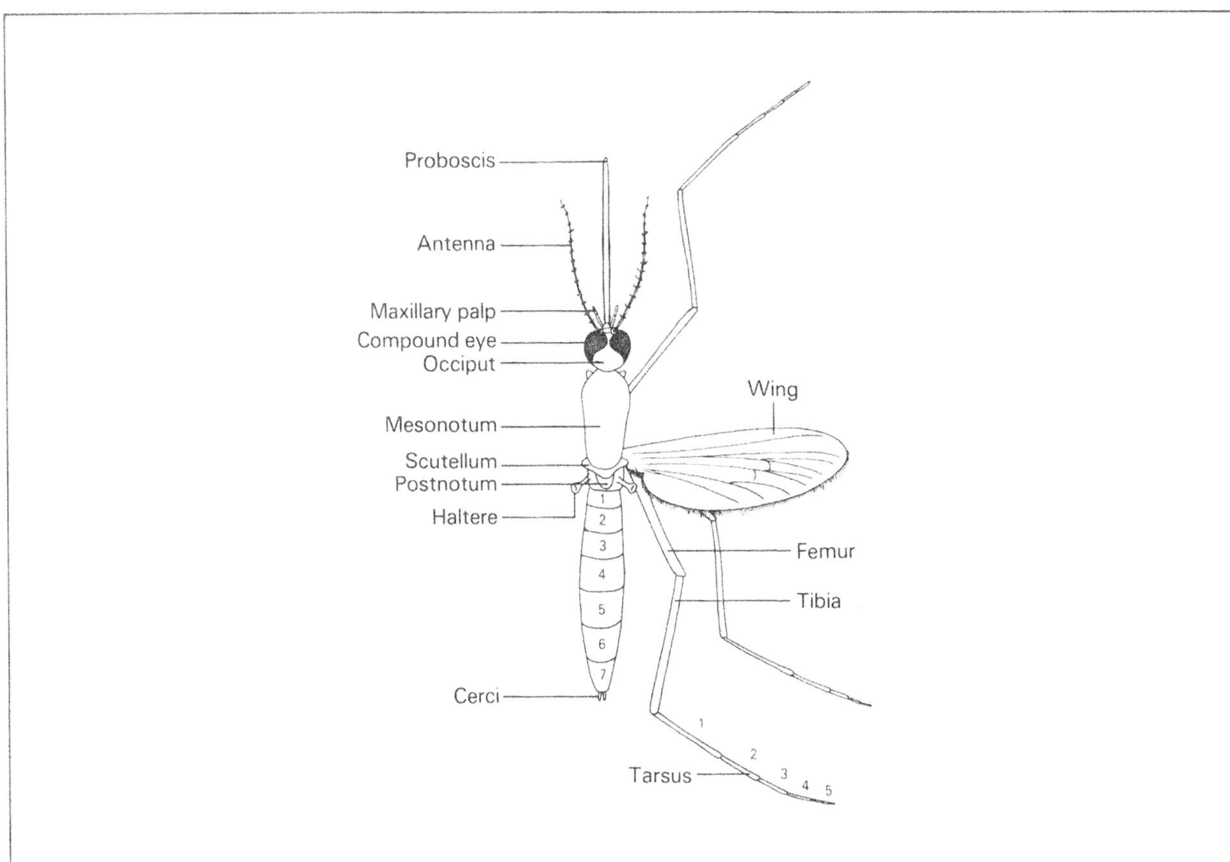


391

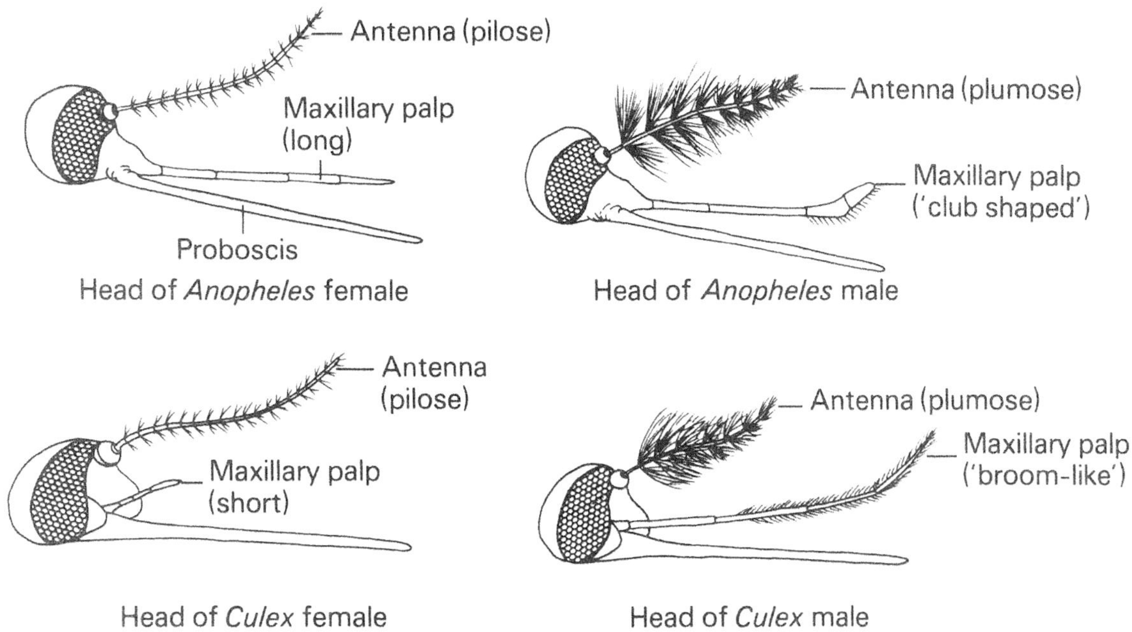
Antenna (pilose)
Maxillary palp (long)
Proboscis
Head of *Anopheles* female
Antenna (plumose)
Maxillary palp ('club shaped')
Head of *Anopheles* male
Antenna (pilose)
Maxillary palp (short)
Head of *Culex* female
Antenna (plumose)
Maxillary palp ('broom-like')
Head of *Culex* male

392

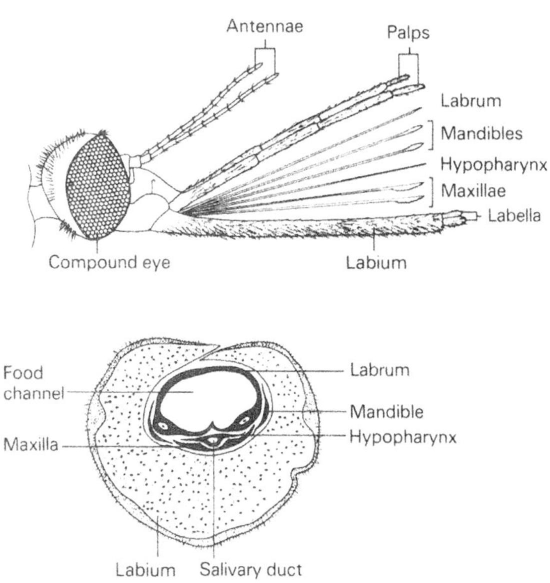
Antennae
Palps
Labrum
Mandibles
Hypopharynx
Maxillae
Labella
Compound eye
Labium
Food channel
Labrum
Mandible
Maxilla
Hypopharynx
Labium
Salivary duct

393

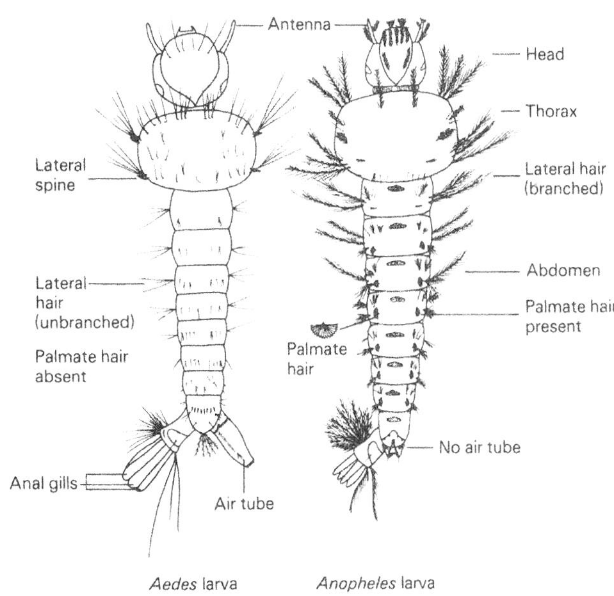
Antenna
Head
Thorax
Lateral spine
Lateral hair (branched)
Abdomen
Lateral hair (unbranched)
Palmate hair present
Palmate hair absent
Palmate hair
No air tube
Anal gills
Air tube
*Aedes* larva
*Anopheles* larva

394

**395 Aedes aegypti egg. Larva beginning to emerge.** The egg breaks near the anterior end due to the pressure exerted by the larva. Interference contrast. X50. Enlarged by 5.4.

**396 Ae. aegypti larva during the process of emergence.** Interference contrast. X50. Enlarged by 5.4.

**397 Ae. aegypti first stage larva, after emergence.** Note the presence of gills and siphon at the posterior end. Interference contrast. X50. Enlarged by 5.4.

**398 Eggs of Mansonia attached to Pistia (water lettuce).** The two dark bodies at the margin of the central leaf are the clumps of eggs.

**399 Larvae of Mansonia attached to roots of Eichhornia (water hyacinth).** The larvae are attached by their siphon and obtain their oxygen requirements through the plant cells (Courtesy of Dr I. Polunin).

**400 Siphon of Mansonia larva.** Note the chitinized surface and saw-like margin used for penetrating the plant tissues.

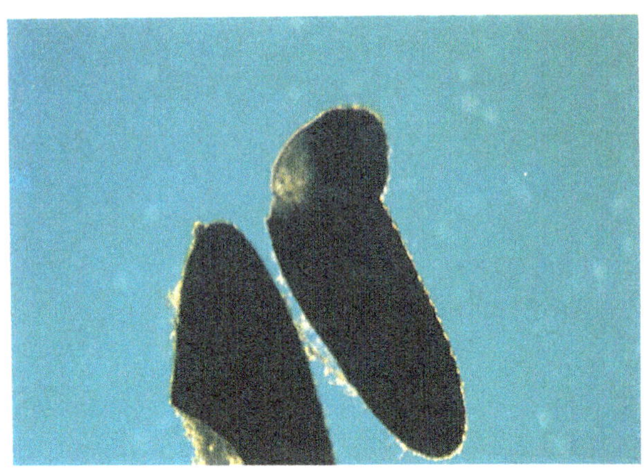

395

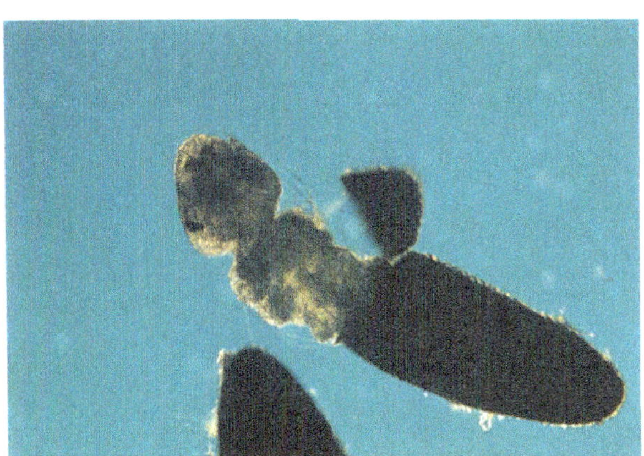

396

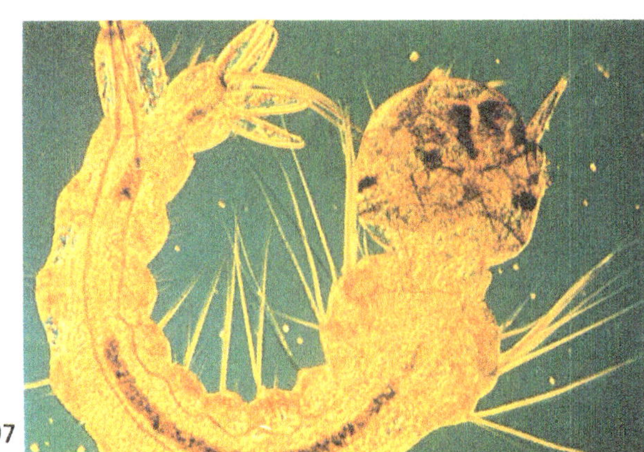

397

398

399

400
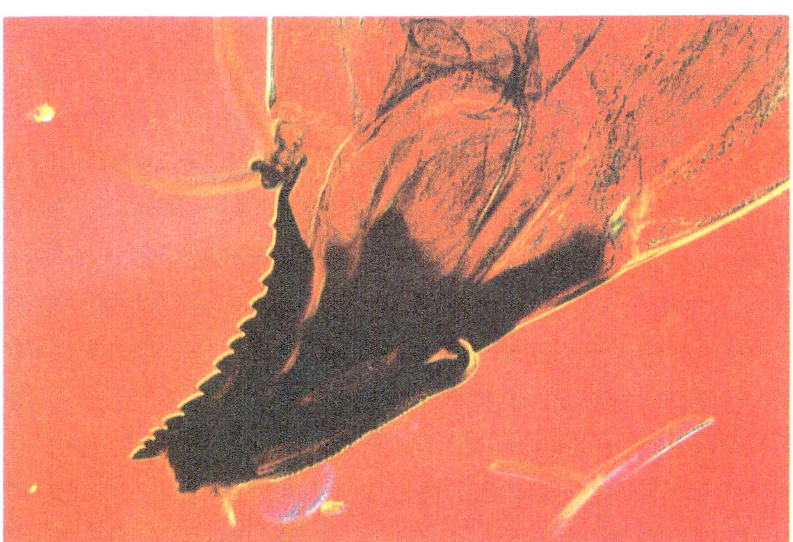

**401 Wing scales of Aedes.** The scales which are attached to veins are thin and elongated. Interference contrast. X100. Enlarged by 5.4.

**402 Wing scales of Mansonia.** The scales are larger and lie close to the wing surface. This character differentiates it from other culicine mosquitoes. Interference contrast. X100. Enlarged by 5.4.

**403 Male genitalia of mosquitoes.** Note the 2 angulated structures with fine teeth. These are the claspers used for holding the female during copulation.

**404 Culicoides sp. is one of the smallest biting insects.** Note the distinct jointed antennae. Interference contrast. X100. Enlarged by 5.4.

**405 An adult house dust mite (Dermatophagoides sp.).** Note the presence of suckers on the distal end of all 4 legs. Interference contrast. X100. Enlarged by 5.4.

**406 A trombiculid mite larva.** Note the presence of 3 legs and fine hairs which cover the body and integuments. Interference contrast. X100. Enlarged by 5.4.

**407 An Ixodid tick larva.** Note the presence of capitulum at the anterior end and 3 legs. The dark area in the body is ingested blood. Interference contrast. X100. Enlarged by 5.4.

**408 Hypostome of a tick.** Note the presence of inverted teeth on the bulbous anterior end. Interference contrast. X100. Enlarged by 5.4.

**409 Section of a tick attached to skin.** Note the hypostome is embedded in the dermis. H and E. X100. Enlarged by 5.4.

**410 Section of a mite attached to skin.** Note the chelicerae (sharply pointed structures) are embedded in the dermis. H and E. X100. Enlarged by 5.4.

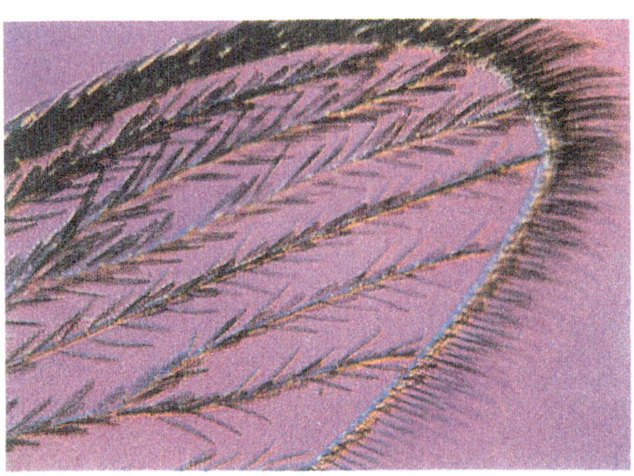

401

402

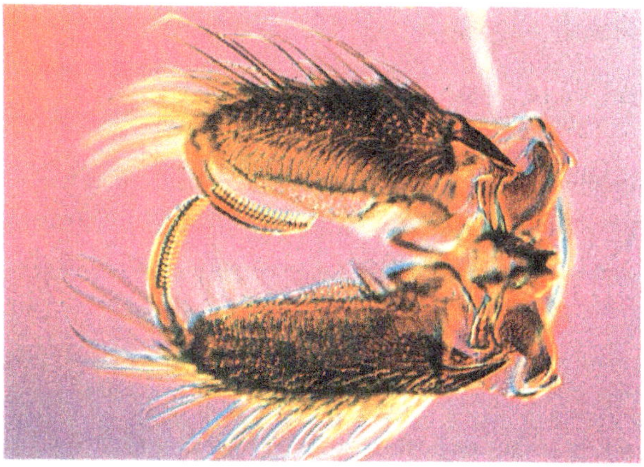
403

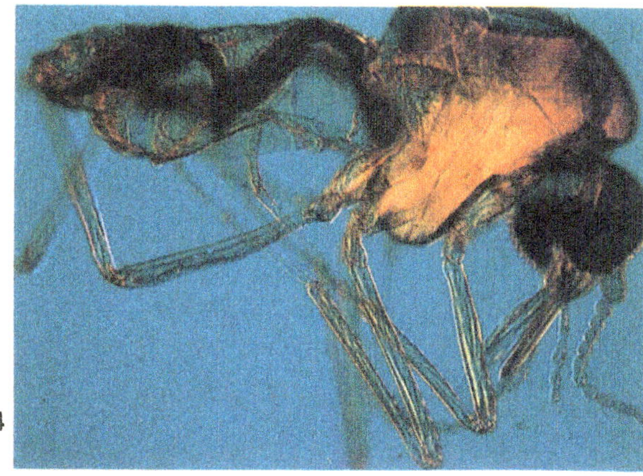
404

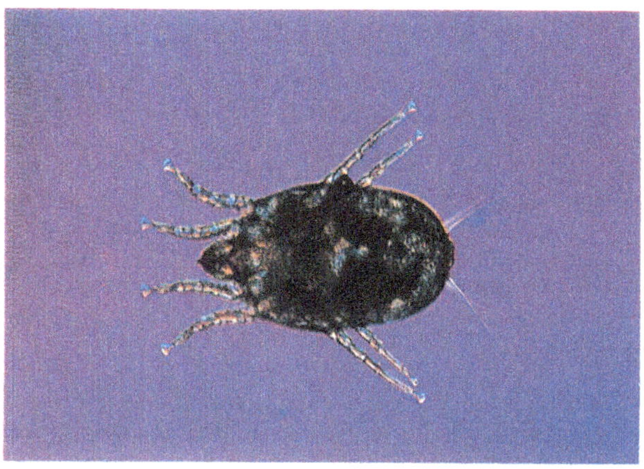
405

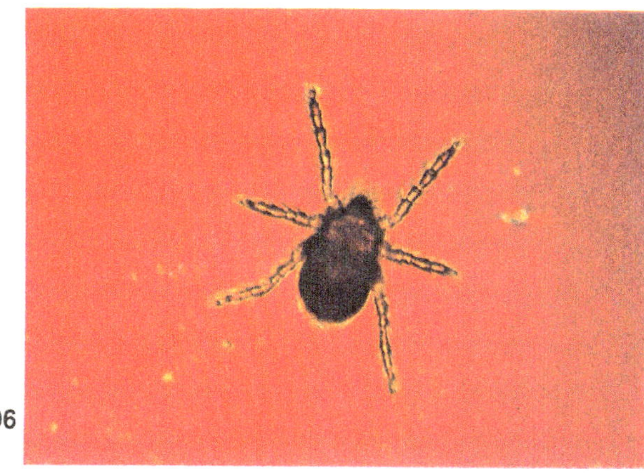
406

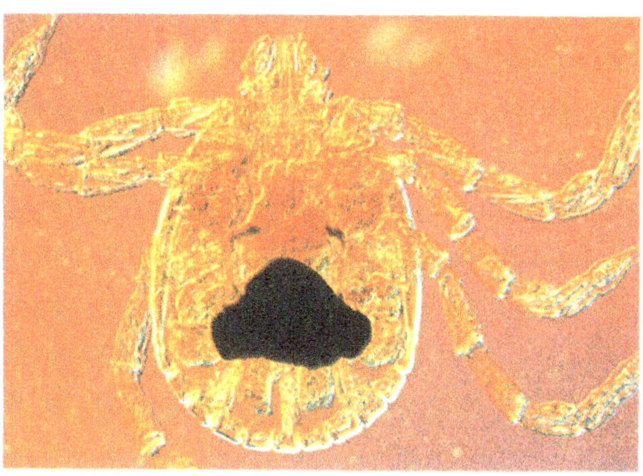
407

408

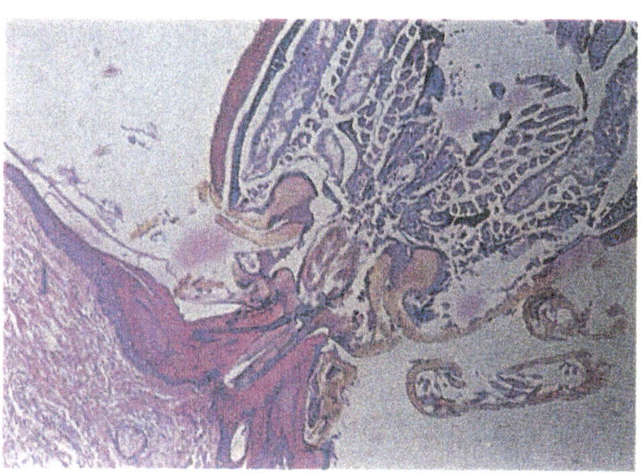
409

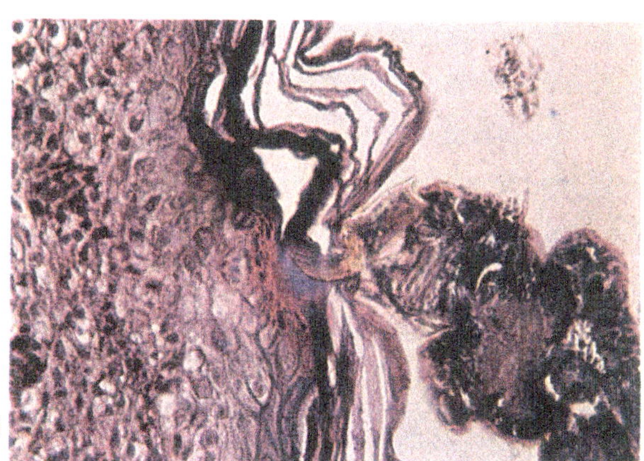
410

**411 Demodex sp. (Follicular mite).** These are characterised by an elongated abdomen and 4 pairs of stubby legs at the anterior end. Interference contrast. X100. Enlarged by 5.4.

**412 Section of skin with a Sarcoptes scabiei.** The mite is lying in a small tunnel in the epidermis.

**413 Culicoides larva.** Anterior end with eye spots and a segmented body. Interference contrast. X100. Enlarged by 5.4.

**414 Sandfly larva.** Posterior end with caudal bristles and feathered hairs projecting from the body. Interference contrast. X100. Enlarged by 5.4.

**415 Simulium larva.** Anterior end showing one of the fan-like feeding brushes which is also known as a cephalic fan. X100. Enlarged by 5.4.

**416 Mouth hook of a muscoid fly larva.** Large hooks such as these are seen in the anterior segment of all muscoid fly larvae. They are supported on a cephalopharyngeal skeleton. Interference contrast. X100. Enlarged by 5.4.

**417 Posterior spiracles of a muscoid fly larva.** In this case the D-shaped spiracles are of *Musca domestica.* The spiracular openings are sinuous and there are 3 on each plate. Interference contrast. X100. Enlarged by 5.4.

**418 Posterior spiracles of Chrysomyia sp.** The spiracular openings are slit-like and there are 3 on each side.

**419 Hooks on the anterior end of a Porocephalus larva.** There are 5 hooks out of which only 1 is completely visible. The shape and the large size of the hooks is an important diagnostic feature. Interference contrast. X100. Enlarged by 5.4.

**420 Eggs of head louse (nits).** The eggs are firmly cemented to the hair when they are laid. In this case 2 eggs are visible, 1 contains a nymph (dark in colour) and the other is empty. Interference contrast. X100. Enlarged by 5.4

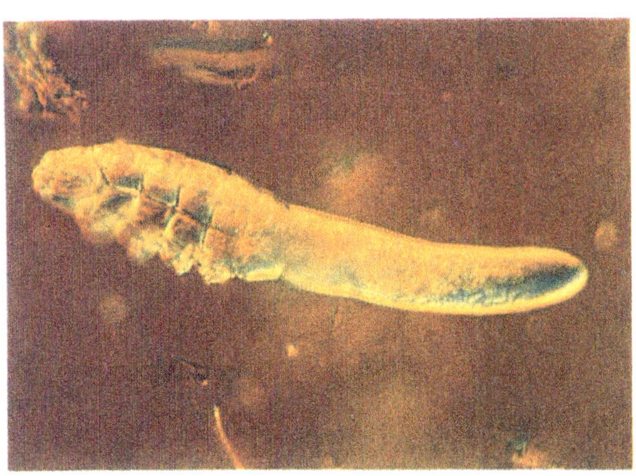

411

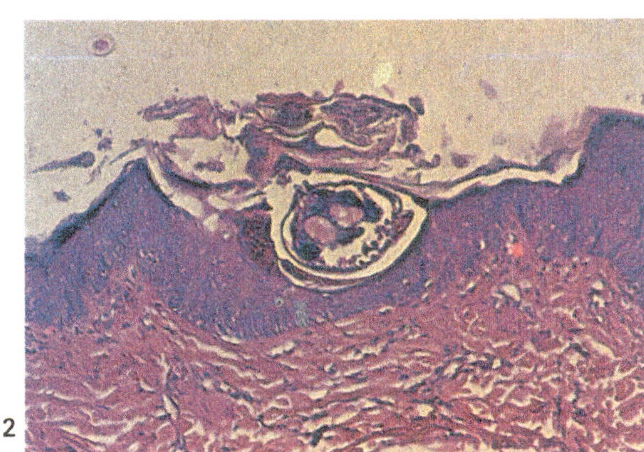

412

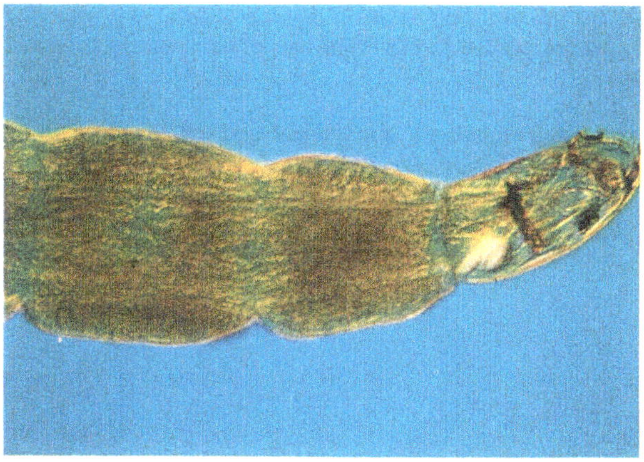

413

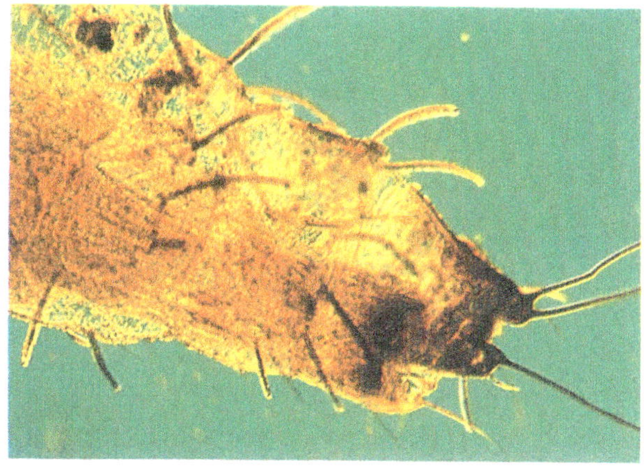

414

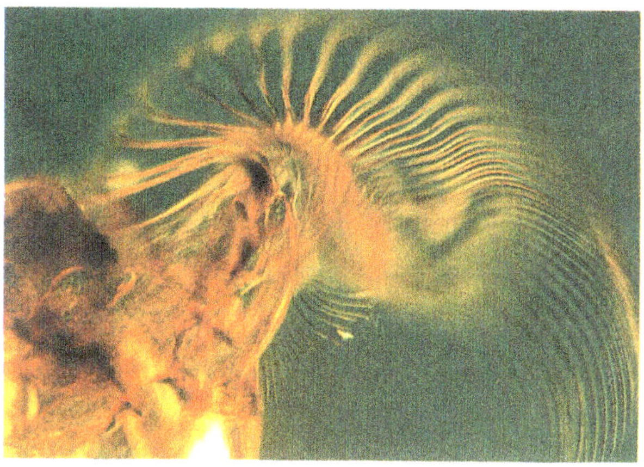

415

416

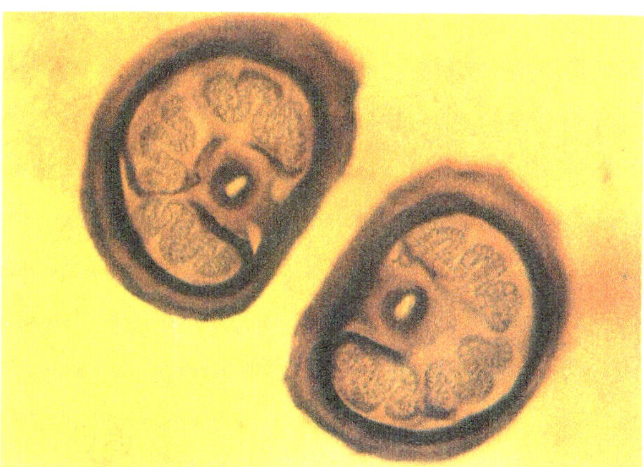

417

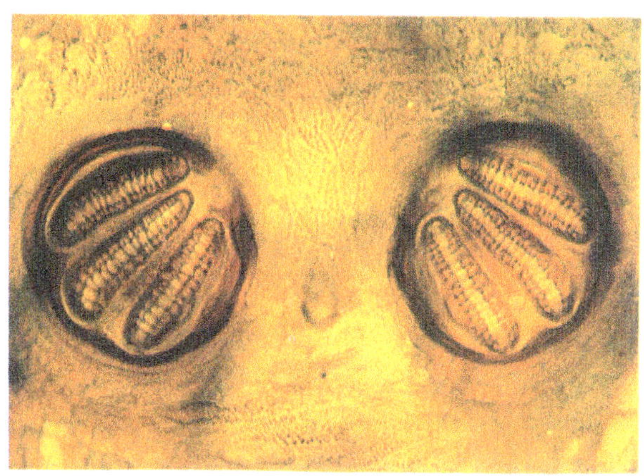

418

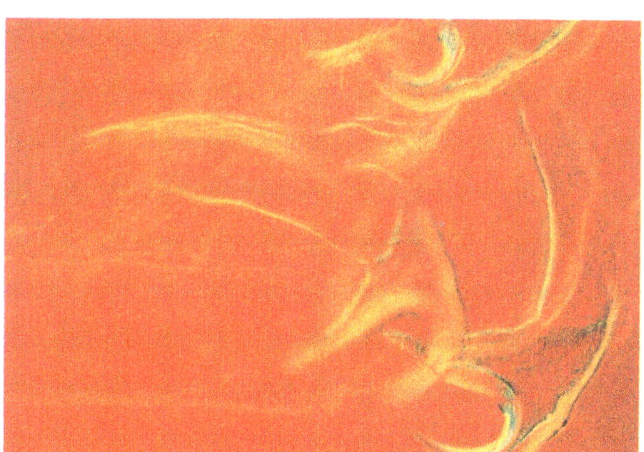

419

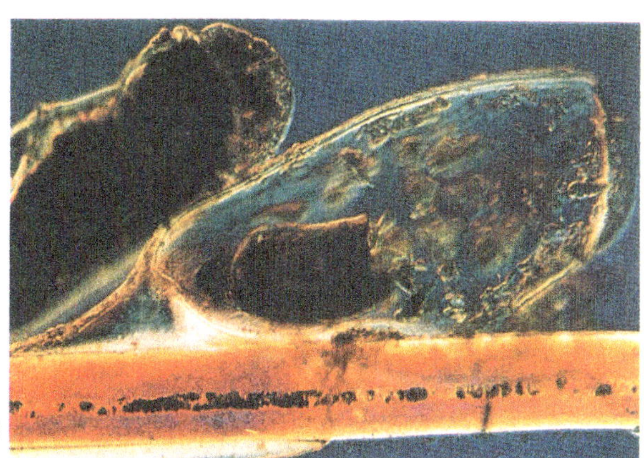

420

**421 Mouthparts of a house fly (Musca domestica)**, showing labella with pseudotracheae. Fluid and food particles are sucked through the pseudotracheae. Interference contrast. X100. Enlarged by 5.4.

**422 Scorpion, showing the distinct pedipalps** at the anterior end and the sting at the posterior end.

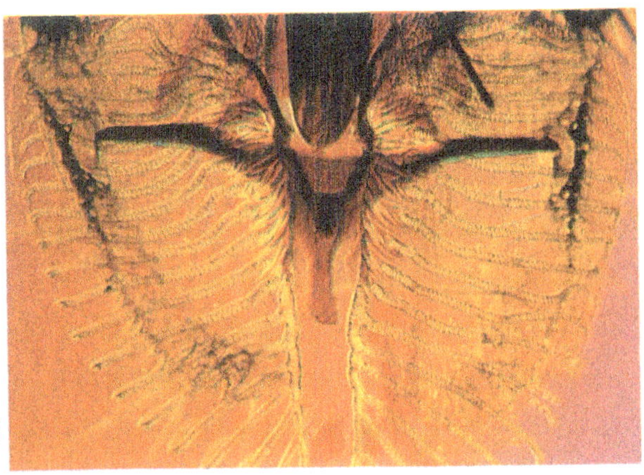

421

422

**423 Ventral surface of the head of a centipede.** The poison claws are curved objects with dark anterior ends. These are not visible from the dorsal surface. Note also the long jointed antennae. X7. Enlarged by 5.4.

**424 Child with hypersensitivity to mosquito bites.** Mosquito bites can produce a severe allergic reaction. In this case the child displayed both the immediate and the delayed (tuberculin) type reactions. The scars on the forearms are due to necrotic changes which occurred during the delayed type reaction.

**425 An eschar of scrub typhus (R. tsutsugamushi)** on the dorsum of the foot. The eschar forms at the site of attachment of the infected mite. (Courtesy of Dr G.W. Brown).

**426 Scabies, showing typical early papular lesion** on the web of fingers. (Courtesy of Dr V.S. Rajan).

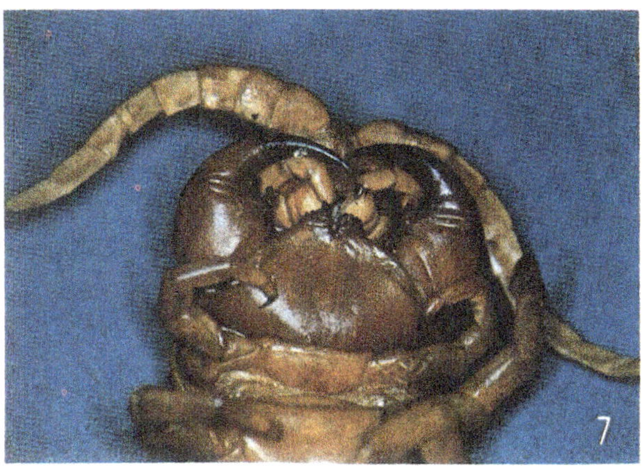

423

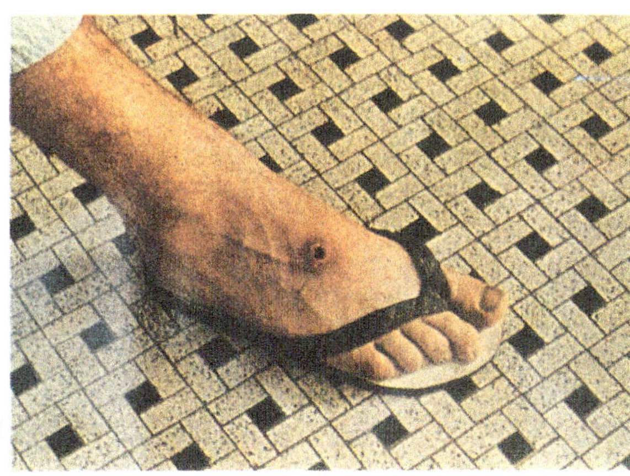

425

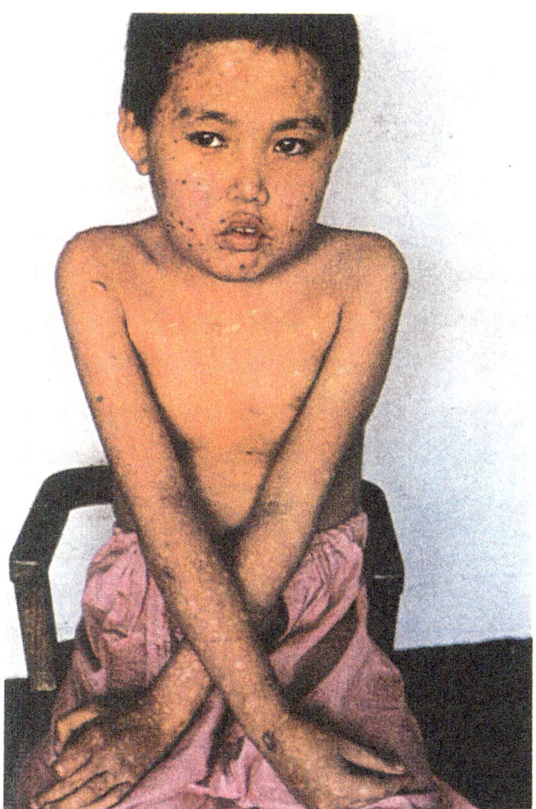

424

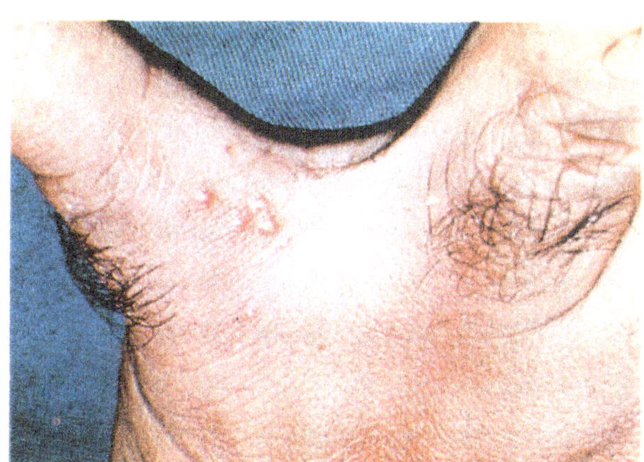

426

**427 Scabies, ulceration and secondary bacterial infection** on the web of fingers. (Courtesy of Dr V.S. Rajan).

**428 Scabies, showing papular lesions** on the dorsal surface of the forearm.

**429 Scabies, pustule formation** and severe secondary bacterial infection involving most of the body.

**430 Scabies, chronic infection** of the scrotum resulting in nodule formations in the skin. (Courtesy of Dr V.S. Rajan).

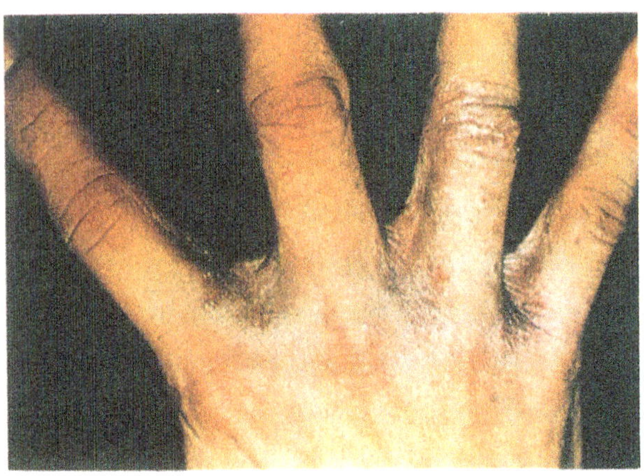

427

428

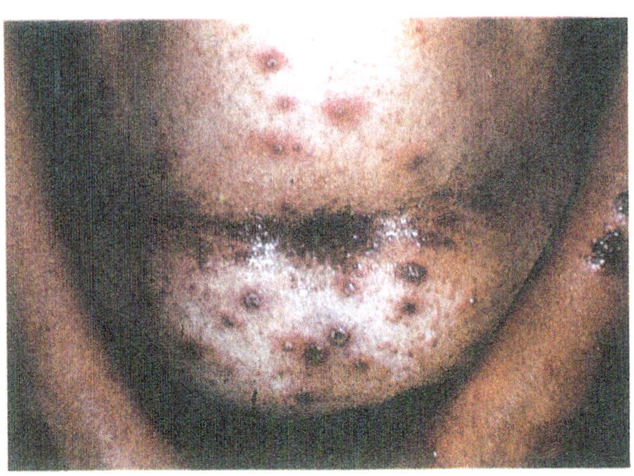

429

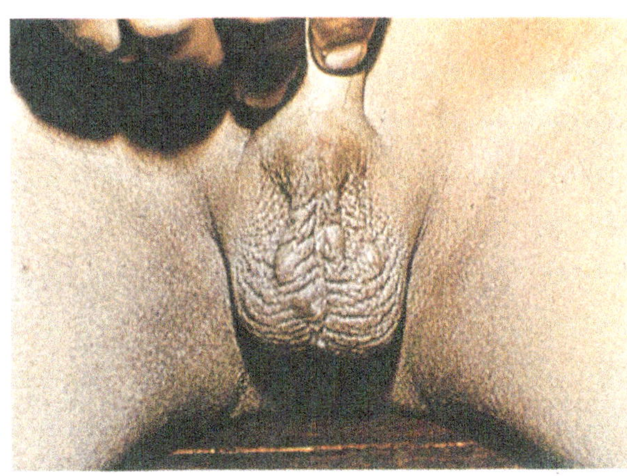

430

# Subject Index

Figures refer to page numbers. Figures in parentheses refer to the illustration number.


**A**

Acanthamoeba sp 28 (38)
- cyst 30 (40b), 32 (45)
- trophozoite 30 (40a), 32 (44)

Aedes aegypti
- egg (395)
- larva (396,397)
- Wing scales (401)

Amoebae infecting man
- classification 4
- terms used 3

Angiostrongylus cantonensis 208 (286), 209 (287), 246 (348,349)
Anisakis sp 212, 247 (350)
Arthropods
- classification 259
- general characteristics 263
- terms used 260, 261

Ascaris lumbricoides
- clinical aspects 210
- cross-section 288 (298,299, 300,301)
- diagnosis 210
- fertilised egg 228 (302), 230 (304,305,306)
- hepatic ascariasis 210, 230 (308,309)
- life cycle 210 (288)
- unfertilised eggs 228 (307)
- 3rd stage larva 230 (307)

**B**

Balantidium coli
- binary fission 75 (99)
- clinical aspects 74
- conjugation 75 (98)
- cyst 76 (100), 78 (106)
- cytostome 77 (104)
- diagnosis 74
- infection 78 (108)
- life cycle 74 (97)
- macronucleus 77 (102), 77 (103)
- trophozoite 76 (101), 78 (105, 107)

Biting midges 271 (404)
Bugs 268 (388)
Brugia malayi
- clinical aspects 202
- diagnosis 202
- larva 250 (360)
- life cycle 202
- microfilaria 249 (355,356,357)

**C**

Calcareous corpuscles 153 (195), 156 (198)
Capillaria hepatica 216, 236 (326,327)
Capillaria phillippinensis 216, 236 (325)
Cestodes
- classification 131
- terms used 132

Charcot Leyden crystals 18 (13)
Chilomastix mesnili 37 (46a), 42 (52)
Clonorchis sinensis
- egg 192 (272,273)
- fish host 192 (274)
- life cycle 180 (237)
- location 182 (238), 190 (269)
- testes 190 (270)

Cysticercus bovis 158 (206)
Cysticercus cellulosa 134, 158 (211,212)

**D**

Demodex sp 278 (411)
Dicrocoelium dendriticum 180
Dientamoeba fragilis 22 (23)
Differential characters of microfilaria 204
Dipetalonema perstans 206, 250 (359)
Dipetalonema streptocerca 206
Diphyllobothrium latum 150 (193)
- coracidium 160 (223)
- egg 160 (220,221,222)
- reproductive organs 152 (194)
- scolex 156 (199)

Dipylidium caninum 148 (192)
- scolex 156 (204,205)

Dirofilaria sp 207, 252 (367,368)
Disseminated cutaneous leishmaniasis 56 (68,71,72)
Dracunculus medinensis 207, 247 (351,352)

**E**

Echinococcus granulosus 138 (186)
- brood capsule 158 (208,209, 210)

Echinococcus multilocularis 138
Elephantiasis 256 (381,382,383, 384,385)
Endolimax nana 22 (24,25)
Entamoeba coli 21 (20,21,22)
Entamoeba hartmanni 22 (28)
Entamoeba histolytica
- clinical aspects 6, 8
- complications 9 (2)
- cyst 18 (15,16,17), 20 (18,19)
- diagnosis 6
- fluorescent antibody test 24 (36)
- food ingestion 10 (3), 11 (4), 12 (5)
- life cycle 4 (1)

liver abscess 24 (35)
trophozoite 17 (11,12), 18 (14)
ulceration of gut 24 (29,30,31, 32,33,34)
ultrastructure 13 (6), 14 (8), 16 (10)
Entamoeba polecki 22 (27)
Enterobius vermicularis
adult 233 (313)
anal smear 233 (315)
clinical aspects 220
diagnosis 220
in appendix 233 (314)
life cycle 220 (293)
Eosinophilic lung 202

F

Fasciola gigantica 180
Fasciola hepatica
cross-section 192 (279)
life cycle 178 (236)
miracidia 192 (278)
Fasciolopsis buski
clinical aspects 176
diagnosis 176
life cycle 176 (235)
water chestnut 192 (280)
Fleas 270 (390)

G

Gastrodiscoides hominis 178, 192 (277)
Giardia lamblia 36, 37, 38
cyst 40 (48), 42 (55)
trophozoite 40 (49,50), 42 (53,54,56)
Gnathostoma spinigerum 220, 244 (344,345,346,347)

H

Haemoflagellates 45
Heterophyes heterophyes 178
Hookworms
buccal capsule 238 (330,331)
clinical aspects 222
diagnosis 222
differentiation 224 (295), 226 (296,297)
egg 238 (333,334)
in section 238 (332)
infective larva 238 (335)
life cycle 222 (294)
House fly 270, 278 (416,417), 280 (421)
Hydatid cyst 140 (187), 154 (196,197)
Hymenolepis diminuta 144 (190), 146 (191)
egg 158 (215)
Hymenolepis nana 142 (189)
cysticercoid 158 (207)
egg 158 (214)
scolex 156 (201,202)

I

Intestinal and genital flagellates
list of species 36
terms used 25
Iodamoeba butschlii 22 (26)
Isospora belli 126 (180)

L

Larval tapeworms 141 (188)
Leishmania
binary fission 52 (61)
diagnosis 49
epidemiology 47
life cycle 45 (57)
morphological forms 52 (60)
sandfly 54 (62), 271
species 45
temperature chart 48
Leishmania donovani 54 (63)
Leishmania tropica 50 (58,59), 52 (60,61), 54 (64,65), 56 (66,69,70)
Lice 269 (389), 278 (420)
Loa loa 206
adult 252 (373)
microfilaria 250 (359)

M

Malaria parasites
anopheline 90 (118), 92 (120,121,122)
brain in *P. falciparum* 96 (138,139,142)
clinical aspects 82
course of infection 85 (112)
culicine 90 (119)
diagnosis 82
differential diagnosis 82, 86, 87 (113)
fever pattern 83 (110)
liver in *P. falciparum* 96 (137)
oocyst 88 (117), 92 (124)
ookinete 92 (123)
*P. falciparum* life cycle 80 (109)
*P. falciparum* morphology 94 (127,128,129,130)
*P. malariae* morphology 94 (136)
*P. vivax* life cycle 80
*P. vivax* morphology 94 (131,132,133,134,135)
pigment 88 (115,116)
placenta in *P. falciparum* 98 (140)
pre-erythrocytic schizont 92 (126)
relapse 84 (111)
species 80
sporozoite 92 (125)
spleen in *P. falciparum* 98 (141)
splenomegaly syndrome 98 (143)
terms used 79
Mansonella ozzardi 206
Mansonia sp
egg (398)
larva (399,400)
wing scales (402)
Metagonimus yokogawi 178, 192 (275)
Mites 266 (387a,387b), 276 (405, 406,410)
Mosquitoes 275
Multiceps multiceps 138

N

Naegleria fowleri
brain infection 32 (41,42,43)
clinical aspects 26
comparison with *Acanthamoeba* 28
comparison with *Entamoeba* 26
cyst 30 (39b)
diagnosis 26
life cycle 26 (37)
trophozoite 30 (39a)
ultrastructure 13 (7)
Nematodes
body wall 200 (283a)
classification 195
internal structure 200 (282)
morphology 198 (281)

muscle cell 200 (283b)
terms used 196

O

Onchocerca volvulus
adult section 252 (366)
dermatitis 254 (374,375)
keratitis 254 (376,377,378)
microfilaria 249 (358)
skin snip 252 (372)
Opisthorcis viverrini 180, 192 (271)

P

Paragonimus westermanii
crab host 190 (265)
cross-section 190 (263-264)
egg 190 (267)
life cycle 182 (239)
location 184 (240)
miracidium 190 (268)
morphology 190 (262)
Pneumocystis carinii 128 (181,182)
Post Kala-Azar dermal leishmaniasis 56 (73)
Protozoa
classification 1

R

Reduviid bug 68 (90,91,92), 88 (68), 89 (68)

S

Sarcocystis and Isospora
cyst 124 (170), 126 (173,174)
cystozoite 124 (171), 126 (175,176)
differential characters 121
*Isospora* stages 125 (172)
life cycle 122 (169)
oocyst 126 (177)
Scabies 278 (412), 281 (426), 282 (427,428,429,430)
Schistosoma haematobium 186 (245, 253, 254)
Schistosoma japonicum 186 (246,247,255,257,258,259)
Schistosoma mansoni 185 (243,244,251,252)
Schistosome
cercaria 186 (249,250)
clinical aspects 176
cross-section 185 (242)
diagnosis 176
differential characters 172 (230)
fluorescent-antibody test 188 (260,261)
life cycle 171 (229)
location 172 (231,232)
Scrub typhus 281 (425)
Snail 186 (248)
Strongyloides stercoralis 214 (290), 240 (336,337,338)

T

Taenia saginata 134 (183)
egg 158 (216), 160 (217,218)
Taenia solium 134 (184)
scolex 156 (200)
Ticks 265 (386), 276 (407,408, 409)
Toxocara sp 232 (310,311,312)
Toxoplasma gondii
cat intestine 114 (155,156, 157)
clinical aspects 102,103
cyst 108 (147,148), 116 (158), 117 (159,160,161,162,163)
endodyogeny 106 (145)
endozoite 118 (165,166,167)
fluorescent antibody test 112 (152)
fundus 118 (168)
laboratory diagnosis 104
life cycle 100 (144)
merozoite 110 (149)
oocysts 112 (153,154)
pseudocyst 118 (164)
terms used 99,100
Trematodes
alimentary system 166 (225a,225b)
classification 163
comparative size 169 (227)
eggs 170 (228)
execretory system 166 (225c)
larval stages 168 (226)
nervous system 166 (225d)
reproductive system 166 (224)
terms used 164
Trichinella spiralis
clinical aspects 212
diagnosis 212
larva in muscles 237 (328,329)
life cycle 212 (289)
Trichomonas sp
clinical aspects 38
diagnosis 38
life cycle 38 (47)
trophozoite 42 (51)
Trichostrongylus sp 214,240 (339)
Trichuris trichiura
clinical aspects 216
cross-section 234 (319,320)
diagnosis 216
egg 234 (321,322)
in rectum 234 (323,324)
life cycle 216 (291)
morphology 234 (316,317,318)
Trypanosoma cruzi 66 (87), 70 (93,94,95)
Trypanosomes infecting man
clinical aspects 60
diagnosis 60
glossina 63 (77,78,79,80,81,82), 271
life cycle 58, 59 (74)
ultrastructure 61 (75), 62 (76)
Trypanosoma rangeli 70 (96)
Trypanosoma rhodesiense 65 (83,84,85,86)

V

Venomous insects 271, 280 (422), 281 (423)
Visceral larva migrans 218 (292)

W

Wuchereria bancrofti
clinical aspects 202
cross-section 252 (365)
larva 251 (361,362,363,364)
life cycle 202 (284)
genital involvement 256 (379,380)
microfilaria 248 (353,354)

CPSIA information can be obtained at www.ICGtesting.com
Printed in the USA
LVOW05s1252080315

429659LV00016B/863/P